Plant Conservation Biotechnology

This book is dedicated, with love and thanks to my late father, Irwin Benson, and my mother Ella.

List of Figures

Foreword

Plant genetic resources (the genetic material which determine the characteristics of plants and hence their ability to adapt and survive) are the biological basis of world food security. Directly or indirectly they support the livelihoods of every person on the Earth. Whether used by farmers or by plant breeders, plant genetic resources are a reservoir of genetic adaptability that acts as a buffer against potentially harmful environmental and economic change. However, genetic erosion is occurring all around the world at an alarming rate due to changes in land use, rising population pressure and industrial development. Many millions of hectares of forest, including tropical forests, are lost every year in some of the most diverse ecosystems in the world. In agriculture, diversity is threatened by the replacement of traditional landacres by high-yielding crop varieties and the move to cash crops. Agricultural production is also affected by globalization of the world economy.

At the end of the twentieth century, access to food around the world is not secure. 800 million people are still inadequately fed. In the next 30 years, the world population is expected to grow by 2 500 million to reach 8 500 million. Eighty percent of this growth will take place in developing countries already affected by poverty and undernourishment. To eradicate hunger in existing populations and to feed their children and grandchildren will require a rate of increase in food production never before achieved. More intensive high and medium-input farming methods will help to increase productivity in parts of the world. However, increased food production in many developing countries will have to come from improved low-input agriculture under difficult environmental conditions. Threats to forests will have to be controlled and the productivity of forest cropping systems improved. Genetic resources will be fundamental in achieving these new levels of production but making that production sustainable will require prudent conservation and use strategies for the genetic resources, supported by effective technologies.

Traditional approaches to germplasm storage in seed and field genebanks make a major contribution to the secure conservation of genepools, as do *in situ* and on-farm conservation. However, recognizing that these approaches are not without drawbacks, efforts have been made in recent years to develop new conservation methods based on biotechnology. The complementary application of these new approaches alongside traditional ones is already having a substantial impact on the conservation and use of

plant genetic resources. Building upon the foundation of success in *in vitro* propagation
and medium-term storage of cultures in slow growth, cryopreservation is being used for
storage of the germplasm of species that cannot be conserved as seed – it is likely to
become increasingly available for a wide range of species in the next few years. New
molecular techniques offer opportunities to make substantial advances in our knowledge
of the diversity of some of the most important crop and forest species. A range of
molecular markers is being used to determine the extent and distribution of genetic
diversity and to support conservation decisions.

The Global Plan of Action for the Conservation and Sustainable Utilization of Plant
Genetic Resources for Food and Agriculture was formally adopted by representatives of
150 countries during the Fourth International Technical Conference on Plant Genetic
Resources which was held in Leipzig, Germany, in June 1996. The importance of using
and developing biotechnologies for improving *in situ* and *ex situ* conservation, as well as
for the utilization of plant genetic resources, was highlighted in this document.

This book seeks to make a contribution to a most worthwhile objective, the implementation of the Global Plan of Action. It provides an overview of the latest biotechnological
methods, techniques and procedures developed for the conservation and exchange of
plant genetic resources. It also presents examples of the current state of their application
to a wide range of plant species, from algae to tropical forest tree species, drawing upon
experiences in research institutes and national and international genetic resources conservation centers located in developed and developing countries.

Florent Engelmann
In Vitro Conservation Officer
IPGRI

Lyndsey A. Withers
Assistant Director General
IPGRI

Preface

Plant Conservation Biotechnology is an interdisciplinary subject, to which the tools of modern biotechnology are applied for plant conservation. Importantly, these techniques must not replace traditional *ex situ* and *in situ* conservation methods, but, rather, they should provide complementary and enabling means of plant genetic resource management.

A wide range of biotechnological methods are now utilized (including tissue culture techniques, molecular genome analysis, immunological diagnostics and cryopreservation protocols) for the collection, characterization, disease indexing, propagation, patenting, storage, documentation and exchange of plant genetic resources. Thus, biotechnology has a major role in all aspects of plant genetic resource management, conservation, and utilization. Examples of user sectors and industries include: aquaculture, agriculture, agroforestry, forestry, horticulture, and the secondary products industries. Importantly, biotechnology is rapidly gaining importance for the conservation of endangered plant species.

As biotechnology continues to have a key role in the conservation, and sustainable utilization of all types of biodiversity it is essential to chart the progress of the many new and innovative developments, expressly within the context of plant diversity. Thus, the aim of this book is to review 'Plant Conservation Biotechnology' in its broadest sense and explore its use across many fields of application: from the conservation of endangered species to the storage of economically important crop plants and industrial plant cell culture collections. This volume also collectively considers a wide spectrum of plant systems, including, for example, freshwater algal protists and Brazilian rain forests. Interestingly, there is considerable commonality across these diverse areas and where interdisciplinary differences do occur it is hoped that they will stimulate our interest to explore the use of new protocols and methodologies in novel contexts. It is exciting to consider that the simple cryopreservation methods developed to conserve recalcitrant rain forest tree seeds of Malaysia also have potential applications for the conservation of unicellular aquatic plants. The continued development of plant conservation strategies based on biotechnological procedures will greatly benefit from the interfacing of multi- and interdisciplinary areas.

Conservation is an international issue and one of the major aims of this book has been to explore the importance of biotechnology as evidenced by its use in major international plant genetic resource centres located in different parts of the globe. Most agronomically important plant groups have been considered and particular emphasis has been given to those species which have a significant conservation problem such as the vegetatively propagated tuber crops.

The subject fields of Conservation, Biotechnology and Biodiversity are attracting considerable international interest in the tertiary education sector. Undergraduate and postgraduate programmes now include aspects of conservation and biodiversity, even in the most general of curricula. This volume has been composed for biotechnology and environmental resources students and especially for those entering their final year of tertiary education, and for MSc and PhD level postgraduate scholars.

Increasingly, biotechnology is incorporated into advanced training courses and specialized workshops which are targeted at assisting the international transfer of conservation technologies to professional users. Many of these personnel are attached to international genebanks, botanical gardens, culture collections and germplasm repositories. Importantly, the book has been largely targeted at professional scientists with an interest in using biotechnology to assist conservation. The volume is in two major sections, Part I, 'Principles of Plant Conservation Biotechnology: Methods, Techniques and Procedures' has been designed to assist newcomers to the subject, and 'set-the-scene' for Part II. Part I was not intended to be a laboratory manual, as many other excellent texts already provide this information. However, the purpose of this section is to provide the reader with a broad introduction to the subject and an overview of the methods and procedures involved, and where necessary the resources required. Part II, 'Applications of Biotechnology in Plant Diversity Conservation' shows the subject of Conservation Biotechnology 'in action', the depth and breadth of which will provide an information source for well established conservation researchers. It is also hoped that it will assist newcomers in developing their own applications strategies.

In order to assist the reader, the Editor has provided cross-reference points between the two sections and related chapters throughout the book. It is thus hoped that the volume can be viewed as an integrated piece of work, whilst at the same time each individual chapter provides an overview of a specific area in its own right. The contributory chapters are written by international experts who are, in the main, biotechnologists *and* conservationists; they represent many different conservation sectors and geographical regions.

As appropriate, and particularly in Part II, individual authors have considered their work in a wider context and debated their conservation programmes in terms of economic, social, country-specific, regional and global issues. It is perhaps fitting to finally note that one team of contributors to the book (González-Benito *et al.*) poses the two key, conservation questions: 'What to conserve?' and 'How to conserve?' and debates that our capacity for (endangered) species preservation is largely influenced by economic factors. Plants which may not have an immediate economic benefit today, may do so in the future. Plants which do not have commercial value, still however, deserve our consideration and protection. Their 'worth' is indicated, in many countries, by the growing and co-operative activities of professional and amateur botanists who are increasingly working together to save their endangered indigenous floras. Furthermore, the general public is becoming more and more involved in actively campaigning for conservation issues. It is thus imperative that we consider conserving not only economically important crop plants but also their wild relatives and endangered species. It is hoped that the considerable progress made in biotechnology will allow more cost-effective and efficient plant conservation strategies to be implemented and in doing so broaden our overall capacity to conserve the Earth's vast diversity of plant species.

Erica E. Benson

Acknowledgements

The Editor gratefully acknowledges the support and co-operation of the chapter contributors; it has been a pleasure working with you. Many thanks to those colleagues at the University of Abertay Dundee who gave their help and precious time to assist in the production of this book and a special 'big' thank you to Brenda, Sandra and Maggie in the MLS School office; to Shona for advice with artwork and to Kevin on the IT-help desk. Thank you to Ian, in UAD's Copy Shop, for helping me meet my deadlines on time. Thank you Dominique, Linda and Isobel, for keeping the research lab going for me whilst I was busy editing and I am especially grateful to Linda and Mark for sorting out my many computer support and e-mail problems. Finally, many, many thanks to my husband Keith for his encouragement during the times I spent working, at all hours, at home.

Contributing Authors

Dr Erica E. Benson
Plant Conservation Biotechnology Group
School of Science and Engineering
University of Abertay Dundee
UK

Dr Rex M. Brennan
Soft Fruit Genetics Department
Scottish Crop Research Institute
Invergowrie, Dundee
UK

Dr John G. Day
NERC Institute of Freshwater Ecology
Windermere Laboratory
Ambleside
UK

Dr Ali M. Golmirzaie
International Potato Centre (CIP)
Lima, Peru

Dr M. Elena González-Benito
Dpto. De Biología Vegetal
Escuela Universitaria de Ingeniería Tecníca Agricola
Universidad Politecnica de Madrid
Ciudad Universitario
Madrid
Spain

Dr Keith Harding
Crop Genetics Department
Scottish Crop Research Institute
Invergowrie
Dundee
UK

Dr Stephen A. Harris
Department of Plant Sciences
University of Oxford
Oxford
UK

Dr Kim E. Hummer
United States Department of Agriculture
Agricultural Research Service
National Clonal Repository
Corvallis
USA

Dr J.M. Iriondo
Escuela Técnica Superior de Ingenieros Agrónmos
Universidad Politécnica de Madrid
Ciudad Universitario
Madrid
Spain

Dr Baskaran Krishnapillay
Forest Plantations Division
Forest Research Institute of Malaysia
Kepong
Kuala Lumpur
Malaysia

Dr Paul T. Lynch
Division of Biological Sciences
University of Derby
Derby
UK

Dr Binay B. Mandal
National Bureau for Plant Genetic Resources
Pusa Campus
New Delhi
India

Dr Sinclair H. Mantell
Department of Biological Sciences
Wye College
University of London
Ashford
UK

Dr C. Martín
Escuela Técnica Superior de Ingenieros Agrónmos
Universidad Politécnica de Madrid
Ciudad Universitario
Madrid
Spain

Dr Robert R. Martin
United States Department of Agriculture
Agricultural Research Service
National Clonal Repository
Oregon
USA

Dr M. Marzalina
Seed Technology Laboratory
Forest Research Institute of Malaysia
Kepong
Kuala Lumpur
Malaysia

Dr Maria Cristina Mazza
Centro Nacional de Pesquisas Florestais
EMBRAPA
Estrada da Ribeira
Colombo
Brasil

Dr Stephen Millam
Crop Genetics Department
Scottish Crop Research Institute
Invergowrie
Dundee
UK

Dr N.Q. Ng
Plant Tissue Culture Genebank
Tropical Root Crop Improvement Programme
International Institute of Tropical Agriculture (IITA)
Ibadan
Nigeria

Dr S.Y.C. Ng
Plant Tissue Culture Genebank
Tropical Root Crop Improvement Programme
International Institute of Tropical Agriculture (IITA)
Ibadan
Nigeria

Dr Ana Panta
International Potato Centre (CIP)
Apartado 1558
Lima
Peru

Dr Valerie C. Pence
Plant Conservation Section
Centre For Research of Endangered Wildlife (CREW)
Cincinnati Zoo and Botanical Garden
Cincinnati
Ohio
USA

Dr César Pérez
Escuela Técnica Superior de Ingenieros Agrónmos
Universidad Politécnica de Madrid
Ciudad Universitario
Madrid
Spain

Dr Joseph D. Postman
United States Department of Agriculture
Agricultural Research Service
National Clonal Repository
Oregon
USA

Dr Barbara M. Reed
United States Department of Agriculture
Agricultural Research Service
National Clonal Germplasm Repository
Oregon
USA

Dr Heinz Martin Schumacher
Plant Culture Department
German Collection of Microorganisms and Cell Cultures (DSMZ)
Braunschweig
Germany

Dr Judith Toledo
International Potato Centre (CIP)
Lima
Peru

Dr Ana Maria Viana
Plant Physiology and Plant Biotechnology
Departamento de Botânica
Centro de Ciências Biológicas
Universidade Federal de Santa Catarina
Florionapolis
Brazil

Principles of Plant Conservation Biotechnology: Methods, Techniques and Procedures

An Introduction to Plant Conservation Biotechnology

ERICA E. BENSON

1.1 Integrating biotechnology into conservation programmes

The tools of modern biotechnology are being increasingly applied for plant diversity characterization and undoubtedly they have a major role in assisting plant conservation programmes. However, their value is dependent upon ensuring that biotechnological methods are targeted effectively and utilized as complementary and enabling technologies. *Most importantly, they must be applied in the appropriate context.* Biotechnology is advancing so rapidly that it may be sometimes difficult for potential 'conservation' users to assess the value and role of new techniques and procedures within their own specific area. It is important to recognize that the effective integration of biotechnology in conservation programmes requires multi- and interdisciplinary co-operation. Thus, present and future conservation teams should comprise personnel from a broad spectrum of disciplines.

Figure 1.1 outlines the key steps which must be considered when embarking on an existing, or new, conservation strategy which has the potential for incorporating biotechnology. First, the conservation need (Figure 1.1A) must be evaluated judiciously. For example, in the case of certain endangered species, there may be a considerable urgency for conservation and the rapid implementation of already existing strategies may, in the first instance, be the most suitable course of action. An appraisal of existing conservation methods is thus required (Figure 1.1B) and if traditional options are already being used with success (e.g. seed banking) there is little benefit in substituting *in vitro* approaches. Biotechnological options must not displace existing successful methods of conservation (Figure 1.1C) and it is essential, from the onset, to assess the 'fitness of purpose' of a new biotechnological method. In addition, the feasibility of incorporating biotechnological techniques must be considered in terms of resources, expertise and specific training needs, cost, and long-term maintenance. Finally, successful methods must be validated and integrated into a programme on a routine basis (Figure 1.1E) and, as appropriate, it may be useful, and cost-effective, to consider the broader applications of methods to other systems and species. It is also important to regularly re-evaluate the biotechnological approach in the context of existing, *ex situ* and *in situ* conservation options (Figure 1.1F).

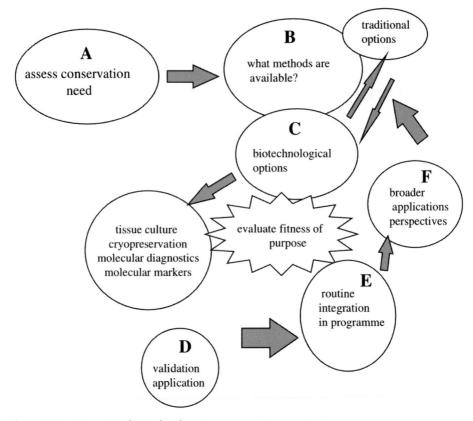

Figure 1.1   Integrating biotechnology in conservation projects

1.2   A general overview: how does biotechnology assist plant conservation?

There are four main areas of biotechnology which can directly assist plant conservation programmes:

A.   Molecular markers technology.
B.   Molecular diagnostics.
C.   Tissue culture (*in vitro* technologies).
D.   Cryopreservation.

In addition, 'information technology' (IT) will have an increasingly important role in facilitating conservation programmes and the interface between IT and biotechnology provides considerable potential for many aspects of plant genetic resource management (for example, see Anderson and Cartinhour, 1997).

For economically important plant species, it is also essential to consider the relationship between conservation and utilization and to recognize that biotechnology can enable sustainability programmes (Callow *et al.*, 1997). Figure 1.2 summarizes the different applications of biotechnology in plant conservation. Molecular biology (Figure 1.2A) and more specifically, marker techniques, have a key role in enabling the assessment of plant

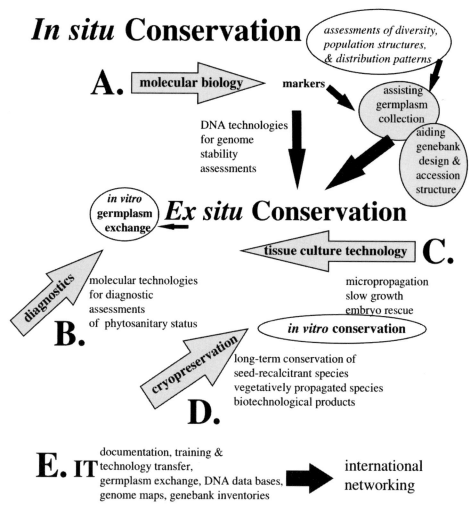

Figure 1.2 Applications summary: the use of biotechnological techniques in plant conservation

diversity at the genomic level (Ayad *et al.*, 1997; Karp *et al.*, 1997). The elucidation of population structures and gene distribution patterns within ecosystems provides information which can be used to support *in situ* conservation programmes. Contrasting examples of techniques include the assessment of restriction fragment length polymorphisms (RFLPs) which permit the detection of specific markers genes (see Harding, Chapter 7, this volume) and polymerase chain reaction-based marker technologies (PCR) used in association with RAPD (randomly amplified polymorphic DNA) analysis. For detailed reviews of molecular marker techniques and their application in plant conservation see Westman and Kresovich (1997) and Harris, Chapter 2, and Harding, Chapter 7, this volume. Direct, practical applications of DNA marker technologies include advising on germplasm collecting missions and genebank design. Importantly, a molecular knowledge of genetic diversity can greatly assist the decision making processes associated with *ex situ* conservation and, more directly, facilitate germplasm collection (see Hummer, Chapter 3, this volume) and genebank management. The duplication of accessions can be costly and

DNA inventories provide an excellent means of ensuring that repositories are well structured and complete. Similarly, DNA marker techniques can be used, by curators, to identify significant omissions in germplasm collections and thus enable them to target, more effectively, the acquisition requirements of future collecting missions.

DNA marker technologies also have an important role in the monitoring of genetic stability in conserved germplasm. It is essential that storage methods can be used with confidence and molecular markers can be used to confirm that conserved germplasm retains genetic fidelity (see Harding, Chapter 7, this volume). This may be especially important for germplasm which is conserved using tissue culture procedures.

Both *in situ* and *ex situ* conservation strategies have a requirement for germplasm transfer and international exchange (Figure 1.2B). Thus, molecular diagnostics based on immunological and molecular DNA methods are applied for the assessment of phytosanitary status (see Martin and Postman, Chapter 5, this volume).

Tissue culture (or *in vitro*) technologies (see Lynch, Chapter 4, this volume) have had a major impact on the *ex situ* conservation of plant genetic resources (Figure 1.2C) and importantly, disease indexed *in vitro*-maintained germplasm provides an excellent means of mediating international germplasm exchange. Micropropagation, using somatic embryo and shoot culture techniques assists many crop plant improvement programmes and increasingly these methods are being used for the conservation of endangered plant species (see Pence, Chapter 15 and González-Benito *et al.*, Chapter 16, this volume). Crop plants which are vegetatively propagated present particular conservation problems as their seeds are not available for banking. Whilst field genebanks provide important conservation options, germplasm maintained in this manner can be at risk from pathogen attack and climatic damage. For vegetatively propagated species, *in vitro* conservation using tissue culture methods is the only reliable, long-term means of preservation. Storage in the active growing state or under reduced (slow) growth provides cost effective, medium-term conservation options. Most major, international germplasm centres use *in vitro* conservation as their method of choice for vegetatively propagated crops (see Ashmore, 1997). Within this volume examples include: the United States Department of Agriculture, National Clonal Germplasm Repository, Oregon (Reed, Chapter 10), the International Potato Centre (CIP), Peru (Golmirzaie *et al.*, Chapter 12), the International Institute of Tropical Agriculture (IITA), Nigeria (Ng *et al.*, Chapter 13) and the National Bureau for Plant Genetic Resources (NBPGR), India (Mandal, Chapter 14).

Maintenance of plant germplasm in the active or slow growth state provides a medium-term storage option, however the long-term conservation of *in vitro*-derived plant germplasm is increasingly achieved using cryopreservation in liquid nitrogen (see Benson, Chapter 6, this volume). Cryo-conservation (Figure 1.2D) is thus applied to plant germplasm which cannot be conserved using traditional seed banking techniques and/or to vegetatively propagated germplasm. In 1980, Withers and Williams highlighted the problems associated with 'difficult to store' and seed-recalcitrant germplasm. It is encouraging to note that during the last decade or more, major advances have been made in the successful application of cryopreservation methods to once termed 'difficult' germplasm types (Ashmore, 1997; Bajaj, 1995; Callow *et al.*, 1997; Normah *et al.*, 1996; Razdan and Cocking, 1997). Particularly significant advances have been in the cryopreservation of the recalcitrant seeds (and excised embryos) derived from tropical agroforesty and plantation crops and rain forest tree species (Normah *et al.*, 1996; see also Marzalina and Krishnapillay, Chapter 17, this volume). Indeed, the next phase of cryo-conservation activity will be the establishment of large scale 'working' cryopreserved genebanks (see the report of the CIAT-IBPGR collaborative project, CIAT-IPGRI, 1994). However, whilst successes have been significant, conservation recalcitrance does still pose a problem for certain plant species and a combined effort, involving both fundamental and applied research, must be maintained.

Advances in biotechnology have only been equalled by the activity of the information science and technology sector. The 'IT revolution' is indeed rapidly changing the way and means in which conservation scientists perform their research and implement their conservation strategies. Figure 1.2E indicates those areas for which the interface between information and (bio)-technologies offers greatest benefit for progressing global conservation initiatives. On a practical basis, IT does, and will, continue to assist all aspects of documentation associated with genetic resource transfer and management, genome mapping, DNA databasing and genebank inventories. However, in the future it will be important to enhance and consolidate the enabling role of IT in international training and technology transfer. Distance learning and electronic networking specifically designed for and targeted at plant conservation programmes will promote the expediency of concerted international conservation activities.

1.3 Conservation biotechnology and the sustainable utilization of plant genetic resources

In 1995, the popular, UK-based, plant conservation journal, *Plant Talk*, produced a communication entitled 'Yew in the fight against cancer: sustainability or pillage?' The article refers, of course, to the use of taxus species for the production of the secondary metabolite taxol, which is used to produce a potent anti-cancer drug. Whilst synthesis of the secondary product has been reported, and indeed, the drug has been launched in the US, the article presents some interesting facts, such that it takes approximately ten Pacific Yew trees to yield enough bark for the 2g of taxol required to treat a single cancer patient. The link between plant conservation, and sustainable utilization (as opposed to exploitation) is indeed of major importance.

Biotechnology can directly *and* indirectly enable conservation strategies, yet at the same time allow economically significant species to be both utilized and protected. This is a major issue for those global areas, rich in biodiversity and for which there is an urgent need for populations to realize the economic potential of their rich biological resources and yet at the same time preserve them for future generations. This is particularly so for the complex ecosystems of tropical rain forests; the relationships between conservation, sustainable management and tropical forest products utilization have been debated elsewhere in this volume (Marzalina and Krishnapillay, Chapter 17 and Viana *et al.*, Chapter 18).

Figure 1.3 outlines the potential involvement of biotechnology in the sustainable utilization of plants. Thus, the germplasm of economically significant species may be preserved, *ex situ* in culture collections and genebanks (see Schumacher, Chapter 9, this volume), circumventing the need to continuously take germplasm from natural environments. Furthermore, tissue cultures of utilizable species can be used as a direct source of natural products. Similarly, through clonal micropropagation, biotechnologically-derived plants or trees can be used as sources of nursery stock for plantations. However, it is essential to consider germplasm acquisition in relation to ownership and patenting rights and this is particularly important for those countries which are the centres of origin for economically important plants (see Hummer, Chapter 3, this volume).

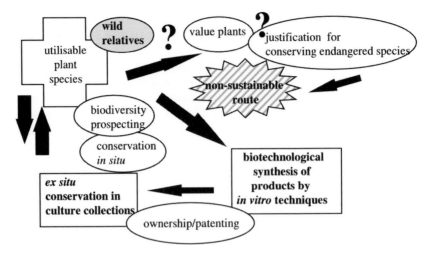

Figure 1.3   Biotechnology and the sustainable utilization of plants

When considering conservation strategies for utilizable plants, it is also important to maintain their wild relatives and indeed the whole habitats from which they were originally derived. *In situ* conservation is essential (Prance, 1997) and the protection of natural ecosystems is justifiable, for without this, natural evolutionary pressures will not be imposed and, in the long-term, this will limit biodiversity prospecting for future generations. That support for both *ex situ* and *in situ* conservation can only be justified by the economic value of plants is questionable and where this value is unknown and especially for endangered species (see Figure 1.3), other factors must be taken into consideration (see Pence, Chapter 15 and González-Benito *et al.*, Chapter 16, this volume). Moreover, for certain groups of plants it may be important to maintain and inter-link *ex situ* and *in situ* conservation strategies for the purpose of environmental monitoring (see Day, Chapter 8, this volume).

1.4   Conclusions and future prospects

Biotechnology is now integrated in all aspects of plant germplasm characterization, acquisition, conservation, exchange, and genetic resource management. Future prospects are highly encouraging in terms of the development and application of new techniques and protocols within the context of germplasm conservation. The sustainable utilization of plant diversity can be greatly assisted by the application of direct and indirect biotechnological procedures. Future prospects and needs must target certain key areas including: the development of appropriate structures for cryopreserved genebanks, the use of *in vitro* methods for the safe transfer of disease-free germplasm, and the application of genetic marker technologies for rationalizing germplasm procurement and genebanking. As *in vitro* and molecular approaches to plant conservation become more amenable it will become more and more important to validate routine operational protocols within and between genebanks and repositories. This must be considered on an international basis and the provision of networking and training infrastructures, with the aid of IT, will assist by enabling cost-effective training, and collaborative communications.

Whilst considerable progress has been made in the application of biotechnology to plant conservation there still remains the requirement to perform fundamental research. Seed recalcitrance, tissue culture recalcitrance, somaclonal variation and cryopreservation injury can be problematic for certain species. Similarly, whilst there has been considerable success in the use of molecular techniques, our current knowledge of the molecular biology of many groups of plants (e.g. temperate woody perennial tropical rain forest trees) is still limited.

Unlike many biotechnological 'applications', conservation biotechnology programmes must be considered with a long-term perspective. Cryopreserved and *in vitro* genebanks, once created, must be maintained in perpetuity. Within an international context there is thus a need for individual governments and regional and global networks to have a commitment to provide sustainable and long-term funding. To date, many advances in plant conservation biotechnology have been largely due to the efforts of specifically targeted projects which have had a short-term remit to solve a particular conservation problem or develop a certain procedure. The successes of many of these programmes are exemplified by the work of the contributors to this volume.

Visionary and sustainable funding policies, organized in concerted action by individual governments and appropriate international organizations will be essential to enable the next phase of conservation. Without such support it will not be possible to capitalize on the achievements to date and use them to implement long-term, and safe, plant diversity conservation programmes.

References

ANDERSON, M.L. and CARTINHOUR, S.W., 1997, Internet resources for the biologist, in CALLOW, J.A, FORD-LLOYD, B.V. and NEWBURY, H.J. (Eds), *Biotechnology and Plant Genetic Resources, Conservation and Use*, Biotechnology in Agriculture Series, No. 19, pp. 281–300, Oxford, UK: CAB International.

ASHMORE, S.E., 1997, *Status Report on the Development and Application of* in vitro *Techniques for the Conservation and Use of Plant Genetic Resources*, Rome: IPGRI.

AYAD, W.G., HODGKIN, T., JARADAT, A. and RAO, V.R., 1997, *Molecular Genetic Techniques for Plant Genetic Resources, Report of an IPGRI Workshop, 9–11 October, 1995, Rome, Italy*, Rome: IPGRI.

BAJAJ, Y.P.S., 1995, *Biotechnology in Agriculture and Forestry, 32, Cryopreservation of Plant Germplasm*, Berlin, Germany: Springer-Verlag.

CALLOW, J.A, FORD-LLOYD, B.V. and NEWBURY, H.J. (Eds), 1997, *Biotechnology and Plant Genetic Resources, Conservation and Use*, Biotechnology in Agriculture Series, No. 19, Oxford, UK: CAB International.

CIAT-IPGRI, 1994, *Report of a CIAT-IBPGR Collaborative Project Using Cassava (*Manihot esculenta, *Crantz) as a Model, Establishment and Operation of a Pilot* in vitro *Active Genebank*, Rome: CIAT-IPGRI.

KARP, A., KRESOVICH, S., BHAT, K.V., AYAD, W.G. and HODGKIN, T., 1997, *Molecular Tools in Plant Genetic Resources Conservation: A guide to the technologies*, IPGRI Technical Bulletin No. 2, Rome: IPGRI.

NORMAH, M.N., NARIMAH, M.K. and CLYDE, M.M. (Eds), 1996, In vitro *Conservation of Plant Genetic Resources, Proceedings of the International Workshop on* in vitro *Conservation of Plant Genetic Resources, July 4–6 1995, Kuala Lumpur, Malaysia*, Kebangsaan, Malaysia: Universiti.

Plant Talk, Plant Conservation Worldwide, 1995, Yew in the fight against cancer: sustainability or pillage?, *Plant Talk*, July, p. 7.

PRANCE, G.T., 1997, The conservation of botanical diversity, in MAXTED, N., FORD-LLOYD, B.V. and HAWKES, J.G. (Eds), *Plant Conservation: the* in situ *Approach*, pp. 3–14, London, UK: Chapman and Hall.

RAZDAN, M.K. and COCKING, E.C., 1997, *Conservation of Plant Genetic Resources* in vitro, *Volume I: General Aspects*, Enfield, New Hampshire, USA: Science Publishers, Inc.

WESTMAN, A.L. and KRESOVICH, S., 1997, Use of molecular marker techniques for description of plant genetic variation, in CALLOW, J.A, FORD-LLOYD, B.V. and NEWBURY, H.J. (Eds), *Biotechnology and Plant Genetic Resources, Conservation and Use*, Biotechnology in Agriculture Series, No. 19, pp. 49–76, Oxford, UK: CAB International.

WITHERS, L.A. and WILLIAMS, J.T., 1980, *Crop Genetic Resources the Conservation of Difficult Material, Proceedings of an International Workshop, Reading, UK*, IUBS Series B42, Paris, France.

2

Molecular Approaches to Assessing Plant Diversity

STEPHEN A. HARRIS

2.1 Introduction

Economic, social, ecological, cultural and aesthetic cases have been made for identification, quantification and understanding the distribution and relationships of biological diversity (Kunin and Lawton, 1996). Biological diversity may be assessed at three different levels: the community, the species and the gene (Frankel *et al.*, 1995). Whilst the importance of ecological and taxonomic diversity is recognized in conservation programmes, the value of genetic diversity is more controversial. The majority of researchers, either implicitly or explicitly, take the view that genetics is an essential component of any conservation programme (Falk and Holsinger, 1991; Hamrick and Godt, 1996), although others argue that organisms go extinct for ecological rather than for genetic reasons (Lande, 1988; Schemske *et al.*, 1994).

Interest in intraspecific genetic variation is primarily for three reasons: (1) the rate of evolutionary change is proportional to the available genetic diversity (Hamrick and Godt, 1996); (2) heterozygosity is positively related to fitness (Allendorf and Leary, 1986); and (3) the global gene pool represents all the information on the planet's biological processes (Wilson, 1992). That is, loss of diversity is likely to decrease the ability of organisms to respond to environmental perturbation and discard anthropocentric biological information (Wilson, 1992).

Within a plant cell there are nuclear (nDNA), chloroplast (cpDNA) and mitochondrial (mtDNA) genomes. As a result of mutation rate variation among these genomes (Wolfe *et al.*, 1987), and their different inheritance patterns (Birky *et al.*, 1989), markers associated with different genomes are suitable for different types of problem. Genome-marker associations place constraints on the questions that may be addressed by a marker system.

Genetic markers are observable traits (the expression of which indicates the presence or absence of certain genes) that are classified into five broad groups: morphological, cytological, chemical, protein and DNA (Szmidt and Wang, 1991). The characteristics of an ideal genetic marker are: detect qualitative or quantitative variation, show no environmental or developmental influences, show simple codominant inheritance, detect silent nucleotide changes, detect changes in coding and non-coding portions of the genome, and detect evolutionary homologous changes (Weising *et al.*, 1995). Such a marker allows the possibility of unambiguously assigning a genotype to a taxon and then using these data either to estimate genetic variation present within and between populations or to compare taxa directly.

Efficient utilization, improvement and conservation of taxa must be based on a sound understanding of: phylogeny; the amount and distribution of genetic variation; and the design of effective sampling and conservation methods. Crucial to the success of long-term taxon management is an understanding of genetics and demography, enabling bio-logically sound strategies to be designed (Falk and Holsinger, 1991). Such data are increasingly important in the development of integrated conservation strategies, com-bining population and taxon management with *in situ* and *ex situ* conservation (Maxted *et al.*, 1997).

Biodiversity assessment has come to mean different things; the breeder is interested in variation within a particular collection or species' geographic range, whilst the evolution-ary biologist is interested in populations and species and understanding the evolutionary bases of diversity patterns. In this paper examples of major marker system types will be described, although technologies continue to develop at a tremendous rate and none of these systems fulfils all of the criteria of an ideal molecular marker system. The applica-tion of molecular markers to issues associated with germplasm, population and system-atic investigations will be considered, followed by the prospects for molecular markers in plant diversity assessment.

2.2 Molecular marker systems

The perceived importance of genetic variation and the availability of powerful marker systems has led to the widespread application of marker technologies to biodiversity issues (Avise, 1994). Molecular marker technologies may be broadly grouped into DNA-based and protein-based techniques (Table 2.1) and numerous publications are available that describe marker techniques in detail (e.g. Karp *et al.*, 1998).

Allozymes are the most widely used and understood of the marker systems currently used for characterizing biological diversity (Butlin and Tregenza, 1998). Continued inter-est in allozyme markers, despite arguments against their use (Newbury and Ford-Lloyd, 1993), is a result of their codominant expression in most species, cost effectiveness and simplicity (Wendel and Weeden, 1990). In addition, considerable information is known about allozymes, and detailed analyses of polyploid speciation are possible (Weeden and Wendel, 1990). However, allozymes only detect low levels of polymorphism in a limited range of water-soluble, nuclear-encoded enzymes, and gene variation is underestimated due to codon redundancy and synonymous nucleotide substitutions (Nei, 1987), although additional polymorphisms are often identified via isoelectric focusing (Sharp *et al.*, 1988). Furthermore, fresh material is needed, and there are problems of environmental and ontogenetic expression with some enzyme systems (Wendel and Weeden, 1990).

Restriction fragment length polymorphisms (RFLP) analysis uses restriction enzymes (REs) to detect variation in primary DNA structure, followed by Southern blotting and a suitable detection method to reveal the variation in any of the plant genomes (Dowling *et al.*, 1996). RFLP analysis measures DNA variation that affects the relative positions of restriction sites and is usually codominant in nDNA (Dowling *et al.*, 1996). Since DNA fragments migrate logarithmically, changes in large fragments are more difficult to detect than similar size changes in small fragments. RFLP analyses require large amounts of DNA, access to radioisotopes and only limited numbers of suitable nDNA markers are available

Table 2.1 Characterization of different molecular marker types used in biodiversity assessment

	Allozymes	RFLP	RAPD	AFLP	PCR-RFLP	SSR	SSCP	Sequencing
Basis	Detection of charged AA distribution differences	Detection of relative RE site positions	Distribution of random primers through genome	PCR of RE fragment subset using modified primers	RE digest of PCR products	PCR of simple sequence repeat regions	ssDNA takes up different structures in non-denaturing gels	Direct sequencing of PCR products
Polymorphism	Charged AA substitutions	Nucleotide substitutions; indels; insertions	Nucleotide substitutions; indels; insertions	Nucleotide substitutions; indels; insertions	Nucleotide substitutions; indels; insertions	Repeat length changes	Nucleotide substitutions; indels	Nucleotide substitutions; indels; insertions
Abundance in genome	Low	High	very high	High	High	Medium	Medium	High
Level of polymorphism	Low	Medium	Medium	Medium	Medium	High	High	Medium
Dominance	Usually codominant	Codominant	Dominant	Codominant/dominant	Codominant	Codominant	Codominant	Codominant
Amount of material	Very little	2–10 µg DNA	10–25 ng DNA	1–2 µg DNA	50–100 ng DNA	50–100 ng DNA	50–100 ng DNA	10–25 ng DNA
Multiplex ratio*	?	1–2	5–10	30–100	Low	1	?	?
Sequence information	No	No	No	No	Yes	Yes	No	Yes
Radioactive detection	No	Yes/No	No	Yes/No	No	No/Yes	No/Yes	No/Yes

Table 2.1 (cont'd)

	Allozymes	RFLP	RAPD	AFLP	PCR-RFLP	SSR	SSCP	Sequencing
Development costs	Low	Medium	Low	Medium	Medium/high	High	Medium	Medium
Start-up costs	Low	Medium/high	Low	Medium	High	High	Medium	High
Applications[†]	Genetic diversity; polyploidy; hybridization; phylogeny; mating system	Genetic diversity; polyploidy; hybridization; phylogeny; mating system	'Fingerprinting'; genetic diversity; polyploidy; hybridization; phylogeny	'Fingerprinting'; genetic diversity	Genetic diversity; polyploidy; hybridization; phylogeny	Genetic diversity; mating system	Genetic diversity; mating system	Genetic diversity; polyploidy; hybridization; phylogeny
Automation	No	Limited	Yes	Yes	Limited	Yes	Limited	Yes
Reproducibility	Medium/high	High	Low	Medium	High	High	Medium	High
Manual	Soltis and Soltis (1990)	Dowling et al. (1996)	Williams et al. (1993)	Matthes et al. (1997)	Dowling et al. (1996)	Morgante et al. (1998)	Jordan et al. (1998)	Hillis et al. (1996)

AA, amino acid; AFLP, amplified fragment length polymorphism; PCR, polymerase chain reaction; PCR-RFLP, polymerase chain reaction-restriction fragment length polymorphism; RAPD, randomly amplified polymorphic DNA; RE, restriction endonuclease; RFLP, restriction fragment length polymorphism; SSCP, single-stranded conformation polymorphism; ssDNA, single-stranded DNA; SSR, simple sequence repeats.

* From Rafalski et al. (1997); "?" indicates that multiplex ratio not calculated.

[†] Application types that markers have been used in; does not necessarily indicate that markers are appropriate for application.

(Dowling *et al.*, 1996). However, as with all other DNA-based methods, dried leaves may be used as a source of DNA.

Randomly amplified polymorphic DNA (RAPD) analysis utilizes single, arbitrary decamer DNA oligonucleotide primers to amplify regions of the genome using the polymerase chain reaction (PCR; Williams *et al.*, 1993). Priming sites are thought to be randomly distributed throughout the plant's genomes and polymorphism results in differing amplification products (Williams *et al.*, 1993). The technique is cheap, simple, requires no sequence information and a large number of putative loci may be screened (Newbury and Ford-Lloyd, 1993). However, the technique has been criticized on technical (Jones *et al.*, 1997) and theoretical (Harris, in press) grounds, and in these respects is similar to the largely abandoned technique of total protein analyses (Crawford, 1990b). Some of these criticisms may be overcome by the development of sequence characterized amplified regions (SCARs; Paran and Michelmore, 1993), but at the cost of reducing the number of products scored, or using very large sample sizes (Furman *et al.*, 1997).

Polymerase chain reaction restriction fragment length polymorphisms (PCR-RFLPs) are similar to RFLPs, except that differences are visualized within specific PCR products (Konieczny and Ausubel, 1993; Rafalski *et al.*, 1997). The technique is cheap and simple once suitable products have been identified, although information content of individual products may be low since short products (<2kb) give the best amplification results and many REs need to be screened to identify suitable polymorphism (Rafalski *et al.*, 1997). Suitable primers may be designed from sequence databases, analysis of low copy number random clones and universal cpDNA, mtDNA and nDNA sequences (Demesure *et al.*, 1995; Dumolin-Lapegue *et al.*, 1997; Rafalski *et al.*, 1997; Strand *et al.*, 1997). Combining sequence and PCR-RFLP analyses is effective for initial polymorphism identification and subsequent screening (Ferris *et al.*, 1995), whilst combined with RAPDs, PCR-RFLPs may prove effective for identification of additional variation (Paran and Michelmore, 1993) or confirmation of RAPD band identity (Rieseberg, 1996).

Amplified fragment length polymorphism (AFLP) analysis involves the selective amplification of an arbitrary subset of restriction fragments generated by single or double RE digestion of DNA (Vos *et al.*, 1995). Prior to amplification fragment ends are modified by addition of double-stranded adapters, and during amplification pairs of end-labelled primers are used that span the adapter, the restriction site and one to three nucleotides of the fragment. Thus only fragments with ends similar to the primer's arbitrary sequence will be amplified. The number of bands generated in an AFLP reaction is determined by the number of bases in the variable part of the amplification primer (Vos *et al.*, 1995). AFLPs are expected to be highly polymorphic, either dominant or codominant (although allelic relations may not be immediately obvious) and requires no prior sequence knowledge (Rafalski *et al.*, 1997). However, the technique requires a high degree of technical skill, large amounts of high quality DNA and methylation insensitive REs.

Microsatellites (simple sequence repeats; SSRs) are short tandem repeats of mono- to tetra-nucleotide repeats which are assumed to be randomly distributed throughout the nDNA, cpDNA and mtDNA and are detected using specifically designed PCR primers (Jarne and Lagode, 1996). SSRs are simple to detect, as either silver-stained or radiolabelled products, once suitable primers have been designed, are easily scored, highly polymorphic and codominant, in the case of nDNA SSRs (Jarne and Lagode, 1996). Unfortunately, it is expensive to identify nDNA SSR primers and these do not generally amplify between species, although organelle SSR primers are easily identified from published sequences and appear to amplify many different taxa (Powell *et al.*, 1996a). Technical modifications, through the use of anchored microsatellites (Charters *et al.*, 1996), and evidence that some nDNA microsatellites amplify across species (Steinkellner *et al.*, 1997; White and Powell, 1997) eliminate some of these disadvantages.

Single-stranded conformation polymorphism (SSCP) analysis is based on the principle that single-stranded DNA molecules have specific sequence-based secondary structures under non-denaturing conditions; molecules with one or a few base differences may form conformations that result in different gel mobilities (Jordan *et al.*, 1998). The method is quick and simple and has great potential for the identification of DNA polymorphism and codominant nDNA fragments in diversity studies (Bodénès *et al.*, 1996). However, it is necessary to test segregation ratios to validate genetic hypotheses and the methodology is sensitive to both sequence composition and the sequence itself (Jordan *et al.*, 1998).

DNA sequence analysis (SA) provides information of nucleotide variation directly, rather than indirectly as other molecular methods do, and with the availability of automated sequencing and high-powered computer facilities SA is likely to become increasingly important and has become the method of choice for phylogenetic studies (Hillis *et al.*, 1996). The method provides very high quality information that may quickly and easily be compared between studies, whilst universal sequence primers make it possible to sequence most taxa with no knowledge of DNA sequence (Baldwin, 1992; Demesure *et al.*, 1995). The method is, however, labour intensive, expensive for general diversity surveys and loci are screened one at a time.

Two criteria have been proposed for technique comparison: information content and multiplex ratio (Rafalski *et al.*, 1997); the greater a marker's information content the easier it is to detect polymorphism, whilst the multiplex ratio indicates the number of loci scored in each experiment. These criteria have been summarized into a single parameter, marker index, which is highly correlated in AFLPs, RFLPs and SSRs (Powell *et al.*, 1996b). However, the assessment of information content, as defined by Rafalski *et al.* (1997), for anonymous markers (e.g. RAPDs, AFLPs) is problematic since locus, and hence allele, identities are generally unknown. All molecular markers have limitations, therefore it is important that technology appropriate to the problem being studied is applied.

2.3   Molecular markers in germplasm characterization

Molecular marker development and testing has been widely associated with germplasm characterization (e.g. seed collections, botanic gardens and field gene banks) and often uses different molecular marker types, sources and collections of material, and scoring and data analysis (e.g. Asemota *et al.*, 1996; Howell *et al.*, 1994). Such studies usually illustrate the ability of a particular marker to detect variation and/or quantify genetic diversity (e.g. Wilde *et al.*, 1992; Yu and Nguyen, 1994).

In allozyme and RFLP analyses diversity is commonly assessed by a range of measures (Nei, 1987), whilst for SSRs diversities are corrected for high mutation rate (Slatkin, 1995). In RAPD and AFLP analyses diversity may be calculated under the assumption of Hardy–Weinberg equilibria, that banding positions correspond to loci or using a phenotypic diversity measure (Clark and Lanigan, 1993; Huff *et al.*, 1993; Lynch and Milligan, 1994). The diversity of data analysis methods means that valid comparisons between studies are difficult and become dependent on intra-taxon and genomic sampling and the method of marker generation and scoring (Müller-Starck, 1991). For example, Shannon's measure is widely applied to RAPD data (Gillies *et al.*, 1997) and is phenotypic, assumes that products at the same position are evolutionarily homologous, and assesses the degree of uncertainty regarding an individual's RAPD products selected at random from a population (Magguran,

1988). Furthermore, the upper limit of this measure is determined by the number of products scored and it is affected by the frequency of the most common products (Pielou, 1966). Thus comparisons based on this measure must include the same number of products and the data must display a similar evenness.

However, studies of germplasm collections have proven useful for the identification of accessions that contain disproportionately large amounts of genetic diversity and in the characterization and 'fingerprinting' of taxa; RAPDs and AFLPs have been particularly useful in the latter case (e.g. Ellis *et al.*, 1997; Hartl and Seefelder, 1998; Wachira *et al.*, 1995). Whilst the intellectual problems of RAPDs and AFLPs are the same, the improved reproducibility and greater multiplex ratio of AFLPs makes this an important method for 'fingerprinting' of cultivars and the protection of Breeders Rights (Jones *et al.*, 1997). High SSR mutation rates (Slatkin, 1995), combined with the relatively low number of loci scored (Jarne and Lagode, 1996), means that these markers may have limited value in 'fingerprinting' studies. However, the large number of alleles per SSR locus means they are ideal for gene flow and potential introgression identification (Jarne and Lagode, 1996); such events may have profound effects on germplasm collections (Ennos and Qian, 1994). Apparently aberrant data sets are often described in terms of hybridization and introgression, although confirmation of this through additional studies is rare (Arnold, 1997).

The greatest number of germplasm-based studies have been conducted in commercially important species. However, such studies are dependent on the quality of germplasm collections in terms of taxonomic and geographic representation and collection sampling and documentation. Germplasm studies have been effective in identifying variation and increasing our understanding of intra-collection diversity patterns. However, the challenge for the future is to find a means of effectively using this variation for breeding or improvement purposes, for the identification and construction of acquisition policies and the identification of core collections and their management (Frankel *et al.*, 1995).

2.4 Molecular markers in systematics and population genetics

Issues of taxon relationships and population structure are of apparently lesser interest to breeders, although molecular markers allow detailed studies of patterns and processes of population and species differentiation in non-model organisms (Avise, 1994).

Population genetic and systematic studies often rely on germplasm collections, in association with good field collections. Investigations, that involve different types of sampling strategies from detailed sampling of localized endemics and widespread taxa, to political sampling of widespread taxa, have revealed levels of allozyme diversity correlated with ecological factors (Hamrick *et al.*, 1992), although such comparisons are difficult and have not been undertaken with DNA-based markers. Millar and Libby (1991) recognized five patterns of genetic variation in temperate conifers. Furthermore, there is a growing interest in phylogeography and Milligan *et al.*'s (1994) arguments regarding the importance of genealogies in conservation, hence the need to use marker systems that allow detailed genealogy construction, e.g. RFLP, PCR-RFLP and sequence analyses of nDNA and cpDNA.

It is in the study of phylogeny, whether at the higher, generic or species levels that molecular markers have had a significant systematic impact (Olmstead and Palmer, 1994). With the implicit assumption that alleles are evolutionary units, and individuals the means of distributing alleles, allozymes, RAPDs, RFLPs and SSRs have been used to delimit taxa (e.g. Crawford, 1990b; van Buren *et al.*, 1994; Zamora *et al.*, 1996). However, such studies represent marker system, rather than species, phylogenies since lineage sorting, polymorphism and hybridization may affect intraspecific marker distribution patterns (Avise, 1994; Doyle, 1992). Furthermore, some marker systems are better suited as phylogenetic markers than others. For example, RAPDs have been criticized as phylogenetic markers (Harris, in press) and their high mutation rate renders SSRs unlikely phylogenetic markers (Jarne and Lagode, 1996). Phylogenetic markers are best found in either restriction site polymorphism, the occurrence of rare structural mutations or in sequence analysis (Baldwin *et al.*, 1995; Olmstead and Palmer, 1994). The low level of allozyme polymorphism renders them poor at resolving phylogenetic signals, although when such polymorphism does occur the information content may be high (Crawford, 1990a).

Molecular markers, often in combination with morphological analyses, have enabled detailed analyses of hybridization and introgression to be undertaken (Arnold, 1997; Hughes and Harris, 1994), whilst combined cpDNA and nDNA analyses have led to detailed understanding of polyploid speciation events (Palmer *et al.*, 1983). Such studies have been instrumental in highlighting the importance of autopolyploidy (e.g. Ness *et al.*, 1989) and the role of hybridization in taxon conservation (Rieseberg, 1991).

Increasingly, molecular studies are being utilized in integrated studies concerned with population dynamics, the role of mating system and the consequences of habitat fragmentation in taxon and landscape conservation (Laurance and Bierregaard, 1997). Allozyme or RFLP markers have been widely used for studies of population structure, with allozymes currently being the most effective marker system for the study of mating system (Brown *et al.*, 1990). However, SSR markers promise to make detailed analysis of mating system and paternity a simpler process in higher plants, as a result of the large number of alleles per locus (Jarne and Lagode, 1996). Unfortunately, detailed SSR studies of widely dispersed, largely inbred populations are probably less useful due to the high mutation rate and the accumulation of too much variation for the effective analysis of interpopulation structure (Jarne and Lagode, 1996).

In many conservation studies it is either endemic taxa or taxa with restricted distributions that have attracted attention (e.g. Travis *et al.*, 1996), whilst there are fewer studies of widespread species across their geographical ranges (e.g. Harris *et al.*, 1997); such studies rarely lead to detailed management recommendations. In Scottish *Pinus sylvestris* the importance of genetic data has been highlighted (Ennos *et al.*, 1998), whilst in North America there is concern over the US Endangered Species Act 'Hybrid Policy' that limits the importance of hybrids in taxon conservation (O'Brien and Mayr, 1991). Some authors (e.g. Ennos *et al.*, 1998) have argued that genetic data must be incorporated into policy and that specific gene pool conservation is as important as taxon conservation; biodiversity maintenance must be undertaken at a wide range of levels, from ecosystems to populations.

2.5 Prospects for molecular markers in biodiversity characterization

The brief history of molecular marker use in biodiversity assessment has been impressive, but such markers must meet new challenges, and be combined with detailed fieldwork and collaboration with other biological disciplines; molecular marker data must be placed within the context of the organism's biology. Molecular markers are of two broad types, high information content for a limited number of putative loci (e.g. SSRs, SA) and low information content for a large number of putative loci (e.g. RAPDs, AFLPs). Generalized marker surveys for the purposes of identifying diversity *per se* are unlikely to prove of great value for biodiversity assessment unless additional research is undertaken to understand the basis of the variation, thus allowing effective comparison of studies within and between taxa. One must, therefore, be in a position to use technology that is best for a particular problem, given the constraints placed on the study. Avise (1994) put this eloquently when he considered that if allozyme methods had been discovered after DNA methodologies then one might have a large number of good biological reasons to use allozymes as the markers of choice.

Two areas need further development: methods of data analysis and the understanding of molecular diversity in relation to quantitative variation. Methods for the analysis of molecular data have not kept up with the sophistication of the methods of data generation. Thus it is common to find sophisticated molecular data (e.g. AFLPs) being analysed using similarity measures derived decades ago (e.g. Jaccard's coefficient). Such problems may reduce the amount of useful information derived from molecular studies of diversity, its partitioning and evolutionary origin, although recently analysis methods have started to be developed for dominant molecular markers (e.g. Clark and Lanigan, 1993; Excoffier *et al.*, 1992; Lynch and Milligan, 1994).

Conservation and utilization of taxa has been seen as the maintenance of evolutionary potential, and genetic diversity is one component of this (Hamrick and Godt, 1996). Diversity studies are made using molecular markers that are assumed to be neutral, even though most useful plant characteristics are quantitative (Butlin and Tregenza, 1998; Lynch, 1996). The correlation between molecular variation and quantitative variation has rarely been studied in detail, but is an issue that must be addressed if studies of genetic diversity are to be used more effectively in biodiversity assessment and conservation (Butlin and Tregenza, 1998; Lynch, 1996). Molecular markers have increased our understanding of spatial and temporal patterns of genetic variation, and of the evolutionary mechanisms that generate and maintain variation. However, the direct benefit of these data to either practical biodiversity conservation or germplasm collection management is equivocal (see Chapters 3, 13 and 16, this volume). Whilst much has been made of molecular markers in the study of germplasm resources it is unlikely that they will influence collecting policies because of the absence of a link between the phenotypic expression of quantitative characters, upon which selection may act, and the assumed neutrality of molecular markers.

Unprecedented molecular marker application, and increased threats to the world's biodiversity have occurred over the past two decades. Can molecular markers provide an effective means of aiding taxon conservation and setting conservation priorities? Is it legitimate to base conservation priorities on markers that are assumed to be neutral? These are issues that must be addressed if molecular markers are to be continued to be studied in economically unimportant taxa. Furthermore, the costs and benefits of such investigations must be considered, since markers currently being developed are unlikely to be widely used, except in commercial taxa.

References

ALLENDORF, F.W. and LEARY, R.F., 1986, Heterozygosity and fitness in natural populations of animals., in SOULE, M.E. (Ed.), *Conservation Biology*, pp. 57–76, Sunderland, Massachusetts: Sinauer Associates.

ARNOLD, M.L., 1997, *Natural Hybridisation and Evolution*, Oxford: Oxford University Press.

ASEMOTA, H.N., RAMSER, J., LOPEZ-PERALTA, C., WEISING, K. and KAHL, G., 1996, Genetic variation and cultivar identification of Jamaican yam germplasm by random amplified polymorphic DNA analysis, *Euphytica*, **92**, 341–351.

AVISE, J.C., 1994, *Molecular Markers, Natural History and Evolution*, London: Chapman & Hall.

BALDWIN, B.G., 1992, Phylogenetic utility of the internal transcribed spacers of nuclear ribosomal DNA in plants: an example from the Compositae, *Molecular Phylogenetics and Evolution*, **1**, 3–16.

BALDWIN, B.G., SANDERSON, M.J., PORTER, J.M., WOJCIECHOWSKI, M.F., CAMPBELL, C.S. and DONOGHUE, M.J., 1995, The ITS region of nuclear ribosomal DNA: a valuable source of evidence on Angiosperm phylogeny, *Annals of the Missouri Botanic Garden*, **82**, 247–277.

BIRKY, C.W., FUEST, P. and MARUYAMA, T., 1989, Organelle gene diversity under migration, mutation and drift: equilibrium expectations, approaches to equilibrium, effects of heteroplasmic cells and comparison of nuclear genes, *Genetics*, **121**, 613–627.

BODÉNÈS, C., LAIGRET, F. and KREMER, A., 1996, Inheritance and molecular variations of PCR-SSCP fragments in pedunculate oak (*Quercus robur* L.), *Theoretical and Applied Genetics*, **93**, 348–354.

BROWN, A.H.D., BURDON, J.J. and JAROSZ A.M., 1990, Isoenzyme analysis of plant mating systems, in SOLTIS, D.E. and SOLTIS, P.S. (Eds), *Isozymes in Plant Biotechnology*, pp. 73–86, Chapman and Hall, London.

BUTLIN, R.K. and TREGENZA, T., 1998, Levels of genetic polymorphism: marker loci versus quantitative traits, *Philosophical Transactions of the Royal Society of London, Series B*, **353**, 187–198.

CHARTERS, Y.M., ROBERTSON, A., WILKINSON, M.J. and RAMSAY, G., 1996, PCR analysis of oilseed rape cultivars (*Brassica napus* L. ssp. *oleifera*) using 5′-anchored simple sequence repeat (SSR) primers, *Theoretical and Applied Genetics*, **92**, 442–447.

CLARK, A.G. and LANIGAN, C.M.S., 1993, Prospects for estimating nucleotide divergence with RAPDs, *Molecular Biology and Evolution*, **10**, 1096–1111.

CRAWFORD, D.J., 1990a, Enzyme electrophoresis and plant systematics, in SOLTIS, D.E. and SOLTIS, P.S. (Eds), *Isozymes in Plant Biology*, pp. 146–164, London: Chapman & Hall.

CRAWFORD, D.J., 1990b, *Plant Molecular Systematics: Macromolecular Approaches*, London: John Wiley & Sons.

DEMESURE, B., SODZI, N. and PETIT, R.J., 1995, A set of universal primers for amplification of polymorphic non-coding regions of mitochondrial and chloroplast DNA in plants, *Molecular Ecology*, **4**, 129–131.

DOWLING, T.E., MORITZ, C., PALMER, J.D. and RIESEBERG, L.H., 1996, Nucleic acids III: Analysis of fragments and restriction sites, in HILLIS, D.M., MORITZ, C. and MABLE, B.K. (Eds), *Molecular Systematics*, pp. 249–320, Sunderland, Massachusetts: Sinauer Associates.

DOYLE, J.J., 1992, Gene trees and species trees: Molecular systematics as one-character taxonomy, *Systematic Botany*, **17**, 144–163.

DUMOLIN-LAPEGUE, S., PEMINGE, M.-H. and PETIT, R.J., 1997, An enlarged set of consensus primers for the study of organelle DNA in plants, *Molecular Ecology*, **6**, 393–397.

ELLIS, R.P., MCNICOL, J.W., BAIRD, E., BOOTH, A., LAWRENCE, P., THOMAS, B. and POWELL, W., 1997, The use of AFLPs to examine genetic relatedness in barley, *Molecular Breeding*, **3**, 359–369.

ENNOS, R.A. and QIAN, T., 1994, Monitoring the output of a hybrid larch seed orchard using isozyme markers, *Forestry*, **67**, 63–74.

ENNOS, R.A., WORRELL, R. and MALCOLM, D.C., 1998, The genetic management of native species in Scotland, *Forestry*, **71**, 1–23.

EXCOFFIER, L., SMOUSE, P.E. and QUATTRO, J.M., 1992, Molecular variance inferred from metric distances among DNA haplocytes: Applications to human mitochondrial restriction data. *Genetics*, **131**, 479–491.

FALK, D.A. and HOLSINGER, K.E., 1991, *Genetics and Conservation of Rare Plants*, Oxford: Oxford University Press.

FERRIS, C., OLIVER, R.P., DAVY, A.J. and HEWITT, G.M., 1993, Native oak chloroplast reveal an ancient divide across Europe, *Molecular Ecology*, **2**, 337–344.

FERRIS, C., OLIVER, R.P., DAVY, A.J. and HEWITT, G.M., 1995, Using chloroplast DNA to trace postglacial migration routes of oaks into Britain, *Molecular Ecology*, **4**, 731–738.

FRANKEL, O.H., BROWN, A.H. and BURDON, J.J., 1995, *The Conservation of Plant Biodiversity*, Cambridge: Cambridge University Press.

FURMAN, B.J., GRATTAPAGLIA, D., DVORAK, W.S. and O'MALLEY, D.M., 1997, Analysis of genetic relationships of Central American and Mexican pines using RAPD markers that distinguish species, *Molecular Ecology*, **6**, 321–331.

GILLIES, A.C.M., CORNELIUS, J.P., NEWTON, A.C., NAVARRO, C., HERNÁNDEZ, M. and WILSON, J., 1997, Genetic variation in Costa Rican populations of the tropical timber species *Cedrela odorata* L., assessed using RAPDs, *Molecular Ecology*, **6**, 1133–1145.

GOLENBERG, E.M., 1994, DNA from plant compression fossils, in HERRMANN, B. and HUMMEL, S. (Eds), *Ancient DNA*, pp. 237–256, Berlin: Springer.

HAMRICK, J.L. and GODT, M.J.W., 1996, Conservation genetics of endemic plant species, in AVISE, J.C. and HAMRICK, J.L. (Eds), *Conservation Genetics. Case Histories from Nature*, pp. 281–304, London: Chapman & Hall.

HAMRICK, J.L., GODT, M.J.W. and SHERMAN-BROYLES, S.L., 1992, Factors influencing levels of genetic diversity in woody plant species, *New Forests*, **6**, 95–124.

HARRIS, S.A., in press, RAPDs in systematics – A useful methodology?, in HOLLINGSWORTH, P., BATEMAN, R. and GORNALL, R.J. (Eds), *Advances in Molecular Systematics*, London: Academic Press.

HARRIS, S.A., FAGG, C.W. and BARNES, R.D., 1997, Isozyme variation in *Faidherbia albida* (Del.) A. Chev. (Leguminosae; Mimosoideae), *Plant Systematics and Evolution*, **207**, 119–132.

HARTL, L. and SEEFELDER, S., 1998, Diversity of selected hop cultivars detected by fluorescent AFLPs, *Theoretical and Applied Genetics*, **96**, 112–116.

HILLIS, D.M., MABLE, B.K., LARSON, A., DAVIS, S.K. and ZIMMER, E.A., 1996, Nucleic acids IV: sequencing and cloning, in HILLIS, D.M., MORITZ, C. and MABLE, B.K. (Eds), *Molecular Systematics*, pp. 321–381, Sunderland, Massachusetts: Sinauer Associates.

HOWELL, E.C., NEWBURY, H.J., SWENNEN, R.L., WITHERS, L.A. and FORD-LLOYD, B.V., 1994, The use of RAPD for identifying and classifying *Musa* germplasm, *Genome*, **37**, 328–332.

HUFF, D.R., PEAKALL, R. and SMOUSE, P.E., 1993, RAPD variation within and among natural populations of outcrossing buffalo grass [*Buchloë dactyloides* (Nutt.) Engelm.], *Theoretical and Applied Genetics*, **86**, 927–934.

HUGHES, C.E. and HARRIS, S.A., 1994, The characterisation and identification of a naturally occurring hybrid in the genus *Leucaena* (Leguminosae: Mimosoideae), *Plant Systematics and Evolution*, **192**, 177–197.

JARNE, P. and LAGODE, P.J.L., 1996, Microsatellites, from molecules to populations and back, *Trends in Ecology and Evolution*, **11**, 424–429.

JONES, C.J., EDWARDS, K.J., CASTAGLIONE, S., WINFIELD, W.O., SALA, F., VAN DE WIEL, C., *et al.*, 1997, Reproducibility testing of RAPD, AFLP and SSR markers in plants by a network of European laboratories, *Molecular Breeding*, **3**, 381–390.

JORDAN, W.C., FOLEY, K. and BRUFORD, M.W., 1998, Single-stranded conformation polymorphism (SSCP) analysis, in KARP, A., ISAAC, P.G. and INGRAM, D.S. (Eds), *Molecular Tools for Screening Biodiversity*, pp. 152–156, London: Chapman & Hall.

KARP, A., ISAAC, P.G. and INGRAM, D.S., 1998, *Molecular Tools for Screening Biodiversity*, London: Chapman & Hall.

KONIECZNY, A. and AUSUBEL, F.M., 1993, A procedure for mapping *Arabidopsis* mutations using co-dominant ecotype-specific PCR-based markers, *Plant Journal*, **4**, 403–410.

KUNIN, W.E. and LAWTON, J.H., 1996, Does biodiversity matter? Evaluating the case for conserving species, in GASTON, K.J. (Ed.), *Biodiversity. A Biology of Numbers and Difference*, pp. 283–308, London: Blackwell Science.

LANDE, R., 1988, Genetics and demography in biological conservation, *Science*, **241**, 1455–1460.

LAURANCE, W.F. and BIERREGAARD, R.O., 1997, *Tropical Forest Remnants. Ecology, Management, and Conservation of Fragmented Communities*, London: The University of Chicago Press.

LYNCH, M., 1996, A quantitative-genetic perspective on conservation issues, in AVISE, J.C. and HAMRICK, J.L. (Eds), *Conservation Genetics. Case Studies from Nature*, pp. 471–501, New York: Chapman & Hall.

LYNCH, M. and MILLIGAN, B.G., 1994, Analysis of population genetic structure with RAPD markers, *Molecular Ecology*, **3**, 91–99.

MAGGURAN, A., 1988, *Ecological Diversity and its Measurement*, London: Croom Helm.

MATTHES, M.C., DALY, A. and EDWARDS, K.J., 1997, Amplified fragment length polymorphism (AFLP), in KARP, A., ISAAC, P.G. and INGRAM, D.S. (Eds), *Molecular Tools for Screening Biodiversity*, pp. 183–190, London: Chapman & Hall.

MAXTED, N., FORD-LLOYD, B.V. and HAWKES, J.G., 1997, *Plant Genetic Conservation. The in situ Approach*, London: Chapman & Hall.

MILLAR, C.I. and LIBBY, W.J., 1991, Strategies for conserving clinal, ecotypic, and disjunct population diversity in widespread species, in FALK, D.A. and HOLSINGER, K.E. (Eds), *Genetics and Conservation of Rare Plants*, pp. 149–170, Oxford: Oxford University Press.

MILLIGAN, B.G., LEEBENS-MACK, J. and STRAND, A.E., 1994, Conservation genetics: beyond the maintenance of molecular marker diversity, *Molecular Ecology*, 3, 423–435.

MORGANTE, M., PFEIFFER, A., JURMAN, I., PAGLIA, G. and OLIVIERI, A.M., 1998, Isolation of microsatellite markers in plants, in KARP, A., ISAAC, P.G. and INGRAM, D.S. (Eds), *Molecular Tools for Screening Biodiversity*, pp. 206–207, London: Chapman & Hall.

MÜLLER-STARCK, G., 1991, Survey of genetic variation as inferred from enzyme gene markers, in MÜLLER-STARCK, G. and ZIEHE, M. (Eds), *Genetic Variation in European Populations of Forest Trees*, pp. 20–37, Frankfurt-am-Main: J.D. Sauerländer's Verlag.

NEI, M., 1987, *Molecular Evolutionary Genetics*, New York: Columbia University Press.

NESS, B.D., SOLTIS, D.E. and SOLTIS, P.S., 1989, Autopolyploidy in *Heuchera micrantha* (Saxifragaceae), *American Journal of Botany*, **76**, 614–626.

NEWBURY, H.J. and FORD-LLOYD, B.V., 1993, The use of RAPD for assessing variation in plants, *Plant Growth Regulation*, **12**, 43–51.

O'BRIEN, S.J. and MAYR, E., 1991, Bureaucratic mischief: recognising endangered species and subspecies, *Science*, **251**, 1187–1188.

OLMSTEAD, R.G. and PALMER, J.D., 1994, Chloroplast DNA systematics: a review of methods and data analysis, *American Journal of Botany*, **81**, 1205–1224.

PALMER, J.D., SHIELDS, C.R., COHEN, D.B. and ORTON, T.J., 1983, Chloroplast DNA evolution and the origin of amphidiploid *Brassica* species, *Theoretical and Applied Genetics*, **65**, 181–189.

PARAN, I. and MICHELMORE, R.W., 1993, Development of reliable PCR based markers linked to downy mildew resistance genes in lettuce, *Theoretical and Applied Genetics*, **85**, 989–993.

PEREZ DE LA ROSA, J., HARRIS, S.A. and FARJON, A., 1995, Noncoding chloroplast DNA variation in Mexican pines, *Theoretical and Applied Genetics*, **91**, 1101–1106.

PIELOU, E.C., 1966, Shannon's formula as a measure of species diversity: its use and misuse, *American Naturalist*, **100**, 463–465.

POWELL, W., MACHRAY, G.C. and PROVAN, J., 1996a, Polymorphism revealed by simple sequence repeats, *Trends in Plant Sciences*, **1**, 215–222.

POWELL, W., MORGANTE, M., ANDRE, C., HANAFEY, M., VOGEL, J., TINGEY, S. and RAFALSKI, A., 1996b, The comparison of RFLP, RAPD, AFLP and SSR (microsatellite) markers for germplasm analysis, *Molecular Breeding*, **2**, 225–238.

RAFALSKI, J.A., VOGEL, J.M., MORGANTE, M., POWELL, W., ANDRE, C. and TINGEY, S.V., 1997, Generating and using DNA markers in plants, in BIRREN, B. and LAI, E. (Eds), *Non-mammalian Genomic Analysis: A Practical Guide*, pp. 75–134, New York: Academic Press.

RIESEBERG, L.H., 1991, Hybridisation in rare plants: insights from case studies in Cercocarpus and Helianthus, in FALK, D.A. and HOLSINGER, K.E. (Eds), *Genetics and Conservation of Rare Plants*, pp. 171–181, Oxford: Oxford University Press.

RIESEBERG, L.H., 1996, Homology among RAPD fragments in interspecific comparisons, *Molecular Ecology*, **5**, 99–105.

SCHEMSKE, D.W., HUSBAND, B.C., RUCKELHAUS, M.H., GOODWILLIE, C., PARKER, I.M. and BISHOP, J.G., 1994, Evaluating approaches to the conservation of rare and endangered plants, *Ecology*, **75**, 574–606.

SHARP, P.J., DESAI, S. and GALE, M.D., 1988, Isozyme variation and RFLPs at the β-amylase loci in wheat, *Theoretical and Applied Genetics*, **76**, 691–699.

SLATKIN, M., 1995, A measure of population subdivision based on microsatellite allele frequencies, *Genetics*, **139**, 457–462.

SOLTIS, D.E. and SOLTIS, P.S., 1990, *Isozymes in Plant Biology*, London: Chapman and Hall.

STEINKELLNER, H., LEXER, C., TURETSCHEK, E. and GLÖSSL, J., 1997, Conservation of (GA)n microsatellite loci between *Quercus* species, *Molecular Ecology*, **6**, 1189–1194.

STRAND, A.E., LEEBENS-MACK, J. and MILLIGAN, B.G., 1997, Nuclear DNA-based markers for plant evolutionary biology, *Molecular Ecology*, **6**, 113–118.

SZMIDT, A.E. and WANG, X.-R., 1991, DNA markers in forest genetics., in MÜLLER-STARK, G. and ZIEKE, G. (Eds), *Genetic Variation in European Populations of Forest Trees*, pp. 79–94, Frankfurt-am-Main: J.D. Sauerländer's Verlag.

TRAVIS, S.E., MASCHINSKI, J. and KEIM, P., 1996, An analysis of genetic variation in *Astragalus cremnophylax* var. *cremnophylax*, a critically endangered plant, using AFLPs, *Molecular Ecology*, **5**, 735–745.

VAN BUREN, R., HARPER, K.T., ANDERSEN, W.R., STANTON, D.J., SEYOUM, S. and ENGLAND, J.L., 1994, Evaluating the relationship of autumn buttercup (*Ranunculus acriformis* var. *aestivalis*) to some close congeners using random amplified polymorphic DNA, *American Journal of Botany*, **81**, 514–519.

VOS, P., HOGERS, R., BLEEKER, M., REIJANS, M., VAN DE LEE, T., HORNES, M., *et al.*, 1995, AFLP: A new technique for DNA fingerprinting, *Nucleic Acids Research*, **23**, 4407–4414.

WACHIRA, F.N., WAUGH, R., HACKETT, C.A. and POWELL, W., 1995, Detection of genetic diversity in tea (*Camellia sinensis*) using RAPD markers, *Genome*, **38**, 201–210.

WEEDEN, N.F. and WENDEL, J.F., 1990, Genetics of plant isozymes, in SOLTIS, D.E. and SOLTIS, P.S. (Eds), *Isozymes in Plant Biology*, pp. 46–72, London: Chapman and Hall.

WEISING, K., NYBOM, H., WOLFF, K. and MEYER, W., 1995, *DNA Fingerprinting in Plants and Fungi*, London: CRC Press.

WENDEL, J.F. and WEEDEN, N.F., 1990, Visualisation and interpretation of plant isozymes, in SOLTIS, D.E. and SOLTIS, P.S. (Eds), *Isozymes in Plant Biology*, pp. 5–45, London: Chapman and Hall.

WHITE, G. and POWELL, W., 1997, Cross-species amplification of SSR loci in the Meliaceae family, *Molecular Ecology*, **6**, 1195–1197.

WILDE, J., WAUGH, R. and POWELL, W., 1992, Genetic fingerprinting of *Theobroma* clones using randomly amplified polymorphic DNA markers, *Theoretical and Applied Genetics*, **83**, 871–877.

WILLIAMS, J.G.K., HANAFEY, M.K., RAFALSKI, J.A. and TINGEY, S.V., 1993, Genetic analysis using random amplified polymorphic DNA markers, *Methods in Enzymology*, **218**, 704–740.

WILSON, E.O., 1992, *The Diversity of Life*, London: Penguin.

WOLFE, K.H., LI, W.-H. and SHARP, P.M., 1987, Rates of nucleotide substitution vary greatly among plant mitochondrial, chloroplast, and nuclear DNAs, *Proceedings of the National Academy of Sciences, USA*, **84**, 9054–9058.

YU, L.-X. and NGUYEN, H.T., 1994, Genetic variation detected with RAPD markers among upland and lowland rice cultivars (*Oryza sativa* L.), *Theoretical and Applied Genetics*, **87**, 667–672.

ZAMORA, R., JAMILENA, M., RUIZ REJÓN, M. and BLANCA, G., 1996, Two new species of the carnivorous genus *Pinguicula*, (Lentibulariaceae) from Mediterranean habitats, *Plant Systematics and Evolution*, **200**, 41–60.

3

Biotechnology in Plant Germplasm Acquisition

KIM E. HUMMER

3.1 Introduction

This chapter describes the procedures involved in the acquisition of seed and clonal genetic resources. A discussion concerning plant acquisition by exploration and exchange is given. Seeds are the most widely collected and conserved form of plant germplasm. Plants are collected or exchanged through vegetative techniques to maintain exact genotypes. New technologies, such as *in vitro* culture, have been applied to plant exploration. Intellectual property rights and quarantine issues have added complexity to plant acquisition, but may provide additional resources for germplasm conservation efforts. Procedures for documenting background information now include ethnobotanical considerations for plant uses. New molecular techniques confirm the botanical and horticultural identity of accessions and help determine gaps and set priorities for new plant acquisitions.

3.2 Plant genetic resource conservation

While the preservation of species diversity may be looked upon as humanitarian, it is, in fact, a global imperative. With the earth's population increasing from 1 billion in 1850, to more than 5 billion in 1989, and projected to be 10 billion by 2030, agricultural production will be key to human survival. Many countries now realize that their strength, if not continued existence, will be based in a very practical way to agricultural production. Only about 5000 plant species have fed the human population since the beginning of agriculture (Wilkes, 1989). This represents less than 1 per cent of the world species of vascular plants. We are depending on a shorter and shorter list containing the most productive plants. About 150 plant species with about 250 000 local races or named cultivars are important globally to feed humanity in the twentieth century. For some crops, genebanks now preserve a greater diversity than does peasant agriculture (Wilkes, 1989).

One of the most significant increases in agricultural production throughout human history occurred in the United States between 1930 and 1980, when the yields of corn, wheat, and potato increased 333 per cent, 136 per cent, and 300 per cent, respectively (National Research Council, 1991). About half of each of these yield increases could be attributed to the acquisition and incorporation of new genes by traditional breeding methods for crop improvement (National Research Council, 1991).

The necessary genes for crop improvement are contained in a wide array of plant materials and forms and are termed 'genetic resources' or 'germplasm'. Sustaining agricultural productivity will require the continued new acquisition with access to a broad diversity of genetic resources. Some countries, who realize this need, have adopted the philosophy that the management of genetic resources is a strategic necessity. The preservation of genetic resources of economically important crops for food, fibre, pharmaceutical, chemical, ornamental, and other uses is supported through governmental, academic, and private institutions throughout the world. Unfortunately, financial support is generally insufficient for adequate preservation.

If second world development continues at the current pace, most tropical lowland forests will be destroyed within 25 years (Raven, 1998). With these forests could perish more than a quarter of the total existing diversity on earth and the rate of species extinction could escalate to more than 100 species per day (Miller *et al.*, 1989). With each lost species a wealth of potentially valuable germplasm may be gone (see Viana *et al.*, Chapter 18, this volume). To avert that loss systems of plant genetic resource conservation have been established. This conservation can be accomplished *in situ*, in place, or *ex situ*, removed from the original location, as described below.

3.2.1   In situ *conservation*

In situ conservation can occur in different forms. Entire biomes containing animal and plant ecosystems can be designated as locations for preservation. This level is extremely important in slowing extinction rates for tropical forests, or controlling timber reserves in temperate forests, but in general has little impact on genetic resources (Wilkes, 1989). *In situ* conservation is carried out under the auspices of national governments through biosphere reserves, national parks, world heritage sites, and other protected areas (Swaminathan, 1997). Lately, joint forest management procedures involving forest dependent communities and the government departments are evolving.

A second type, *in situ* on-farm conservation, is most applicable to agro-ecosystems. Here village farmers manage land races and wild relatives in ecogeographic pockets of genetic diversity (Wilkes, 1989). For example, Mexico has set aside the Sierra de Manantlan for the perennial maize relative *Zea dipioperennis*.

Another concept of *in situ* conservation consists of preservation in gardens where crops and their wild relatives hybridize and the resulting variation is dynamic evolution with pests and pathogens evolving with their host plants (Wilkes, 1989).

An interesting aspect of *in situ* conservation is that the gene pool is not 'frozen'; but rather is under constant genetic change due to both natural and artificial selection. This method preserves a living population, and if established in a large enough region, is not subject to questions of sampling strategy. It conserves genes at the ideal population level (Wilkes, 1989). This mode of conservation is being considered and implemented more frequently than in previous years.

3.2.2   Ex situ *conservation*

The traditional technique for plant genetic resource conservation has involved obtaining seed or plants and removing them to locations under appropriate conditions for long-term storage. This technique has several associated components: exploration, collection, banking, distribution, documentation, and evaluation (Wilkes, 1989). The material must be bred for enhancement and crop improvement to complete the cycle.

Although *ex situ* preservation has been considered the easiest and most preferred conservation method, the collected samples are only as complete as the starting samples. Low frequency alleles (less than 1 per cent) may be missed during the original collection or may be lost later during seed regeneration (Wilkes, 1989). In addition, the documentation aspect of information management is at least as important as the physical arrangements of the collections, and requires considerable research and data entry effort.

Ex situ samples can be preserved in gene banks, agricultural research or germplasm crop centres, clonal repositories, botanical gardens and arboreta. These locations tend to have specific crop assignments of local, regional, national, or international interest. Wilkes (1989) and Williams (1989) summarized designated international base collections of seed crops and field genebanks. The National Plant Germplasm System of the United States was summarized by Shands *et al.* (1989).

3.2.3 Static and dynamic conservation

Bretting and Duvick (1997) have noted some ambiguities in the literature between *in situ* and *ex situ* preservation and have suggested the terms 'static' or 'dynamic' conservation. According to them: 'Static conservation seeks to dramatically alter the original evolutionary trajectories of the plant genetic resources so that a genetic snapshot of sorts is conserved' (Bretting and Duvick, 1997). This conservation safeguards the genes, outside of the evolutionary context in order to minimize the risk of loss while facilitating easy access by researchers for crop improvement (Shands, 1991).

Bretting and Duvick (1997) define dynamic conservation as seeking to conserve or reconstitute the plant associated evolutionary trajectories and the biological, agroecological and human cultural processes that comprise their original evolutionary milieu. This allows the inclusion of crop conservation within traditional agrarian societies. Programmes working with dynamic conservation may be conducted by traditional people themselves, by non-governmental agencies, by government agencies, or by a combination of these. This concept may enable plant genetic resource-rich but capital-poor countries and people to participate more equitably in conservation (Bretting and Duvick, 1997).

3.3 Acquisition procedures

3.3.1 Plant exploration

Since the beginning of the agricultural age, man has moved, used, and domesticated plants for human benefit (Tanksley and McCouch, 1997). One of the first recorded foreign plant collecting expeditions was made by Queen Hatshepsut of Egypt who sent ships to the land of Punt about 1500 BC to obtain the incense tree (Hodge and Erlanson, 1956). When the later explorers traversed the oceans by boat to expand their country's empire, they brought unusual plants with them. Some of these plants have changed the course of crop production. For example, a French soldier, Lieutenant Colonel Amedee Francois Frezier, who was sent to spy on the Spanish colony in Chile in 1711, brought back five strawberry plants (*Fragaria chiloensis* Duchesne) on his return voyage (Darrow, 1966). Several years later, in France, these white-fruited plants eventually hybridized with Virginian strawberries (*F. virginiana* Duchesne) to produce a large fruited hybrid *F.* × *ananassa* Duchesne. This hybrid species is now cultivated throughout the world and has become the large fruited strawberry of commerce. This is one example of international involvement behind the acquisition and development of genetic resources for crop improvement.

Most of the agriculturally important or economically useful crops originated from wild relatives clustered in specific centres of diversity (Vavilov, 1926, 1992). Vavilov enumerated seven main centres which stretch across the tropical and sub-tropical belts of the old and new world. He recognized the need to incorporate phytogeography and plant evolution in plans for plant introduction and conservation (Vavilov, 1997). The first major Russian expedition to obtain sub-tropical plant genetic resources was led by Professor A.N. Krasnov and the agronomist, I.N. Klingin in 1890 (Vavilov, 1997). During that time, plant collecting by United States government personnel was more eclectic (Vavilov, 1997), based on needs and desires to obtain unusual new crops for cultivation (Fairchild, 1944) rather than on an academic desire to examine botanical relationships.

In the late 1890s the United States Congress began appropriating funds for the Foreign Seed and Plant Introduction Section of the Bureau of Plant Industry of the Department of Agriculture. David Fairchild, head of this section, sent Seaman Knapp to Japan to look for new varieties of upland rice; Mark Carleton to Russia to find winter wheats; W.T. Swingle to Algeria and Asia Minor to investigate new crops for the southwest; and Frank N. Meyer to Asia (Cunningham, 1984) to collect useful plants and 'ornamentals when encountered' according to the Department's policy. These early plant explorers provided many crops new to the North American continent, which have become economically important. The United States Department of Agriculture (USDA) continues to annually fund plant exploration expeditions to broaden the pool of genetic resources available for crop development. From 1987 to 1997 the USDA has sponsored more than 114 plant collecting expeditions throughout the world. About a quarter of the collections at the USDA, ARS National Clonal Germplasm Repository at Corvallis have been obtained through plant exploration while three quarters were obtained through plant exchange (Hummer and Reed, 1998).

3.3.2 *Plant exchange*

A recent survey by the Food and Agriculture Organization in conjunction with national genetic resources programmes has tallied more than 6 million accessions stored in approximately 1300 publicly managed genebanks (FAO, 1996). Many countries have established their own genetic resource conservation programmes, although some regional centres exist. International centres have been designated for the major agricultural crops (Wilkes, 1989). International accessions of temperate fruit and tree nuts have been summarized by Bettencourt and Konopka (1989). The United States National Plant Germplasm (NPGS) preserves more than 450 000 accessions representing 1000 genera and 6000 species at sites throughout the country. The NPGS distributes about 150 000 accessions annually to requesters. About 25 per cent of this plant exchange is shipped to requesters from outside the United States.

Seeds are the traditional means of germplasm conservation and plant exchange. However, maintaining the exact genotype is critical for clonally propagated crops so vegetative conservation techniques must be used. The USDA Agricultural Research Service, National Clonal Germplasm Repository at Corvallis preserves *Corylus, Fragaria, Humulus, Mentha,*

Number of Items Shipped

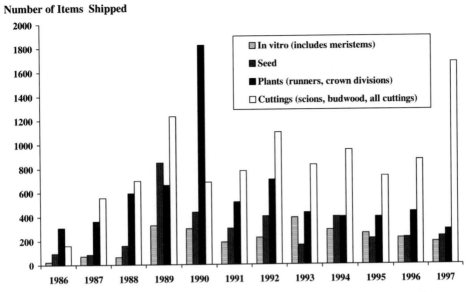

Figure 3.1 Plant distribution from the US Department of Agriculture, Agricultural Research Service, National Clonal Germplasm Repository, Corvallis, Oregon from 1986 to 1997

Pyrus, Ribes, Rubus and *Vaccinium* collections for the NPGS. This facility has shipped more than 25 000 accessions from 1981 to 1997. Although whole plants continued to be significant for clonal exchange, cuttings such as scions, budwood, stem, and roots, remained as the main shipping form (Figure 3.1). *In vitro* culture, though lower in total numbers than other forms, provided almost 70 per cent of foreign requests in 1997.

Quarantine regulations more readily allow *in vitro* cultures rather than whole plants or cuttings, which could harbour insects, mites, or diseases. For example, European and South American quarantines prohibit the entry of fruit trees during the growing season; *in vitro* cultures are acceptable any time of the year. Pear, *Pyrus* L., scionwood is prohibited from countries of the European Community because of *Erwinia amylovora* (Burr.) Winslow *et al.*, the causal agent of fire blight, but virus tested pears in tissue culture can enter these countries. As more plants become available in culture (Reed and Chang, 1997), and as more regional or national quarantines require specific pathogen-free declarations, *in vitro* distribution will increase as a significant form of plant exchange.

3.4 Acquisition planning

Curators for specific crops consult with crop germplasm experts in determining which plants to obtain for collections. Floristic references are consulted for species descriptions, key characters, and geographical ranges. Gaps in the gene pools of existing collections are determined by molecular testing, morphological qualities, and systematic analyses and a plan of plant materials acquisition is developed.

The old explorers worked under an unwritten policy of 'free germplasm exchange' and were not greatly concerned with redistribution of pests or diseases. Those who acquire plants today must first consider the principles of ownership, i.e., intellectual property rights (IPR), and the prevention of spread of diseases or pests before obtaining foreign germplasm.

3.4.1 *Intellectual property rights*

Intellectual property refers to the intangible content of novel products and processes which have been derived from a concept or product of intelligence (Santos and Lewontin, 1997). The value of intellectual property was recognized more than 200 years ago in private competitive industrial production, as in the establishment of the United States Patent and Trademark System in the eighteenth century. The application of IPR to plants and agriculture is more recent and applies to plants or plant parts which are novel or have been improved in some way. These regulations are designed to promote invention and investment in plant breeding.

Plant breeders' and developers' rights

While asexually reproduced plants became protected in the United States by the Plant Patent Act of 1930, seed propagated plants became protected in 1970 under the Plant Variety Protection Act. In 1977, a major change occurred in the Utility Patent Law of the United States. Later judicial reinterpretation of this law extended utility patent protection to novel forms or compositions of natural products including 'man-made', i.e., altered, microorganisms, multi-cellular living organisms, including animals (excluding human beings). These regulations now protect plants, genes, and the novel processes of gene creation and construction.

During the 1960s through the 1980s, the Union for the Protection of New Varieties of Plants (UPOV) established a Plant Variety Protection Act which provided for breeders' rights on newly released varieties (UPOV, 1985). UPOV established that varieties of any plant species could be protected, and recommended a minimum length of protection of exclusive ownership and marketing rights for 20 years with 25 years for woody species. Initially, 28 European countries were signatories, although this number has now increased to nearly 100. Jurisdiction for this protection extends throughout the signatory countries. Some plant breeding programmes are now completely funded through royalties produced by IPR generated from their plant releases.

Farmers' rights

Ethnobotanists have chronicled the invaluable contributions of tribal and rural women and men in the conservation and enhancement of genetic diversity in plants (Swaminathan and Kochhar, 1998). Because these contributions have been financially unrecognized, in contrast with plant breeders' rights, international dialogue was organized in the early 1990s through the Keystone Conferences (Swaminathan, 1997). As a result, the Convention on Biological Diversity (CBD), came into force on 29 December 1993 (Swaminathan, 1997), with eventual ratification by nearly 170 countries. The CBD recognized the 'desirability for sharing equitably benefits arising from the use of traditional knowledge, innovations and practices relevant to the conservation of biological diversity and the sustainable use of its components'. This CBD regulates access to genetic resources as a means of benefit-sharing. Accordingly, access when granted is on mutually agreed terms subject to prior informed consent. The CBD provides undeveloped countries with the opportunity to assert sovereignty over their genetic resources and to ensure that the benefits from the utilization are shared (Swaminathan, 1997). As a result, the legally binding prior informed consent procedures, including material transfer agreements, are an integral part of international plant acquisitions.

Contractual agreements

Contractual agreements are becoming a larger factor in plant genetic resources acquisition. International trade agreements are defining boundaries for the movement of plant genetic resources as well as goods and services. The Economic Community (EC), General Agreement on Tariffs and Trade (GATT), North American Free Trade Agreement (NAFTA), and other international trade alliances affect the movement of genetic resources directly and through the interpretation of quarantine regulations. In some cases private contractual agreements are made which limit the access to specific seed propagated or clonal plant genetic resources. For example, in the case of strawberries (*Fragaria* × *ananassa* Duchesne), raspberries (*Rubus idaeus* L.), and blackcurrants (*Ribes nigrum* L.), cultivar owners have chosen to protect their clones through contractual agreements with specific propagators because IPR protection cannot be obtained for older foreign cultivars through the United States patent system. Unfortunately these contracts can 'bottleneck' developer, grower, or public access to the clones.

3.4.2 *Quarantine regulations*

The movement of genetic resources for plant acquisition involves a risk of accidentally introducing harmful biotic agents along with the host material. In some cases, pathogens may be symptomless within the host although the introduced disease could destroy a commercially important crop if allowed to escape. Plant quarantines are established on the basis of pest risk assessment (Parliman and White, 1985). FAO/IPGRI Technical Guidelines for the safe movement of germplasm, such as that for small (soft) fruit (Diekmann *et al.*, 1994), have been prepared for many crops (see Martin and Postman, Chapter 5, this volume). Governments use these guidelines and other considerations in the development of quarantine regulations. The benefit of plant introduction for the purpose of new crop development must exceed the risks of moving hazardous pests or testing for and eliminating the pest prior to the plant introduction.

Sovereign countries or trading groups have defined plant quarantine regulations. These regulations specify whether plant materials are restricted upon entry into the country. Restricted germplasm in most cases, requires an *a priori* import permit (IP) from the department or ministry of agriculture, or corresponding regulatory agency of the requesting country. The IP specifies diseases or pests that each accession must be certified against, prior to admission. An approved agricultural inspector from the accession's country of origin examines the plant material for the specified pests and prepares a document of phytosanitary certification (PC). Both the IP and the PC are shipped or carried with the plant material for legal entry.

In some countries certain plants can only enter through specific national germplasm quarantine facilities. These facilities are located away from commercial crop production regions and have trained plant pathologists who perform biological, chemical, or molecular assays to test for pathogens. These tests for clonal plant material can include graft inoculation onto indicator plants, serological tests such as enzyme linked immunosorbent assay (ELISA), or sensitive molecular tests which detect viruses or viroids in low titre. Some tests historically may require several years of examination of field grown material. Recent newly developed molecular tests may more accurately detect pathogens within hours rather than the years of inspection previously required (see Martin and Postman, Chapter 5, this volume).

Plum pox has been particularly devastating to stone fruit production in Eastern Europe. This disease is caused by a virus and can now be detected by a polymerase chain reaction procedure (Levy *et al.*, 1994) rather than grafting onto a woody plant indicator and visually rating results. Another example is blackcurrant reversion associated virus which can now be detected using molecular tests (Lemmetty *et al.*, 1997) rather than the three-year visual inspection of leaf and flower symptoms on indicator plants.

Unfortunately, funding for testing plant quarantine accessions has been insufficient in some countries and systems. Canada has chosen to 'privatize' its quarantine station at Saanichton, British Columbia. Fees are also charged at the NRSP-5 in Prosser, Washington, which processes commercially sponsored tree-fruit cultivars through quarantine. Quarantine backlogs occur when resources are insufficient to test and process the requested imported items.

3.5 Methods of acquisition

3.5.1 *Seeds*

Seeds are the most commonly conserved plant genetic resource (Englemann, 1997). Those seeds which are capable of retaining viability after being dried are termed 'orthodox', and are generally long lived (Roberts, 1973). Desiccation-intolerant seeds are termed 'recalcitrant' and may be too short-lived for conservation efforts. Plants with recalcitrant seeds are generally collected and propagated through vegetative techniques.

3.5.2 *Vegetative propagation*

Crown division, cuttings, layering, grafting

Plants are collected and propagated through vegetative means where seeds are either non-viable or not produced, or the exact genotype needs to be maintained, i.e., cloned. Vegetative propagation includes crown division; cuttings such as stem, leaf, or roots; layering; and budding or grafting. These materials cannot survive excessive desiccation or too much moisture. These propagules can easily succumb to rot or wilt diseases. The successful transport of delicate vegetative plant materials over long distances is a challenge. Explorers have reported great losses of plant materials during collecting trips (Cunningham, 1984; Fairchild, 1944; Vavilov, 1997). Knowledge of propagation requirements of new plant species has improved. New techniques, such as *in vitro* culture, have enabled more propagating success with difficult-to-propagate crops.

In vitro *plant collecting*

In vitro plant collecting (see Pence, Chapter 15, this volume) has been somewhat successful for temperate and tropical fruit trees, grasses, and vines (Table 3.1). This technique is

Table 3.1 Crops collected from the wild through *in vitro* means

Taxon	Crop	Reference
Citrus spp.	Oranges, lemons, limes, grapefruit	Withers (1995)
Cocos nucifera L.	Coconut	Withers (1995); Rillo *et al.* (1991); Assy-Bah *et al.* (1987)
Cynodon dactylon (L.) Pers.	Bermuda grass	Rurezdo (1991)
Digitaria pentzii Stent.	Pentz fingergrass	Rurezdo (1991)
Musa spp.	Banana	Withers (1995); INIBAP (1993)
Prunus spp.	Peaches, plums, cherries, almonds	Elias (1988)
Theobroma cacao L.	Cacao	Withers (1995); Yidana *et al.* (1987)
Coffea arabica L.	Coffee	Withers (1995)
Palmae Juss.	Palm	Sossou *et al.* (1987)
Vitis vinifera L.	Grapes	Elias (1988)

inappropriate for collecting most genera, whose plants are readily propagated by seed or cuttings from the field, and would only be considered where the collecting situation and germplasm condition merits the additional effort. Internal plant contaminants, such as soil bacteria and yeasts, can prevent successful *in vitro* establishment. Ashmore (1997) reports that this technique is unimportant for cassava, *Manihot esculenta* Crantz, or potato relatives (*Solanum* spp.).

In vitro collecting is most valuable where seed is unavailable or non-viable, is very large and fleshy and difficult to transport, or may not remain viable during transportation; and where cuttings have poor rootability and remoteness of the collecting site from the laboratory may play a factor. For example coconut, *Cocos nucifera* L., seed are large, fleshy, germinate rapidly, and can easily be contaminated and die during normal field collection. *In vitro* techniques developed for the coconut allow the collection of the relatively small zygotic embryo for transport to the laboratory (Assy-Bah *et al.*, 1987). For cotton, *Gossypium* L., viable seeds were difficult to find at the time plant collection trips were made so that vegetative propagation was considered. Cotton stem cuttings were removed in the field, surface-sterilized, transported to the laboratory, and placed on a simple sterile tissue culture medium containing antibiotics and fungicides (Altman *et al.*, 1990). Internal contamination of these field-collected propagules prevented *in vitro* establishment and survival (Ashmore, 1997). Cacao, *Theobroma cacao* L., germplasm collection is reduced by the rapid deterioration of samples during transit. Many sites for cacao collection are remote. Yidana *et al.* (1987) sterilized cuttings in the field with water-sterilizing tablets, and placed the tissue into prepared culture vials with semi-solid medium containing specifically selected antibiotics and fungicides. Rurezdo (1991) used a similar technique to collect several species of forage grasses. Normally, grasses are collected by shoot cuttings because seed production was poor and the germination percentage very low. However, the shoot cuttings of *Cynodon dactylon* (L.) Pers. and *Digitaria pentzii*

Stent. deteriorate in a short time. *In vitro* culture of material directly in the field was tested to maintain viability of collected material during transit. In addition, these collected cultures more easily satisfied quarantine specifications. Rurezdo (1991) attained success with 75 per cent of the *in vitro* collected material after four weeks.

In vitro collection has been somewhat limited to date but has the potential for wider development and application for a great range of plant species (Ashmore, 1997); see also Pence, Chapter 15, this volume.

3.6   Documentation

Information management is an invaluable aspect of plant acquisition. The background origin information associated with newly acquired accessions is commonly referred to as 'passport' data. When the plant material is collected in the wild, names of the collectors, collection date, site locality information, climatic and ecological conditions are noted (Figure 3.2).

Hand-held geopositioning devices now record accurate geographical coordinates. Ethnobotanical information available from indigenous peoples of the region is also re-corded. If cultivated material is obtained, breeder, pedigree, release date and institution are documented.

As genebanks acquire accessions, a sequential number is assigned. In the United States, 'Plant Inventory' also called 'PI' numbers are designated. Initially, these numbers along with collection information were published in more than 196 volumes (White and Briggs, 1989). Now this information is maintained in publicly accessible databases such as the Germplasm Resource Information Network (GRIN) (Mowder and Stoner, 1989). With the advent of the World Wide Web this information can readily be searched from web-linked sites anywhere in the world.

Unfortunately most plant acquisitions arrive at genebanks without complete passport information. The plant background must be extracted from plant exploration reports, collectors' notes, and research station summaries. For cultivated material, copies of re-lease notices and written summaries of plant descriptions are obtained.

Translation is an important aspect of new acquisitions. Plant material which has passed through intermediary countries may acquire additional synonyms or name changes. For example, the Japanese pear, *Pyrus pyrifolia* (Burm. f.) Nakai cv. Nijisseiki is known as 'Twentieth Century' in the United States and as 'Er Shi Shinge' in China. A Finnish gooseberry, *Ribes uva-crispa* L. cv. Hinnonmaen keltainen, has come to be known in the United States, and some other European countries as 'Hinnomakki gold'. Genetic resource managers must constantly evaluate new acquisitions to determine potential synonymy with known cultivars.

3.7   Identity confirmation

The confirmation of the identity of accessions of genetic resources is a relentless challenge for genetic resource curators. Confirmation is generally done in three ways: morphological comparisons with prior written descriptions; consultation with taxonomists or crop specialists; and molecular analysis (see Harris, Chapter 2, this volume).

Accessions may arrive with a paucity of background information. In-depth research must be performed to obtain as much information as is known on the accession. The

Plant Collection Field Notes

Date: **Collectors:**

Field Collection # _____   Repository Local # _____   PI # _____

Genus _____ species _____ subtax. _____ Auth. _____

Local Clone Name   _____Local Type Name _____

      in English   _____   in English _____

Site Information

Country: Prov./State _____ Dist./Co. _____

Town _____   Km./Direction from _____

Property Owner _____Site Details _____

                                    _____

                                    _____

Latitude _____ Longitude _____ Altitude (m.)_____

GPS •   Map •

Exposure _____   Slope _____   Aspect _____

General Topography _____

Soil Texture _____Drainage (1 poor - 5) _____

Associated Vegetation _____

                       _____

                       _____

                       _____

                       _____

Collection Information

Propagule: **seed plant cutting scion rhizone** Propagules (#) ____ Plants Sampled (#) _____

Improvement status: **wild cultivated cultivar landrace clone breeding**

• Photograph   Abundance (1 few - 5) _____

• Herbarium Sample   Variability (1 little - 5) _____

Other   Comments:

_____

_____

_____

Figure 3.2   Sample genebank form for plant acquisition information

minimal labelling received with the plant propagule may be written in a foreign language and intuitive interpretation by the recipient may be necessary. In clonal germplasm preservation, the genotype is of decisive importance. Collaborators unfamiliar in working with vegetatively propagated crops may unwittingly distribute seedlings of the clone; sometimes a sport or subclone may be propagated producing an 'off-type'. Identity confirmation must begin upon receipt of the plant material. For wild collected material, voucher specimens obtained during the collecting expedition validate species designations. For clonal material, written descriptions of the accession are invaluable. The morphology and phenology of the new accession must be compared with the written description to assist in identity confirmation. Curators must develop familiarity with crop history and references. Connections with current crop research are also imperative. Taxonomists and crop specialists can be consulted to confirm accession identity on a whole plant basis.

Many different molecular tests assist in identity confirmation (see Harris, Chapter 2, this volume). Plant products, such as phenolics (Challice and Westwood, 1973) or essential oils identify specific clones. Proteins such as isozymes were frequently used in the 1980s (Soltis and Soltis, 1989). Nucleic acid assays such as hybridization, restriction analyses, minisatellite sequences, and nucleotide sequencing are more direct fingerprinting techniques (Avise, 1994; Karp *et al.*, 1997).

In clonal germplasm acquisition management the guiding procedure should be to:

- Acquire genotypes from the original, pathogen-free, certified source wherever possible.
- Confirm the botanical and horticultural identity through morphological evaluation, nomenclatural expertise, and molecular techniques.
- Acquire the genotype from additional sources, if the identity of the initially received clone seems questionable.
- Eliminate the duplicate or incorrectly identified material from multiple sources upon confirmation of a primary clone.
- Recheck genotype identity periodically.

Re-propagation and routine maintenance can cause identity difficulties. The curator must maintain a constant vigil concerning the identity of the collections.

3.8 Summary

Eloquent arguments (Avise, 1994; Tanksley and McCouch, 1997; Wilkes, 1989) have spoken to the need for genetic resource conservation. With the advent of clonal germplasm conservation facilities, not only the species genes but specific genotypes are now preserved for humanity. These plants can be acquired by seed or vegetative propagation techniques. New technologies of *in vitro* collection and maintenance and cryogenic preservation have enhanced our ability to preserve significant clones. New molecular pathogen detection methods enable rapid testing and certification of pathogen-negative plant material for distribution. Nucleic acid assays such as hybridization, restriction analyses, minisatellite sequences, and nucleotide sequencing now assist as fingerprinting techniques for confirmation of botanical or horticultural identity. New interpretations of intellectual property rights issues have changed the world from 'free germplasm exchange' to exchange with permission and potential future financial reimbursement. Although these concepts impose complexity on distribution, if managed appropriately, they could generate resources to ensure long-term conservation efforts throughout the world.

References

ALTMAN, D.W., FRYXELL, P.A., KOCH, S.D. and HOWELL, C.R., 1990, *Gossypium* germplasm conservation augmented by tissue culture techniques for field collecting, *Economic Botany*, **44**, 106–113.

ASHMORE, S.E., 1997, *Status Report on the Development and Application of* in vitro *Techniques for the Conservation and Use of Plant Genetic Resources*, Rome: International Plant Genetic Resources Institute.

ASSY-BAH, B., DURAND-GASSELIN, T. and PANNETIER C., 1987, Use of zygotic embryo culture to collect germplasm of coconut (*Cocos nucifera* L.), FAO/IBPGR *Plant Genetic Resources Newsletter*, **71**, 4–10.

AVISE, J.C., 1994, *Molecular Markers, Natural History and Evolution*, NY: Chapman and Hall.

BETTENCOURT, E.J. and KONOPKA, J., 1989, *Directory of Germplasm Collections. 6. II. Temperate Fruits and Tree Nuts*, Rome: International Board for Plant Genetic Resources.

BRETTING, P. and DUVICK D., 1997, Dynamic conservation of plant genetic resources, *Advances in Agronomy*, **61**, 1–51.

CHALLICE, J.S. and WESTWOOD, M.N., 1973, Numerical taxonomic studies of the genus *Pyrus* using both chemical and botanical characters, *Botanical Journal of the Linnean Society*, **67**, 121–148.

CUNNINGHAM, I.S., 1984, *Frank N. Meyer, Plant Hunter in Asia*, Ames: Iowa State University Press.

DARROW, G.M., 1966, *The Strawberry, History, Breeding, and Physiology*, New York: Holt Rinehart and Winston.

DIEKMANN, M., FRISON, E.A. and PUTTER, T. (Eds), 1994, *FAO/IPGRI Technical Guidelines for the Safe Movement of Small Fruit Germplasm*, Food and Agriculture Organization of the United Nations, Rome: International Plant Genetic Resources Institute.

ELIAS, K., 1988, *In vitro* culture and plant genetic resources. A new approach: *in vitro* collecting. *Lettere d'Informazione*, Valenzano, Italy: Istituto Agronomico Mediterraneo, **3**, 33–34.

ENGLEMANN, F., 1997, Importance of desiccation for the cryopreservation of recalcitrant seed and vegetatively propagated species, *Plant Genetic Resources Newsletter*, **112**, 9–18.

FAIRCHILD, D., 1944, *The World Was My Garden*, New York: Charles Scribner's Sons.

FAO, 1996, *State of the World Report on Plant Genetic Resources for Food and Agriculture*, Rome: Food and Agriculture Organization of the United Nations.

HODGE, W.H. and ERLANSON, C.O., 1956, Federal plant introduction – a review, *Economic Botany*, **10**, 299–334.

HUMMER, K. and REED, B., 1998, Establishment and operation of a temperate clonal field genebank, *Consultation on the Management of Field and* in vitro *Genebanks*, Colombia: CIAT.

INIBAP, 1993, *Annual Report 1992*, Montpellier, France: International Network for the Improvement of Banana and Plantain.

KARP, A., KRESOVICH, S., BHAT, K., AYAD, W. and HODGKIN, T., 1997, Molecular tools in plant genetic resources conservation: a guide to the technologies, *IPGRI Technical Bulletin No. 2*, Rome: International Plant Genetic Resources Institute.

LEMMETTY, A., LATVALA, S., JONES, A.T., SUSI, P., McGAVIN, W.J. and LEHTO, K., 1997, Purification and properties of a new virus from black currant, its affinities with nepoviruses, and its close association with black currant reversion disease, *Phytopathology*, **87** (4), 404–413.

LEVY, L., LEE, I.M. and HADIDI, A., 1994, Simple and rapid preparation of infected plant tissue extracts for PCR amplification of virus, viroid and MLS nucleic acids, *Journal of Virological Methods*, **49**, 295–304.

MILLER, D.R., ROSSMAN, A.Y. and KIRKBRIDE, J., 1989, Systematics, diversity and germplasm, in *Biotic Diversity and Germplasm Preservation, Global Imperatives*, edited by L. KNUTSON and A.K. STONER, The Netherlands: Kluwer Academic Publishers, pp. 3–11.

MOWDER, J.D. and STONER, A.K., 1989, Plant germplasm information systems, in *Biotic Diversity and Germplasm Preservation, Global Imperatives*, edited by L. KNUTSON and A.K. STONER, The Netherlands: Kluwer Academic Publishers, pp. 419–426.

National Research Council (US) Committee on Managing Global Genetic Resources: Agricultural Imperatives, 1991, *Managing Global Genetic Resources: The US National Plant Germplasm System*, Washington, DC: National Academy Press, p. 171.

PARLIMAN, B.J. and WHITE, G.A., 1985, The Plant Introduction and Quarantine System of the United States, in *Plant Breeding Rev*, Vol. 3, edited by J. JANNICK, AVI Pub. Co., pp. 361–434.

RAVEN, P., 1988, Our diminishing tropical forests, in *Biodiversity*, edited by E.O. WILSON and F.M. PETER, Washington, DC: National Academy Press, pp. 119–122.

REED, B.M. and CHANG, Y., 1997, Medium- and long-term storage of *in vitro* cultures of temperate fruit and nut crops, in *Conservation of Plant Genetic Resources In Vitro*, Vol. 1, edited by M.K. RAZDAN and E.C. COCKING, Enfield, NH, USA: Science Publishers, Inc., pp. 67–105.

RILLO, E.P., BELEN, M. and PALOMA, F., 1991, Storage and transport of zygotic embryos of *Cocos nucifera* L. for *in vitro* culture, *FAO/IBPGR Plant Genetic Resources Newsletter*, **86**, 1–4.

ROBERTS, H.F., 1973, Predicting the viability of seeds, *Seed Science and Technology*, **1**, 499–514.

RUREZDO, T.J., 1991, A minimum facility method for *in vitro* collection of *Digitaria eriantha* ssp. *penizii* and *Cynodon dactylon*, *Tropical Grasslands*, **25**, 56–63.

SANTOS, M. DE M. and LEWONTIN, R.C., 1997, Genetics, plant breeding and patents: conceptual contradictions and practical problems in protecting biological innovations, *FAO/IBPGR Plant Genetic Resources Newsletter*, **112**, 1–8.

SHANDS, H., 1991, Complementarity of *in situ* and *ex situ* germplasm conservation from the standpoint of the future user, *Israeli Journal of Botany*, **40**, 521–528.

SHANDS, H.L., FITZGERALD, P.J. and EBERHART, S.A., 1989, Program for plant germplasm preservation in the United States: the US National Plant Germplasm System, in *Biotic Diversity and Germplasm Preservation, Global Imperatives*, edited by L. KNUTSON and A.K. STONER, The Netherlands: Kluwer Academic Publishers, pp. 97–115.

SOLTIS, D.E. and SOLTIS, P.S., 1989, Isozymes in plant biology, *Advances in Plant Sciences*, Vol. 4, Portland: Diosccorides Press.

SOSSOU, J., KARUNARATNE, S. and KOVOOR, A., 1987, Collecting palm: *in vitro* explanting in the field, *FAO/IBPGR Plant Genetic Resources Newsletter*, **69**, 7–18.

SWAMINATHAN, M.S., 1997, Implementing the benefit-sharing provisions of the convention on biological diversity: challenges and opportunities, *FAO/IBPGR Plant Genetic Resources Newsletter*, **112**, 19–27.

SWAMINATHAN, M.S. and KOCHHAR, S.L., 1998, *Plants and Society*, London: Macmillan.

TANKSLEY, S.D. and MCCOUCH, S.R., 1997, Seed banks and molecular maps: unlocking genetic potential from the wild, *Science*, **277**, 1063–1066.

UPOV, 1985, International Convention for the Protection of New Varieties of Plants of December 2, 1961; Additional Act of November 10, 1972; and Revised Text of October 23, 1978. *Document 293(E)*, Geneva: UPOV.

VAVILOV, N.I., 1926, Tsentry proiszhokhdeniya kultur'nykh rasteniy [Centres of origin of cultivated plants]. *Trudy po prikl. botan., genet. i. selek. [Papers on applied botany, genetics and plant breeding]*, 16(2).

VAVILOV, N.I., 1992, *Origin and Geography of Cultivated Plants*, London: Cambridge University Press.

VAVILOV, N.I., 1997, *Five Continents*, Rome, Italy: International Plant Genetic Resources Institute.

WHITE, G.A. and BRIGGS, J.A., 1989, Plant germplasm acquisition and exchange, in *Biotic Diversity and Germplasm Preservation, Global Imperatives*, edited by L. KNUTSON and A.K. STONER, The Netherlands: Kluwer Academic Publishers, pp. 405–417.

WILKES, G., 1989, Germplasm preservation: objectives and needs, in *Biotic Diversity and Germplasm Preservation, Global Imperatives*, edited by L. KNUTSON and A.K. STONER, The Netherlands: Kluwer Academic Publishers, pp. 13–41.

WILLIAMS, J.T., 1989, Plant germplasm preservation: a global perspective, in *Biotic Diversity and Germplasm Preservation, Global Imperatives*, edited by L. KNUTSON and A.K. STONER, The Netherlands: Kluwer Academic Publishers, pp. 81–96.

WITHERS, L.A., 1995, Collecting *in vitro* for genetic resources conservation, in *Collecting Plant Genetic Diversity*, Technical Guidelines, edited by L. GUARINO, V. RAMANATHA RAO and R. REID, Wallingford, Oxon: CABI, pp. 511–526.

YIDANA, J.A., WITHERS, L.A. and IVINS, J.D., 1987, Development of a simple method for collecting and propagating cocoa germplasm *in vitro*, *Acta Horticulturae*, **212**, 95–98.

4

Tissue Culture Techniques in
In Vitro Plant Conservation

PAUL T. LYNCH

4.1   Introduction

The historic development of *in vitro* plant cell and tissue culture has undoubtedly been a major factor in the advancement of our knowledge of cell biology, physiology, biochemistry (Bhojwani and Razdan, 1996) and more recently, molecular biology (Raghaven, 1997). However, its exploitation for applied purposes could be argued as being of even greater consequence. Plant biotechnology utilizes a range of *in vitro* techniques to manipulate plant germplasm, including clonal multiplication, generation of novel variants and the production of genetically modified plants through somatic hybridization and genetic transformation (Vasil and Thorpe, 1994). In addition, *in vitro* culture can also have an important role in the conservation of plant germplasm. The use of *in vitro* germplasm storage in plant biotechnology programmes has a growing significance, as it improves the efficiency of research activities and secures the valuable products of such activities for both scientific and commercial purposes (Lynch, 1999). Conservation of plant germplasm can itself be the goal of *in vitro* plant cell and tissue culture programmes (Feijoo and Iglesias, 1998; Prance, 1997), by the use of techniques including micropropagation (Edson *et al.*, 1994) and embryo rescue (Dixon, 1994). Tissue culture approaches have been vital in the re-establishment of endangered plant species, for example the lady's slipper orchid (*Cypripedium calceolus* L.) (Ramsay and Stewart, 1998). However, *in vitro* plant germplasm conservation requires an understanding and appreciation of the inherent problems of biotechnology and of the specific culture requirements of different plant species. In common with most *in vitro* plant cell and tissue manipulations, an overriding concern is the maintenance of the genetic fidelity of the stored germplasm. Thus, with time the phenotype and genotype of *in vitro* plant cultures change (Jahne *et al.*, 1991; Wang *et al.*, 1993). Such changes constitute the basis of somaclonal variation (Larkin and Scowcroft, 1981), the significance of which to *in vitro* germplasm conservation is reviewed by Harding (Chapter 7, this volume).

4.2   *In vitro* propagation

Micropropagation is a general term which describes a variety of routes (as shown in Figure 4.1), for the propagation of selected germplasm using *in vitro* techniques. Although the

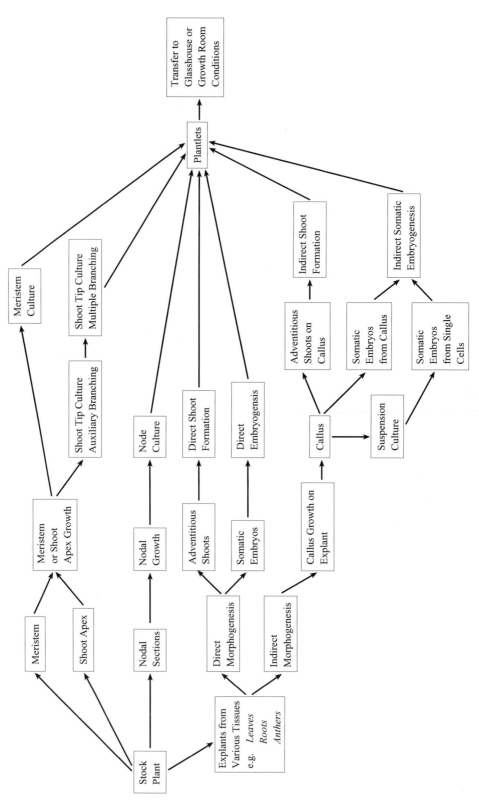

Figure 4.1 The principle methods of micropropagation

details of the different routes vary significantly in the mechanisms of plant multi-plication, the five basic stages for successful micropropagation are comparable. The stages are:

- Stage 0: Preparative stage, involving germplasm selection.
- Stage 1: Establishment stage, involving the production of axenic, viable cultures.
- Stage 2: Multiplication stage, during which the number of propagules is increased.
- Stage 3: Plantlet production, involving the development of germplasm of sufficient size and quality for transfer to *in vivo* conditions.
- Stage 4: Establishment under *in vivo* conditions, involving the acclimatization of plantlets to glasshouse conditions.

Notable features of the *in vitro* propagation process are discussed in the following sections.

4.2.1 *Germplasm acquisition*

The first step towards *in vitro* conservation is germplasm acquisition (see Hummer, Chapter 3, this volume). For source plants in managed environments, collection can be reasonably straightforward. However, even for this germplasm it is important to consider factors, such as seasonal effects, which can significantly affect the ability to establish *in vitro* cultures (Enjalric *et al.*, 1988). Before germplasm collection from either managed or non-managed habitats, the associated legal issues dealing with aspects of ownership, sovereignty and intellectual property rights have to be considered (Guarino *et al.*, 1995). It is also vital to ensure that target plants have been correctly identified and tissue samples can be collected from sufficient individuals to maximize the diversity of the collection, without endangering the remaining natural population.

In situations where germplasm is difficult to transport, or there is a significant risk of loss of viability during transit, *in vitro* collection provides a potential means of overcoming some of these difficulties (see Pence, Chapter 15, this volume). *In vitro* collection involves the placing of surface-sterilized tissue on to pre-prepared sterile culture medium in the field, prior to transport to the laboratory. To reduce the problem of *in vitro* culture contamination, pre-prepared culture medium is often supplemented with antimicrobial agents (Yidana *et al.*, 1987). Techniques used in the field are kept as simple as possible to ensure that only rudimentary equipment needs to be transported. This approach has been successfully used for a number of species, including coconut, cotton and cacao (Ashmore, 1997; Withers, 1995).

4.2.2 *Selection of tissue for* in vitro *culture*

It has long been recognized that there are a number of factors which can significantly influence the *in vitro* behaviour of an explant (Murashige, 1974), and which should be considered when selecting tissue for collection; these include:

- The genotype of the source plant (Brown and Atanassov, 1985; Lynch *et al.*, 1991).
- The explant tissue type (Kameya and Widholm, 1981).
- The physiological age of the explant (Atree and Fowke, 1991; von Arnold *et al.*, 1996).

- The season in which the explant was obtained (Cassells and Minas, 1983).
- Explant size (Reeves *et al.*, 1983, 1995).
- The health and vigour of the source plant (Oertel and Breuel, 1987).

For some plant species a specific explant may be required for successful *in vitro* plant regeneration, for example immature and mature embryos are among the few explant types from which embryogenic cereal callus, a prerequisite for plant regeneration, can be initiated (Lynch *et al.*, 1991; Vasil and Vasil, 1986). The general consensus is that larger explants give better survival and regeneration rates as compared with smaller explants (Al Mazrooei *et al.*, 1997). However, the use of smaller explants does have the advantage of increasing the chance of virus elimination from subsequent cultures (Kartha, 1986).

To reduce the significance of some of the above factors, appropriate pretreatment of the donor plant can be important. For example, the growth of rice anther and immature embryo cultures is influenced by the light regime the donor plants were grown under (Lee *et al.*, 1988). In plants with strong apical dominance, removal of the terminal meristem can improve the *in vitro* response of lateral buds (Hasegawa, 1979). Control of glass-house or growth room conditions in which donor plants are maintained, for example by using irrigation systems which avoid soil splash, can reduce subsequent microbial contamination of *in vitro* cultures (Debergh and Read, 1991). Similarly, field-collected explants have been shown to carry more microbial contamination during the wet season as compared with the dry season (Enjalric *et al.*, 1988).

In selecting explant tissue for *in vitro* culture initiation for conservation purposes it is important to consider the influence of different explant types on the occurrence and frequency of somaclonal variation in regenerants (Harding, Chapter 7, this volume). In general terms the older and/or more specialized the explant the greater the potential for variation in derived plants (Karp, 1995). For example, plants regenerated from cultured petals of *Chrysanthemum* had higher frequencies of abnormalities compared with plants derived from pedicels (Bush *et al.*, 1976; De Jong and Custers, 1986). Such effects relate to changes in the genome, including endopolyploidy and DNA sequence amplification (D'Amato, 1989).

4.2.3   *Microbial contamination and disease indexing*

Plant surfaces provide habitats for many types of microorganism (Campell, 1985). Additionally, plants may also have endophytic micro floras (Cassells, 1991). Depending on the explant used, epiphytic and/or endophytic microorganisms can be carried over to *in vitro* culture. Surface microorganisms are normally removed by surface sterilization of the explant prior to culture initiation. The precise sterilization procedures are dependent on the explant type and plant species. Explants such as mature seeds and nodal stem pieces can be directly treated with sterilizing agents. However, tissue such as immature ovules and embryos tends to be easily damaged by disinfectants, therefore the ovary or ovule can be disinfected and the desired explants removed by dissection under sterile conditions. Although a range of chemicals have been successfully used to surface sterilize plant material (Table 4.1) most laboratories use diluted commercial sodium hypochlorite bleaches, such as Domestos and Clorox. These preparations contain wetting agents which increase the effectiveness of the active compounds, they are easily available, cheap and relatively safe. After sterilization it is vital to remove the surface sterilizing agent by washing, normally with sterile water, to minimize toxic effects to the plant tissue.

Table 4.1 Examples and effectiveness of surface sterilizing agents

Disinfectant	Concentration used (%)	Duration of treatment (min)	Effectiveness
Calcium hypochlorite	9–10	10–45	Very good
Sodium hypochloride	0.9–2.0	10–45	Very good
Commercial bleach	5–30	10–45	Very good
Mercuric chloride	0.1–1	2–10	Good
Ethanol	70	2–3	Good
Antibiotics	4–150mg l^{-1}	30–60	Quite good

Many of the endophytic microorganisms present within plant tissue will be capable of growing on plant culture medium, although some may be inhibited by the high salt or sucrose concentration, or pH (Cassells *et al.*, 1988). Some types of endophytic micro-organism will overgrow and kill slower growing plant tissues, while less adapted species can proliferate in the tissues of the explant utilizing nutrients from dead and damaged cells; this can also lead to the death of an explant. Latent or subliminal microorganisms may overgrow plant tissue after transfer to fresh media, especially if the medium has reduced salt or sucrose concentrations (Cassells, 1988). Even where plant cultures are not killed by overgrowth of subliminal microorganisms such contamination can affect the vigour of *in vitro* and *ex vitro* plants (Long *et al.*, 1988).

The ability to utilize *in vitro* cultures as a means of virus elimination (see Martin and Postman, Chapter 5, this volume) is an important advantage of *in vitro* plant germplasm conservation (Villalobos *et al.*, 1991). The technique of virus elimination pioneered by Morel and Martin (1952) was based on the principle that meristematic domes may be free of viral particles. For example in sweet potato there is a direct correlation between explant size and elimination of the sweet potato yellow dwarf virus (Green and Lo, 1989). However, not all viruses can be eliminated this way (Theiler-Hedtrich and Baumann, 1989; see also Martin and Postman, Chapter 5, this volume). Heat treatment (thermotherapy) in conjunction with meristem culture has been used to improve the success of virus elimination (Brown *et al.*, 1988). The combination of meristem culture and thermotherapy has been used with a 97 per cent success rate in an *in vitro* cassava genebank at the Centro Internacional de Agricultura Tropical (CIAT) (IPGRI/CIAT, 1994). Chemotherapy has also been assessed as a means of virus elimination, but only limited success has been reported, and the role of antiviral agents in virus suppression or inactivation has not been clear (Cassells, 1987).

It is vital to define and maintain the phytosanitory status of *in vitro* cultures being used for germplasm conservation, especially those cultures that are to be involved in international exchange programmes (Martin and Postman, Chapter 5, this volume). Therefore the availability of rapid, effective and sensitive procedures for disease indexing is vital. A number of approaches have been used, including enzyme-linked immunosorbent assay (ELISA) (Greno *et al.*, 1990) and nucleic acid hybridization (Fuchs *et al.*, 1991).

4.2.4 *Tissue culture media*

Plant tissue culture medium usually consists of major inorganic salts, trace elements, a carbon source, vitamins, growth regulators and a gelling agent. The systematic study of

Table 4.2 Inorganic and organic compounds of four major plant culture media (mg l^{-1})

Component	Murashige & Skoog (MS)	Gamborg's B5	White's	Nitsch's
$KHPO_4$	170			68
KNO_3	1900	2527.5	80	950
KCl			65	
Na_2SO_4			200	
$NaH_2PO_2·H_2O$		150	19	
$Na_2MoO_4·2H_2O$	0.25	0.25		0.25
$NaEDTA·2H_2O$	37.3	37.3		37.3
$CaCl_2·2H_2O$	440	150		
$CaCl_2$				166
$Ca(NO_3)_2·4H_2O$			300	
$MgSO_4·7H_2O$	370	246.5	750	185
$ZnSO_4·7H_2O$	8.6	2.0	3.0	10
KI	0.83	0.75	0.75	
$NiCl_2·6H_2O$			0.03	
$Fe(SO_4)_3$			2.5	
$FeSO_4·7H_2O$	27.8	27.8		27.8
NH_4NO_3	1650			720
$(NH_4)_2SO_4$		134		
$CuSO_4·5H_2O$	0.025	0.025	0.001	0.025
H_3BO_3	6.2	3.0	1.5	10
MoO_3			0.001	
$MnSO_4·4H_2O$	22.3		5.0	25
$CoCl_2·6H_2O$	0.025	0.025		
$MnSO_4·H_2O$		10.0		
Myo-inositol	100	100		100
Pyridoxine-HCl	0.5	1.0	0.01	0.5
Thiamine-HCl	0.1	10.0	0.01	0.5
Nicotinic acid	0.5	1.0	0.05	5.0
Glycine	2.0		3.0	2.0
Folic acid				0.5
Biotin				0.05
Sucrose	30 000	20 000	20 000	20 000

the nutrient requirements of plant tissue under *in vitro* conditions has led to the development of a range of nutrient formulations (Gamborg *et al.*, 1968; Linsmaier and Skoog, 1965; Murashige and Skoog, 1962; Nitsch, 1969; White, 1963) which have now become the basis of commercial preparations used by most plant tissue culture laboratories. Examples of the components of such media are detailed in Table 4.2. Optimization of *in vitro* growth is normally achieved by the modification of standard media formulae, which has resulted in numerous reports for a vast range of plant species (George, 1993). Significant factors in culture media formulation are discussed in the following subsections.

Sources and concentration of nitrogen

There are varied reported effects of the nitrogen composition of plant culture media. Explants from some species cannot tolerate high levels of nitrogen in the culture media, e.g. immature embryos of *Impatiens platypetala* (Han and Stephens, 1992). In rice anther culture medium the ratio of NH_4^+ and NO_3^- is critical for cell proliferation and plant regeneration (Grimes and Hodges, 1990), while the presence of organic nitrogen in culture media has been shown to enhance plant regeneration in, for example *Agrostis alba* (Shetty and Asano, 1991). Hence most culture media contain a mixture of inorganic and organic nitrogen (Shetty and Asano, 1991). The form and concentration of nitrogen in culture media have been shown to induce genetic variation in cultured plant cells. The number of albino plants regenerated from anther callus cultures of wheat is influenced by the potassium nitrate concentration in the culture medium (Feng and Ouyang, 1988). Furner *et al.* (1978) showed that the haploid *Datura innoxia* cell lines retained their ploidy in medium containing both inorganic and organic nitrogen, while cultures in medium containing only organic nitrogen were composed of diploid and tetraploid cells.

Carbon source

Sucrose, at concentrations of 2–5 per cent w/v, is the most commonly used carbon source in plant tissue and cell culture media. Higher concentrations have been utilized to induce embryogenesis (Lu *et al.*, 1983) and bulblet development in *Allium* sp. (Zel, *et al.*, 1997). Increasing the sucrose concentration also provides a means of reducing tissue growth and has been the basis of several slow growth *in vitro* storage protocols for plant germplasm (Bonnier and van Tuyl, 1997). Different types of sugar in culture media have been shown to enhance, for example plant regeneration (Jain *et al.*, 1997) and seed germination (Foley, 1992). Interestingly the maintenance of micropropagated plants under photoautotrophic conditions can result in the promotion of plantlet growth *in vitro* and *ex vitro*, while the resulting simplification of the micropropagation process can reduce labour costs (Kozai, 1991).

Gelling agent

The most commonly used gelling agents for *in vitro* plant culture are agar, agarose and gellan gums, such as Gelrite (Bhojwani and Razdan, 1996). The physical and chemical properties of gelling agents have been shown to vary significantly between source and batch (Scholten and Pierik, 1998). As a result there are a growing number of reports indicating that media gelling agents influence culture characteristics, such as somatic embryo formation (Tremblay and Tremblay, 1991), shoot elongation (Barbas *et al.*, 1993), shoot multiplication (Podwyszynska and Olszewski, 1995), and hyperhydricity (Franck *et al.*, 1998).

Plant growth factors

There are several groups of plant growth regulator substances, specifically, auxins, cytokinins, gibberellins, ethylene and absisic acid. *In vitro* plant growth and morphogenesis are regulated by the interaction and balance between the growth regulators supplied in the medium, and those produced endogenously. Arguably the most important of the plant growth regulator groups are the auxins and cytokinins. Some typical responses include:

1   Promotion of auxiliary shoot formation by the presence of very low concentrations of auxins in combination with high concentrations of cytokinins, for example in *Populus* sp. (Lubrano, 1992).

2   Plant regeneration in monocotyledons is often promoted by transferring callus tissue to medium without auxins, or by replacing active auxins such as 2,4-dichlorophenoxyacetic acid (2,4-D) with indole-3-acetic acid (IAA) or naphthalene acetic acid (NAA) (Lynch *et al.*, 1991).

3   Promotion of adventitious shoot formation by the presence of low concentrations of auxins in combination with high concentrations of cytokinins, for example in *Rubus* species (Mezetti *et al.*, 1997).

Although gibberellic acid (GA$_3$) tends to prevent the formation of organized root and shoot meristems in callus cultures, it can promote the further growth and development of pre-existing organs. For example, the presence of GA$_3$ in culture medium has been reported to improve the growth of potato meristems and inhibit callus proliferation (Novak *et al.*, 1980), and increase the number of shoots produced from seedling and tuber shoot internodes of *Apios americana* (Wickremesinhe *et al.*, 1990).

However, it must be remembered that the effect of growth regulators can be significantly influenced by the basal culture medium. For example 0.5 mg l^{-1} 2,4-D in MS medium resulted in the production of approximately two plants from each embryo-derived callus of *Oryza sativa*, whereas on N6 medium (Chu *et al.*, 1975) with the same auxin concentration, up to nine plants per embryo-derived callus were produced (Koetje *et al.*, 1989).

Where *in vitro* plant culture is to be used for germplasm conservation it is also important to consider the role on the culture environment of the frequency of somaclonal variation (see Harding, Chapter 7, this volume). There is considerable evidence to indicate that somaclonal variation is influenced particularly by the type and concentration of plant growth regulators (Karp, 1992), and that growth regulators can act as mutagens. For example, 2,4-D has been shown to increase the frequency of blue to pink mutations in the *Tradescantia* stamen hair (Dolezel and Novak, 1984). Phenotypic variants in palms and African plantains have been attributed to the use of high concentrations of auxins and cytokinins respectively (Corley *et al.*, 1986; Vuylsteke and Swennen, 1990). 2,4-D has been shown to be more genetically 'damaging' than other auxins. For example, Jha and Roy (1982) showed that in suspension cultures of *Vigna sinensis* the maximum ploidy of cells grown in medium with NAA was hexaploid, compared with octaploid in cells in medium with 2,4-D. This phenomenon is particularly important in plants which are vegetatively propagated, apomictic and/or have long lived species, which would not normally or readily go through a sexual cycle which would eliminate 'epigenetic modifications'.

It is therefore vital to undertake a comprehensive literature search prior to starting *in vitro* culture to assist in the targeting of significant media components for study. Information from related plant species can often provide useful indicators for species for which *in vitro* culture has not been previously reported.

4.2.5   *Problems of culture establishment*

After excision from the donor plant, transportation to the laboratory, trimming and surface sterilization, the explant which is finally placed on the culture medium will inevitably contain stressed, damaged and dying cells. This results, particularly in many woody plant species, in the production of deleterious polyphenol and tannins (Collins and Symons, 1992), and lipid peroxidation products (Benson and Roubelakis-Angelakis, 1994), which results in browning blackening of explant tissue and often cell death. Browning of tissues and the medium may be prevented or reduced, by the addition, for example, of glutamine (Bergmann *et al.*, 1997) or activated charcoal (Kikkert *et al.*, 1996) to the culture medium, and/or preculture explant washing with antioxidant solutions (Block and Lankes, 1996).

Immediately after transfer to fresh culture medium, plant tissues release a range of substances, including alkaloids, amino acids, enzymes, growth factors and vitamins, into the culture medium (Street, 1969). If the cells are inoculated at a low population density, the concentration of essential substances in the cells and in the medium can become inadequate for culture development, at which point the cultured cells can undergo programmed cell death (McCabe *et al.*, 1997). There is a minimum size of explant, or quantity of cells per unit volume of culture medium, for successful culture initiation. The size of the initial explant/cell inoculum also affects the initial rate of *in vitro* cell growth. The minimum inoculum density varies with plant species, explant type and cultural conditions.

4.2.6 *Propagule multiplication (morphogenesis)*

Once the explant tissue has been placed into culture, there are a series of different routes to the production of plantlets as summarized in Figure 4.1. The route which normally results in the most rapid multiplication of propagules is adventitious callogenesis. However, the involvement of a callus stage can result in a greater proportion of non-true-to-type plantlets as compared with propagation routes in which shoot regeneration does not involve callus (Karp, 1995). Hence this approach is the least suitable method for *in vitro* conservation. The use of shoot or meristem (shoot tip) cultures is preferred as the genetic fidelity of the germplasm is more likely to be maintained. However, not all shoots arising from shoot cultures originate from axillary buds. Adventitious shoots can frequently arise directly from shoot tissue or indirectly from callus at the base of the shoot mass. Direct and indirect shoot development has been observed in apple shoot cultures (Nasir and Miles, 1981). The occurrence of such shoots is increased when cytokinin concentrations are greater than the optimum concentration.

Indirect or direct morphogenesis occurs either by organogenesis or somatic embryogenesis. Although these are distinctly different processes, both depend on the ability to redirect cells and tissues which are mitotically quiescent, or already committed to some function or pathway of development, to a meristematic state, i.e. exhibit totipotency, that is the ability of an individual cell to regenerate into a whole organism. The development of adventitious shoot meristems is usually from the periphery of callus or explants, but cells from any histogen or cell layer can be involved (Litz and Gray, 1992). For many years these meristems were thought to have a multicellular origin (Thorpe, 1994). However, studies of plant regeneration from leaf discs of periclinal chimeras indicated that shoot organogenesis usually has a single cell origin (Marcotrigiano, 1986). Two distinct types of somatic embryogenesis have been recognized. The first is direct embryogenesis, in which a single cell, or cell group commences meristematic activity and all the progeny of this cell form part of the embryo. This is a rare event and occurs, for example in citrus nucellular tissue (Barlass and Skene, 1986). More common is indirect somatic embryogenesis in which somatic embryos originate from proembryonic cell masses, of single cell origin.

Whichever pattern of plant regeneration is being followed by an *in vitro* plant culture an event must initially occur that involves a change in the development of certain cells within the culture (Christianson and Warnick, 1988). Cells must acquire competence which allows the expression of organogenic or embryogenic potential. This change is referred to as an inductive event. The limited number of regeneration competent cells in an explant is illustrated by the reported difficulties of the production of transgenic grapevines. Grapevine leaves after co-cultivation with *Agrobacterium* expressed β-glucuronidase activity at the cut surface, in vascular bundles, or in inner cortical cells of the petiole, but none of these regions produce adventitious shoots (Colby *et al.*, 1991). It is also important to remember that the ability of a culture to sustain morphologically competent cells declines with time in culture as a result of the effects of somaclonal variation (Wang and Marshall, 1996).

The pattern of morphogenic development is determined by medium constituents and by genetic and epigenetic factors. Regeneration via standardized protocols can be restricted to specific cultivars; James *et al.* (1984) demonstrated that *Malus* rootstocks M25 and M27 regenerated by organogenesis from stem segments, but M9 and M26 did not regenerate. Genes from several plant species involved in plant regeneration *in vitro* have been identified in several plant species, including rice (Jung *et al.*, 1998) and orchard grass (Taran and Bowley, 1997). Difficulty in regenerating certain genotypes could be a result of many possibly inter-related factors, including acquisition of competence, induction and differentiation. Each may be mediated in a different manner that requires separate investigation.

4.2.7 Plantlet development

Basically there are only two options at this stage in the micropropagation sequence, to produce plantlets or cuttings for transfer to the *in vivo* environment. The alternative routes are shown in Figure 4.2. For most plantlets, shoot elongation, whether combined with rooting or not, is still necessary and is usually achieved by transferring shoots or shoot clusters to cytokinin-free medium.

4.3 Acclimatization of *in vitro* germplasm to *in vivo* conditions

The quality of germplasm to be transferred to *in vivo* conditions is critical to the success of this procedure. Visual evaluation should be used to select out plantlets exhibiting disorders such as hyperhydricity or apex necrosis. The term vitrification was proposed by Debergh (1988) to encompass any disorder that could check the survival of *in vitro* plantlets once transferred to greenhouse or field conditions. It is common for substantial numbers of micropropagated plants not to survive the transfer from *in vitro* conditions. Losses of micropropagated plants are due to the lower humidities, higher light levels and non-sterile conditions of the *in vivo* environment. To ensure that plants survive and grow vigorously when transferred to compost an acclimatization process is required (Figure 4.2). This process normally requires the provision of a high relative humidity, by fogging, misting etc., shading, manipulation of the photoperiod and in some cases bottom heat. As the plants adapt to the *in vivo* environment the conditions can be gradually adjusted to the 'normal' levels. To assist rooting, plantlets are often given an *in vivo* auxin treatment (Edson *et al.*, 1994). The acclimatization process has been reviewed by Preece and Sutter (1991).

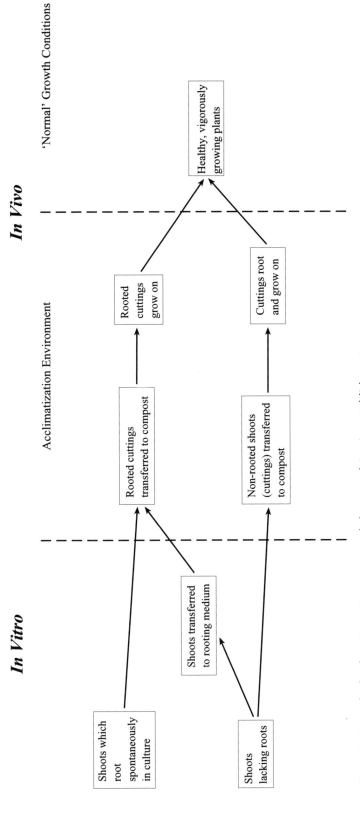

Figure 4.2 Methods of rooting micropropagated shoots and *in vivo* establishment

4.4   *In vitro* culture recalcitrance

Species of woody plants, cereals, grasses and some legumes are difficult to establish in culture, and recover viable regenerated shoots from, and are generally regarded as recalcitrant to tissue culture. Common problems include morphological recalcitrance, often in relation to tissue maturity (Pulido *et al.*, 1990), tissue browning (Li *et al.*, 1998) and hyperhydricity (Marga *et al.*, 1997). This has resulted in a range of empirical approaches in an attempt to overcome these difficulties, from the 'traditional', for example modification of growth factors types and concentrations in the culture medium (Heloir *et al.*, 1997), gelling agent (Zimmerman *et al.*, 1995), type of explant (Hsia and Korban, 1996), and the use of silver nitrate to control ethylene accumulation (Lentini *et al.*, 1995) to more novel approaches, including the use of ventilated culture containers (Majada *et al.*, 1997), bottom cooling (Piqueras *et al.*, 1998) and liquid raft culture (Teng, 1997). However, potentially of more significance have been the fundamental studies of the biochemistry and molecular biology of recalcitrant *in vitro* plant systems to try and understand the mechanisms of these problematic tissue culture responses. For example, cell ageing has been correlated with methylation of DNA sequences (Palmgren *et al.*, 1991) and demethylation implicated in the rejuvenation of woody perennials (Harding *et al.*, 1996). The oxidative stress status of cultures appears to fundamentally affect *in vitro* plant development (Benson *et al.*, 1997). It is from these studies that less empirical developments in plant tissue culture may be possible, with the aim of overcoming current recalcitrant responses. Such studies may also provide markers to tissue culture recalcitrance (Cazaux and Dauzac, 1995). However, the development of new approaches, which, if they are to be adopted by curators of *in vitro* genebanks, cannot be significantly more costly than existing procedures.

4.5   Embryo rescue

Embryo culture or rescue involves the dissection of embryos from seeds and their 'germination' *in vitro* to provide one plant per explant. This technique has been used to overcome post-zygotic incompatibility which would otherwise hamper the production of desired hybrids. For example, Bodanesezanettini *et al.* (1996) used embryo rescue to recover hybrids between Brazilian soybean lines and wild perennial *Glycine* species. Embryo rescue can also provide a means of recovering seedlings from genotypes that have low and/or rapidly lost seed viability or protracted dormancy period (Dilday *et al.*, 1994; Ganguli and Senmandi, 1995; Mian, *et al.*, 1995) and has proved to be of practical importance in the conservation of recalcitrant tree seed germplasm (see Marzalina and Krishnapillay, Chapter 17, this volume).

     To control microorganisms, it is normally sufficient to surface sterilize the fruits or seeds, after which the embryo can be dissected out under aseptic conditions. To ease dissection, hard seeds can be soaked in water to soften them. The culture medium on which the excised embryo germinates is usually of a simple formulation, consisting of inorganic salts and sucrose. Use of immature embryos can result in the recovery of a greater number of seedlings as compared with mature embryos. However, they are more difficult to excise and have additional media ingredient requirements, such as amino acids, vitamins and growth regulators. Seedlings derived from embryo rescue are acclimatized to *in vivo* conditions in a similar manner to micropropagated germplasm. The methods used in embryo rescue have been reviewed by Sharma *et al.* (1996).

4.6   Use of plant tissue culture for germplasm storage

4.6.1   *Slow (minimal) growth*

Reducing the growth rate of *in vitro* plant cultures can provide a convenient option for short- to medium-term germplasm storage. However, it is not suited to long-term pro- grammes, because of risks of selection due to stresses imposed on the cultures during storage (Withers, 1991). A variety of approaches have been used separately and in com- bination to reduce the growth rates of *in vitro* plant tissue. Probably the most successful strategies have involved temperature reductions, but responses vary significantly both between and within species. For example, cold tolerant species such as strawberry and *Prunus* sp. have been successfully stored at 0°C to 4°C (Reed, 1992; Wilkins *et al.*, 1988), but *Musa* plantlets cannot be stored below 15°C (Banerjee and DeLanghe, 1985). *Coffea arabica* plantlets can be maintained at 27°C and only require sub-culture every 12 months, but *Coffea racemosa* plantlets have to be transferred every six months (Bertrand-Desbrunais and Charrier, 1990). A reduction in light intensity can be used in combination with temperature reductions, for example with banana cultures (Banerjee and DeLanghe, 1985).

Modifications to the culture medium, including addition of osmotically active com- pounds such as mannitol (Staritsky and Zandvoort, 1985), reduction of the growth factor concentrations (Dussert *et al.*, 1994), the reduction of the medium's nutrient status (Malaurie *et al.*, 1993), and the use of growth retardants (Jarret and Gawel, 1991a) have all been reported to permit the maintenance of cultures in slow growth. Such changes to the culture medium have also been used in combination with reduced incubation tem- perature (Paulet and Glaszmann, 1994). The addition of activated charcoal to culture medium has also been reported as beneficial in minimal growth conditions. Roca *et al.* (1984) noted that its addition to cassava culture medium reduced defoliation, and limited chlorophyll degradation and root browning.

A reduction in growth of *in vitro* tissue can also be achieved by the lowering of the available oxygen levels. The simplest way of achieving this is to cover the culture with mineral oil. Overlaying callus cultures with mineral oil as a means of germplasm conser- vation was reported as early as 1959 by Caplin, who maintained 30 plant cell lines under mineral oil, with subculturing every 3–5 months, for longer than three years without apparent change in growth characteristics after transfer to normal growth conditions. It has been noted that mineral oil overlay could provide a useful means of short-term plant germplasm conservation in low-tech situations (Constabel and Shyluk, 1994; Engelmann, 1991).

Slow growth is used as a short- to medium-term storage approach in many laborat- ories, international/regional centres, including Centro Internacional de la Papa (CIP) and CIAT (Engelmann, 1991). However, even with increased time between subcultures, man- agement of large *in vitro* collections is problematic. There is also continued concern about the level of somaclonal variation under slow growth conditions (Jarret and Gawel, 1991b). However, cassava stored under slow growth conditions for 10 years has been shown to remain genetically stable (Angel *et al.*, 1996).

4.6.2   *Recovery of germplasm after storage*

Cryopreservation of plant tissue is the only viable option for the long-term storage of germplasm from species which are vegetatively propagated or produce recalcitrant seeds.

Cryopreservation will be discussed at length in other chapters in this volume, but it is important to remember that cryopreservation depends for its success to a significant degree on appropriate tissue culture approaches. Germplasm to be cryopreserved should be healthy and disease-free, therefore an understanding and appreciation of the *in vitro* requirements of germplasm to be cryopreserved is important. On thawing, plant tissue inevitably suffers damage and stress. The composition of culture medium has been shown to significantly affect post-thaw cellular cryoinjury, for example lipid peroxidation (Benson *et al.*, 1992). Therefore the use of specific post-thaw culture media can be important in the initiation of cell regrowth. For example, the use of ammonium ion-free medium (Kuriyama *et al.*, 1989) and supplementing the culture medium with activated charcoal (Scrijnemakers and van Iren, 1995) and the iron chelating agent desferrioxamine (Benson *et al.*, 1995) have all been shown to improve post-thaw cell regrowth.

4.7   Facilities for plant tissue culture

There are many descriptions in the literature of the requirements of laboratory and facility design, such as Mageau (1991), Bhojwani and Razdan (1996), which can be extremely elaborate and expensive both in terms of capital expenditure and running costs and can be unnecessary for the tissue culture being undertaken. At the early planning stages it is important to ensure that the basic infrastructure, such as continuity of power and supply of appropriate quality water are in place, and that local back-up support services are available for the repair and servicing of equipment. Key requirements of any plant tissue culture facility are:

- An area for media preparation, washing up and sterilization, equipped with standard laboratory equipment for preparation of solutions such as pH meters and stirrers, an autoclave and glassware washing and storage facilities.
- A transfer area for manipulation of the *in vitro* culture, in which prepared sterile media and equipment can also be stored. This room should be equipped if possible with sterile transfer hoods or UV light boxes for manipulating *in vitro* cultures.
- A growth room with an air-conditioning unit and open shelves, to aid air flow, preferably of metal construction with fluorescent tubes running under the shelves. It is preferable that the growth room is situated near the transfer area, in a room without windows and if possible away from outside walls to reduce the effects of external temperature fluctuations.

It is vital that the design of the transfer area and growth room is such that both areas can be easily kept clean. This is an important consideration in avoiding loss of cultures due to microbial contamination.

4.8   Conclusions

Tissue culture provides a potentially important means in itself for the *in vitro* conservation of plant germplasm and also vital supportive technology for long-term cryopreservation-based storage. However, its success requires understanding and an appreciation of the shortcomings of plant tissue culture, particularly in terms of minimizing the effects of somaclonal variation. No one tissue culture approach is universally appropriate for the storage of all plant germplasm. Therefore the development of integrated approaches will be important, such as that discussed by Dixon (1994). Finally the cost implications of *in vitro* approaches to plant germplasm conservation have to be compared with more traditional approaches. In a recent cost analysis study of the field and *in vitro* parts of the cassava collection at CIAT, it was shown that the *in vitro* collection, including the isozyne laboratory for genotype characterization was 53 per cent more expensive than field-based collections (Epperson *et al.*, 1997). However, the authors felt that the annual cost of preserving the world's most complete collection of cassava germplasm (Roca *et al.*, 1992) was not excessive considering the importance of the crop as a major food source.

References

AL MAZROOEI, S., BHATTI, M.H., HENSHAW, G.G., TAYLOR, N.J. and BLAKESLEY, D., 1997, Optimisation of somatic embryogenesis in fourteen cultivars of sweet potato [*Ipomoea batatas* (L) Lam], *Plant Cell Reports*, **16**, 710–714.

ANGEL, F., BARNEY, V.E., TOHME, J. and ROCA, W.M., 1996, Stability of cassava plants at the DNA level after retrieval from 10 years of *in vitro* storage, *Euphytica*, **90**, 307–313.

ASHMORE, S.E., 1997, *Status Report on the Development and Application of In vitro Techniques for the Conservation and Use of Plant Genetic Resources*, Rome: IPGRI.

ATREE, S.M. and FOWKE, L.C., 1991, Micropropagation through somatic embryogenesis in conifers, in BAJAJ, Y.P.S. (Ed.), *Biotechnology in Agriculture and Forestry, Vol. 17, High-Tech and Micropropagation*, pp. 53–70, Berlin: Springer-Verlag.

BANERJEE, N. and DELANGHE, E., 1985, A tissue culture technique for rapid clonal propagation and storage under minimal growth conditions of *Musa* (banana and plantain, *Plant Cell Reports*, **4**, 351–354.

BARBAS, E., JAYALLEMAND, C., DOUMAS, P., CHAILLOU, S. and CORNU, D., 1993, Effects of gelling agents on growth, mineral composition and naphthoquinone content of *in vitro* explants of hybrid walnut tree (*Juglans regia* × *Juglans nigra*), *Annales des Sciences Forestieres*, **50**, 177–186.

BARLASS, M. and SKENE, K.G.M., 1986, Citrus (*Citrus* species), in BAJAJ, Y.P.S. (Ed.), *Biotechnology in Agriculture and Forestry, Trees I*, pp. 207–219, Heidelberg: Springer-Verlag.

BENSON, E.E. and ROUBELAKIS-ANGELAKIS, K.A., 1994, Fluorescent lipid peroxidation products and antioxidant enzymes in tissue cultures of *Vitis vinifera* L., *Plant Science*, **84**, 83–90.

BENSON, E.E., LYNCH, P.T. and JONES, J., 1992, The detection of lipid peroxidation products in cryoprotected and frozen rice cells: consequences for post-thaw survival, *Plant Science*, **85**, 107–114.

BENSON, E.E., LYNCH, P.T. and JONES, J., 1995, The use of the iron chelating agent desferrioxamine in rice cell cryopreservation: a novel approach for improving recovery, *Plant Science*, **110**, 249–258.

BENSON, E.E., MAGILL, W.J. and BREMNER, D.H., 1997, Free radical processes in plant tissue cultures: implications for plant biotechnology programmes, *Phyton-Annales rei Botanicae*, **37**, 31–38.

BERGMANN, B.A., SUN, Y.H. and STOMP, A.M., 1997, Harvest time and nitrogen source influence *in vitro* growth of apical buds from Fraser fir seedlings, *HortScience*, **32**, 125–128.

BERTRAND-DESBRUNAIS, A. and CHARRIER, A., 1990, Conservation des ressources genetiues cafeieres, in, *Proceedings of the Colloquium Internation*, pp. 438–446, ASIC, Paipu, Colombia.

BHOJWANI, S.S. and RAZDAN, M.K., 1996, *Plant Tissue Culture: Theory and Practice, a Revised Edition*, London: Elsevier.

BLOCK, R. and LANKES, C., 1996, Measures to prevent tissue browning of explants of the apple rootstock M9 during *in vitro* establishment, *Gartenbauwissenschaft*, **61**, 11–17.

BODANESEZANETTINI, M.H., LAUXEN, M.S., RICHTER, S.N.C., CALALLIMOLINA, S., LANGE, C.E., WANG, P.J. and HU, C.Y., 1996, Wide hybridisation between Brazilian soybean cultivars and wild perennial relatives, *Theoretical and Applied Genetics*, **93**, 703–709.

BONNIER, F.J.M. and VAN TUYL, J.M., 1997, Long-term *in vitro* storage of lily: effects of temperature and concentration of nutrients and sucrose, *Plant Cell Tissue and Organ Culture*, **49**, 81–87.

BROWN, D.C.W. and ATANASSOV, A.I., 1985, Role of genetic background in somatic embryogenesis, in *Medicago, Plant Cell Tissue and Organ Culture*, **4**, 111–122.

BROWN, C.R., KWIATKOWSKI, S., MARTIN, M.W. and THOMAS, P.E., 1988, Eradication of PVS from potato clones through excision of meristems from *in vitro*, heat treated shoot tips, *American Potato Journal*, **65**, 633–638.

BUSH, S.R., EARLE, E.D. and LANGHANS, R.W., 1976, Plantlets from petal epidermis and shoot tips of the periclinal chimera *Chrysanthemum moriloium* 'Indianapolis', *American Journal of Botany*, **63**, 729–737.

CAMPBELL, R., 1985, *Plant Microbiology*, London: Arnold.

CAPLIN, S.M., 1959, Mineral oil overlay for conserving plant tissues, *American Journal of Botany*, **46**, 324–329.

CASSELLS, A.C., 1987, *In vitro* induction of virus-free potatoes by chemotherapy, in BAJAJ, Y.P.S. (Ed.), *Biotechnology in Agriculture and Forestry, Vol. 3, Potato*, pp. 40–50, Berlin: Springer-Verlag.

CASSELLS, A.C., 1988, Bacterial and bacteria-like contaminants of plant tissue cultures, *Acta Horticulturae*, **225**, 225.

CASSELLS, A.C., 1991, Problems of tissue culture contamination, in DEBERGH, P.C. and ZIMMERMAN, R.H. (Eds), *Micropropagation, Technology and Application*, pp. 31–44, Dordrecht: Kluwer Academic.

CASSELLS, A.C. and MINAS, G., 1983, Plant and *in vitro* factors influencing the micropropagation of *Pelargonium* cultivars by bud-tip culture, *Scientific Horticulture*, **21**, 53–65.

CASSELLS, A.C., HARMEY, M.A., CARNEY, B.F., McCARTHY, E. and McHUGH, A., 1988, Problems posed by cultivable bacterial endophytes in the establishment of axenic cultures of *Pelargoniun × domesticum*: the use of *Xanthomonas pelargonii*-specific ELISA, DNA probes and culture indexing in the screening of antibiotic treated and untreated donor plants, *Acta Horticulturae*, **225**, 153–162.

CAZAUX, E. and DAUZAC, J., 1995, Explanation for the lack of division of protoplasts from stems of rubber tree (*Hevea braziliensis*), *Plant Cell Tissue and Organ Culture*, **41**, 211–219.

CHRISTIANSON, M.L. and WARNICK, D.A., 1988, Organogenesis *in vitro* as a developmental process, *Hortscience*, **23**, 515–519.

CHU, C.C., WANG, C.C., SUN, C.S., HSU, C., YIN, K.C., CHU, C.Y. and BI, F.Y., 1975, Establishment of an efficient medium for the anther culture of rice, through comparative experiments on the nitrogen sources, *Scientia Sinica*, **18**, 659–668.

COLBY, S.M., JUNCOSA, M. and MEREDITH, C.P., 1991, Cellular differences in *Agrobacterium* susceptibility and regeneration capacity restrict the development of transgenic grapevines, *Journal of the American Horticultural Society*, **116**, 351–361.

COLLINS, G.G. and SYMONS, R.H., 1992, Extraction of nuclear DNA from grapevine leaves by a modified procedure, *Plant Molecular Biology Reporter*, **10**, 233–235.

CONSTABEL, F. and SHYLUK, J.P., 1994, Initiation, nutrition and maintenance of plant cell and tissue cultures, in VASIL, I.K. and THORPE, T.A. (Eds), *Plant Tissue Culture*, pp. 3–15, Dordrecht: Kluwer Academic Publishers.

CORLEY, R.H.V., LEE, C.H., LAW, L.H. and WONG, C.Y., 1986, Abnormal flower development in oil palm clones, *Planter*, **62**, 233–240.

D'AMATO, F., 1989, Polyploidy in cell differentiation, *Caryologia*, **42**, 183–211.

DEBERGH, P.C., 1988, Improving mass propagation of *in vitro* plantlets, in Organising Committee of the International Symposium on High Technology in Protected Cultivation (Eds), *Horticulture in a High Technological Era – Special Lectures*, pp. 47–57, Japan.

DEBERGH, P.C. and READ, P.E., 1991, Micropropagation, in DEBERGH, P.C. and ZIMMERMAN, R.H. (Eds), *Micropropagation, Technology and Application*, pp. 1–13, Dordrecht: Kluwer Academic Publishers.

DE JONG, J. and CUSTERS, J.B.M., 1986, Induced changes in growth and flowering of chrysanthemums after irradiation and *in vitro* culture of pedicels and petal epidermis, *Euphytica*, **35**, 137–148.

DILDAY, R.H., YAN, W.G., LEE, F.N., HELMS, R.S. and HUANG, F.H., 1994, Application of embryo rescue in recovering rice (*Oryza sativa* L.) germplasm, *Crop Science*, **34**, 1636–1638.

DIXON, K.W., 1994, Towards integrated conservation of Australian endangered plants – the Western Australian model, *Biodiversity and Conservation*, **3**, 148–159.

DOLEZEL, J. and NOVAK, 1984, Effect of plant tissue culture medium on the frequency of somatic mutations in *Tradescantia* stamen hairs, *Z. Pflanzenphysiolgia*, **114**, 51–58.

DUSSERT, S., CHABRILLANGE, N., RECALT, C., BOURSOT, M. and HAMON, S., 1994, ORSTOM germplasm collections: origins and management, in *Rapport d'Activite, Laboratoire de Ressources Genetiques et Amelioration des Plantes Tropicales*, pp. 42–43.

EASTMAN, P.A.C., WEBSTER, F.B., PITEL, J.A. and ROBERTS, D.R., 1991, Evaluation of somaclonal variation during somatic embryogenesis of interior spruce (*Picea glauca–engelmannii* complex) using culture morphology and isozyme analysis, *Plant Cell Reports*, **10**, 425–430.

EDSON, J.L., WENNY, D.L. and LEEGERBRUSVEN, A., 1994, Micropropagation of pacific dogwood, *Hortscience*, **29**, 1355–1356.

EDSON, J.L., LEEGE-BRUSVEN, A.D., EVERETT, R.L. and WENNY, D.L., 1996, Minimising growth regulators in shoot culture of an endangered plant *Hackelia venusta* (Boraginaceae), *In Vitro Cellular and Developmental Biology – Plant*, **32**, 267–271.

ENGELMANN, F. 1991, *In vitro* conservation of tropical plant germplasm – a review, *Euphytica*, **57**, 227–243.

ENJALRIC, F., CARRON, M.P. and LARDET, L., 1988, Contamination of primary cultures in tropical areas: the case of *Hevea brasiliensis, Acta Horticulturae*, **225**, 57–65.

EPPERSON, J.E., PACHICO, D.H. and GUEVARA, C.L., 1997, A cost analysis of maintaining cassava plant genetic resources, *Crop Science*, **37**, 1641–1649.

FEIJOO, M.C. and IGLESIAS, I., 1998, Multiplication of an endangered plant: *Gentiana lutea* L. Subsp. *Aurantiaca* Lainz, using *in vitro* culture, *Plant Tissue Culture and Biotechnology*, **4**, 87–94.

FENG, G.H. and OUYANG, J., 1988, The effects of potassium nitrate concentration in callus induction medium for wheat anther culture, *Plant Cell Tissue and Organ Culture*, **12**, 3–12.

FOLEY, M.E., 1992, Effect of soluble sugars and gibberellic acid in breaking dormancy of excised wild oat (*Avena fatua*) embryos, *Weed Science*, **40**, 208–214.

FRANCK, T., CREVECOEUR, M., WUEST, J., GREPPIN, H. and GASPAR, T., 1998, Cytological comparison of leaves and stems of *Prunus avium* L. shoots cultured on a solid medium with agar and Gelrite, *Biotechnology and Histochemistry*, **73**, 32–43.

FUCHS, M., PINCK, M., ETIENNE, I., PINCK, I. and PALTER, B., 1991, Characterisation and detection of grapevine fan leaf virus by using cDNA probes, *Phytopathology*, **81**, 559–565.

FURNER, I.J., KING, J. and GAMBOURG, O.L., 1978, Plant regeneration from protoplasts isolated from a predominantly haploid suspension culture of *Datura innoxia* (Mill.), *Plant Science Letters*, **11**, 169–176.

GAMBORG, O.L., MILLER, R.A. and OJIMA, K., 1968, Nutrient requirements of suspension cultures of soybean root cells, *Experimental Cell Research*, **50**, 151–158.

GANGULI, S. and SENMANDI, S., 1995, Embryo rescue from non-viable seeds for preservation of wheat (*Triticum aestivum*) germplasm, *Indian Journal of Agricultural Sciences*, **65**, 781–784.

GEORGE, E.F., 1993, *Plant Propagation by Tissue Culture, Part 1, The Technology*, 2nd edition, Edington: Exegetics Ltd.

GREEN, S.K. and LO, C.Y., 1989, Elimination of sweet potato yellow dwarf virus (SPVDV) by meristem culture and heat treatment, *Journal of Plant Disease*, **96**, 464–469.

GRENO, V., CAMBRA, M., NAVARRO, L. and DURAN-VILA, N., 1990, Effect of antiviral chemicals on the development and virus content of citrus buds cultured *in vitro*, *Scientifica Hortica*, **45**, 75–87.

GRIMES, H.D. and HODGES, T.K, 1990, The inorganic NO_3 NH_4^+ ratio influences plant-regeneration and auxin sensitivity in primary callus derived from immature embryos of Indica rice (*Oryza-sativa* L.), *Journal of Plant Physiology*, **136**, 362–367.

GUARINO, L., RAMANATHA RAO, V. and REID R., 1995, *Collecting Plant Genetic Diversity: Technical Guidelines*, Wallingford: CAB International.

HALPERIN, W., 1966, Alternative morphological events in cell suspensions, *American Journal of Botany*, **53**, 443–453.

HAN, K.C. and STEPHENS, L.C., 1992, Carbohydrate and nitrogen-sources affect respectively *in vitro* germination of immature ovules and early seedling growth of *Impatiens platypetala* L., *Plant Cell Tissue and Organ Culture*, **31**, 211–214.

HARDING, K., BENSON, E.E. and ROUBELAKIS-ANGELAKIS, K.A., 1996, Methylated DNA changes associated with the initiation and maintenance of *Vitis vinifera in vitro* shoot and callus culture: A possible mechanism for age-related changes, *Vitis*, **35**, 79–85.

HASEGAWA, P.M., 1979, *In vitro* propagation of rose (*Rosa hybrida* L.), *HortScience*, **14**, 610–612.

HELOIR, M.C., FOURNIOUX, J.C., OZIOL, L. and BESSIS, R., 1997. An improved procedure for the propagation *in vitro* of grapevine (*Vitus vinifera* cv. Pinot noir) using auxillary bud microcuttings, *Plant Cell Tissue and Organ Culture*, **49**, 223–225.

HSIA, C.N. and KORBAN, S.S., 1996, Factors affecting *in vitro* establishment and shoot prolifera-tion of *Rosa hybrida* L. and *Rosa chinensis minima, In Vitro Cellular and Developmental Biology – Plant*, **32**, 217–222.

IPGRI/CIAT, 1994, Establishment and operation of a pilot *in vitro* active genebank, Report of a CIAT–IPGRI Collaborative Project using Cassava (*Manihot esculenta* Crantz) as a model, Rome: IPGRI.

JAHNE, A., LAZZERI, P.A., JAGER-GUSSEN, M. and LORZ, H., 1991, Plant regeneration from embryogenic cell suspensions derived from anther cultures of barley (*Hordeum vulgare* L.), *Theoretical and Applied Genetics*, **82**, 74–80.

JAIN, R.K., DAVEY, M.R., COCKING, E.C. and WU, R., 1997, Carbohydrate and osmotic require-ments for high-frequency plant regeneration from protoplast-derived colonies of Indica and Japonica rice varieties, *Journal of Experimental Botany*, **48**, 751–758.

JAMES, D.J., PASSEY, A.J. and MALHOTRA, S.B., 1984, Organogenesis in callus derived from stem and leaf tissues of apple and cherry rootstocks, *Plant Cell Tissue and Organ Culture*, **3**, 333–341.

JARRET, R.L. and GAWEL, N., 1991a, Absisic acid induced growth inhibition of sweet potato (*Ipomoea batatas* (L.) Lam) *in vitro, Plant Cell Tissue and Organ Culture*, **24**, 13–18.

JARRET, R.L. and GAWEL, N., 1991b, Chemical and environmental growth regulation of sweet potato (*Ipomoea batatas* (L.) Lam) *in vitro, Plant Cell Tissue and Organ Culture*, **25**, 153–159.

JHA T.B. and ROY, S.C., 1982, Effect of different hormones on *Viigna* tissue culture and its chromosome behaviour, *Plant Science Letters*, **24**, 219–224.

JUNG, B.K., PYO, J.H., KIM, W.S., NAM, B.H., HWANG, S.J. and HWANG, B., 1998, Cloning of genes specifically expressed in rice embryogenic cells, *Molecules and Cells*, **8**, 62–67.

KAMEYA, T. and WIDHOLM, J., 1981, Plant-regeneration from hypocotyl sections of *Glycine* species, *Plant Science Letters*, **21**, 289–294.

KARP, A., 1992, The role of growth regulators in somaclonal variation, *British Society for Plant Growth Regulation Annual Bulletin*, **2**, 1–9.

KARP, A., 1995, Somaclonal variation as a tool for crop improvement, *Euphytica*, **85**, 295–302.

KARTHA, K.K., 1986, Production and indexing of disease free plants, in WITHERS, L.A. and ANDERSON, P.G. (Eds), *Plant Tissue Culture and its Agricultural Applications*, pp. 219–238, London: Butterworths.

KIKKERT, J.R., HEBERTSOULE, D., WALLACE, P.G., STRIEM, M.J. and REISCH, B.I., 1996, Transgenic plantlets of chancellor grapevine (*Vitis* sp.) from biolistic transformation of embryogenic-cell suspensions, *Plant Cell Reports*, **15**, 311–316.

KOETJE, D.S., GRIMES, H.D., WANG, Y.C. and HODGES, T.K., 1989, Regeneration of Indica rice (*Oryza sativa* L.) from primary callus from immature embryos, *Journal of Plant Physiology*, **13**, 184–190.

KOZAI, T., 1991, Micropropagation under photoautotrophic conditions, in DEBERGH, P.C. and ZIMMERMAN, R.H. (Eds), *Micropropagation: Technology and Application*, pp. 447–469, Dordrecht: Kluwer Academic.

KURIYAMA, A., WATANABE, K., UENO, S. and MITSUDA, H., 1989, Inhibitory effect of ammonium ions on recovery of cryopreserved rice cells, *Plant Science*, **64**, 231–235.

LARKIN, P.J. and SCOWCROFT, W.R., 1981, Somaclonal variation – a novel source of variability from cell culture for plant improvement, *Theoretical and Applied Genetics*, **60**, 197–214.

LEE, S.Y., LEE, M.T. and LEE M.S., 1988, Studies on the anther culture of *Oryza sativa* L. 3, Growing environment of the donor plant for anther culture, effects of photoperiod and light intensity, *Research Reports of the Rural Development Administration, Suweon, Korea*, **30**, 7–12.

LENTINI, Z., REYES, P., MARTINEZ, C.P. and ROCA, W.M., 1995, Androgenesis of highly recalcitrant rice genotypes with maltose and silver-nitrate, *Plant Science*, **110**, 127–138.

LI, X.Y., HUANG, F.H. and GBUR, E.E., 1998, Effect of basal medium, growth regulators and Phytogel concentration on initiation of embryogenic cultures from immature zygotic embryos of lobolly pine (*Pinus taeda* L.), *Plant Cell Reports*, **17**, 298–301.

LINSMAIER, E.M. and SKOOG, F., 1965, Organic growth factor requirements of tobacco tissue cultures, *Physiologia Plantarum*, **8**, 100–127.

LITZ, R.E. and GRAY, D.J., 1992, Organogenesis and somatic embryogenesis, in HAMMERSCHLAG, F.A. and LITZ, R.E. (Eds), *Biotechnology of Perennial Fruit Crops*, pp. 3–34, Wallingford: CAB International.

LONG, R.D., CURTIN, T.F. and CASSELLS, A.C., 1988, An investigation of the effects of bacterial contaminants on potato nodal cultures, *Acta Horticulturae*, **225**, 83–92.

LU, C.Y., VASIL, V. and VASIL, I.K., 1983, Improved efficiency of somatic embryogenesis and plant regeneration in tissue culture of maize (*Zea mays* L.), *Theoretical and Applied Genetics*, **66**, 285–289.

LUBRANO, L., 1992, Micropropagation of Poplars (*Populus* sp.), in BAJAJ, Y.P.S. (Ed.), *Biotechnology in Agriculture and Forestry, Vol. 18, High-Tech. Micropropagation II*, pp. 151–179, Berlin: Springer-Verlag.

LYNCH, P.T., 1999. Applications of cryopreservation to the long-term storage of dedifferentiated plant cultures, in RAZDAN, M.K. and COCKING, E.C. (Eds), *Applications of In Vitro Plant Germplasm Conservation*, Vol. 2, Oxford: Oxford Press.

LYNCH, P.T., FINCH, R.P., DAVEY, M.R. and COCKING E.C., 1991, Rice tissue culture and its applications, in KHUSH, G.S. and TOENNIESSEN, G.H. (Eds), *Rice Biotechnology; Biotechnology, Agriculture*, No. 6, pp. 135–156, Wallingford: CAB International.

MAGEAU, O.C., 1991, Laboratory design, in DEBERGH, P.C. and ZIMMERMAN, R.H. (Eds), *Micropropagation, Technology and Application*, pp. 15–29, Dordrecht: Kluwer Academic.

MAJADA, J.P., FAL, M.A. and SACHEZ-TAMES, R., 1997, The effect of ventilation rate on proliferation and hyperhydricity of *Dianthus carophyllus* L., *In Vitro Cellular and Developmental Biology – Plant*, **33**, 62–69.

MALAURIE, B., PUNGU, O. and TROUSLOT, M.F., 1993, The creation of an *in vitro* germplasm collection of yam (*Dioscorea* sp.) for genetic resource preservation, *Euphytica*, **65**, 113–122.

MARCOTRIGIANO, M., 1986, Synthesizing plant chimeras as a source of new phenotypes, *Conference Proceedings of the International Plant Propagation Society*, **35**, 582–586.

MARGA, F., VEBRET, L. and MORVAN, H., 1997, Agar fractions could protect apple shoots cultured in liquid media against hyperhydricity, *Plant Cell Tissue and Organ Culture*, **49**, 1–5.

McCABE, P.F., LEVINE, A., MEIJER, P.-J., TAPON, N.A. and PENNELL, R.I., 1997, A programmed cell death pathway activated in carrot cells cultured at low density, *The Plant Journal*, **12**, 267–280.

MEZETTI, B., SAVINI, G., CARNEVALI, F. and MOTT, D., 1997, Plant genotype and growth factor interaction affecting *in vitro* morphogenesis of blackberry and raspberry, *Biologia Plantarum*, **39**, 139–150.

MIAN, M.A.R., SKIRVIN, R.M., NORTON, M.A. and OTTERBACHER, A.G., 1995, Dryness interferes with germination of blackberry (*Rubus* sp.) seeds *in vitro*, *HortScience*, **30**, 124–126.

MOREL. G. and MARTIN, C. 1952, Guerison de dahlias atteints d'une maladie à virus. *Compt. Rend. Acad. Sci., Paris*, **235**, 1324–1325.

MURASHIGE, T., 1974, Plant propagation through tissue culture, *Annual Review of Plant Physiology*, **25**, 135–166.

MURASHIGE, T. and SKOOG, F., 1962, A revised medium for rapid growth requirements of tobacco tissue cultures, *Physiologia Plantarum*, **15**, 473–497.

NASIR, F.R. and MILES, N.W., 1981, Histological origin of EMLA 26 apple shoots generated during micropropagation, *HortScience*, **16**, 417.

NITSCH, J.P., 1969, Experimental androgenesis in *Nicotiana*, *Phytomophology*, **19**, 389–404.

NOVAK, P.J., ZADINA, J., HORACKOVA, V. and MASKOVA, I., 1980, The effect of growth regulators on meristem tip development and *in vitro* multiplication of *Solanum tuberosum* L., plants, *Potato Research*, **23**, 155–160.

OERTEL, C. and BREUEL, K., 1987, Virus freie Klonpflanzen fur die Digitaliszuchtung, *Pharmazie*, **42**, 217.

PALMGREN, G., MATTSSON, O. and OKKELS, F.T., 1991, Specific levels of DNA methylation in various tissues, cell lines, and cell-types of *Daucus carota, Plant Physiology*, **95**, 174–178.

PAULET, F. and GLASZMANN, J.C., 1994, Biotechnological support for varietal extension of sugarcane, *Agriculture et Development, Special Issue*, Dec. 1984, 47–53.

PIQUERAS, A., HAN, B.H., VAN HUYLENBROECK, J.M. and DEBERGH, P.C., 1998, Effect of different environmental conditions *in vitro* on sucrose metabolism and antioxidant enzyme activities in cultured shoots of *Nicotiana tabacum* L., *Plant Growth Regulation*, **25**, 5–10.

PODWYSZYNSKA, M. and OLSZEWSKI, T., 1995, Influence of gelling agents on shoot multiplication and the uptake of macroelements by *in vitro* culture of rose, cordyline and homalomena, *Scientia Horticulturae*, **64**, 77–84.

PRANCE, G.T., 1997, The conservation of botanic diversity, in, MAXTED, N., FORD-LLOYD, B.V. and HAWKES, J.G. (Eds), *Plant Genetic Conservation*, London: Chapman and Hall.

PREECE, J.E. and SUTTER, E.G., 1991, Acclimatisation of micropropagated plants to the greenhouse and field, in DEBERGH, P.C. and ZIMMERMAN, R.H. (Eds), *Micropropagation, Technology and Application*, pp. 71–93, Dordrecht: Kluwer Academic.

PULIDO, C.M., HARRY, I.S. and THORPE, T.A., 1990, *In vitro* regeneration of plantlets of Canary Island pine (*Pinus canariensis*), *Canadian Journal of Forest Research*, **20**, 1200–1211.

RAGHAVEN, V., 1997, *Molecular Embryology of Flowering Plants*, Berlin: Springer-Verlag.

RAMSAY, M.M. and STEWART, J., 1998, Re-establishment of the lady's slipper orchid (*Cypripedium calceolus* L.) in Britain, *Botanical Journal of the Linnean Society*, **126**, 173–181.

REED, B.M., 1992, Cold storage of strawberries *in vitro*: A comparison of three storage systems, *Fruit Varieties Journal*, **46**, 98–102.

REEVES, D.W., HORTON, B.D. and COUVILLON, G.A., 1983, Effect of media and media pH on *in vitro* propagation of Nemaguard peach rootstock, *Scientific Horticulture*, **21**, 353–357.

REEVES, D.W., COUVILLON, G.A. and HORTON, B.D., 1985, Effect of gibberellic acid (GA$_3$) on elongation and rooting of St Julian A rootstock *in vitro*, *Scientific Horticulture*, **26**, 253–259.

ROCA, W.M., RODRIGUES, J.A., MAFLA, G. and ROA, J., 1984, Procedures for recovering cassava clones distributed *in vitro*, CIAT, Cali, Colombia.

ROCA, W.M., HENRY, G., ANGEL, F. and SARRIA, R., 1992, Biotechnology research applied to cassava improvement at the Centro Internacional de Agricultura Tropical (CIAT), *AgBiotech News and Information*, **4**, 303N-308N.

SCHOLTEN, H.J. and PIERIK, R.L.M., 1998, Agar as a gelling agent: chemical and physical analysis, *Plant Cell Reports*, **17**, 230–235.

SCHRIJNEMAKERS, E.W.M. and VAN IREN, F., 1995, A two-step or equilibrium freezing pro-
cedure for the cryopreservation of plant cell suspensions, in DAY, J.G. and MCLELLAN,
M.R. (Eds), *Cryopreservation and Freeze-Drying Protocols, Methods in Molecular Biology*,
Vol. 38, pp. 103–111, Totowa, USA: Humana Press.

SHARMA, D.R., KAUR, R. and KUMAR, K., 1996, Embryo rescue in plants – a review, *Euphytica*,
89, 325–337.

SHETTY, K. and ASANO, Y., 1991, Specific selection of embryogenic cell-lines in *Agrostis alba* L.
using the proline analogue thioproline, *Plant Science*, **79**, 259–263.

STARITSKY, G. and ZANDVOORT, E.A., 1985, *In vitro* propagation and genetic conservation of
tropical woody crops, *Acta Botanica Neerlandica*, **34**, 238.

STREET, H.E., 1969, Growth in organised and unorganised systems – knowledge gained by culture
of organs and tissue explants, in STEWARD, F.C. (Ed.), *Plant Physiology – A Treatise*, 5B,
pp. 3–224, New York: Academic Press.

TARAN, B. and BOWLEY, S.R., 1997, Inheritance of somatic embryogenesis in orchard grass,
Crop Science, **37**, 1497–1502.

TENG, W.L., 1997, Regeneration of *Anthurium* adventitious shoots using liquid or raft culture,
Plant Cell Tissue and Organ Culture, **49**, 153–156.

THEILER-HEDTRICH, R. and BAUMANN, G., 1989, Elimination of apple mosaic-virus and raspberry bushy dwarf virus from infected red raspberry (*Rubus idaeus* L.) by tissue-culture,
Journal of Phytopathology Phytopathologische Zeitschrift, **127**, 193–199.

THORPE, T.A., 1994, Morphogenesis and regeneration, in VASIL, I.K. and THORPE, T.A. (Eds),
Plant Cell and Tissue Culture, pp. 17–36, Dordrecht: Kluwer Academic Publishers.

TREMBLAY, L. and TREMBLAY, F.M., 1991, Effects of gelling agents, ammonium nitrate and
light on the development of *Picea mariana* (Mill) BSP (Black Spruce) and *Picea rubens* sarg
(Red Spruce) somatic embryos, *Plant Science*, **77**, 23–242.

VASIL, I.K. and THORPE, T.A., 1994, *Plant Cell and Tissue Culture*, Dordrecht: Kluwer Academic Publishers.

VASIL, I.K. and VASIL, V., 1986, Regeneration of cereal and other grass species, in Vasil, I.K.
(Ed.), *Cell Culture and Somatic Cell Genetics of Plants*, Vol. 3, pp. 121–150, New York:
Academic Press.

VILLALOBOS, V.M., FERREIRA, P. and MORA, A., 1991, The use of biotechnology in the conservation of tropical germplasm, *Biotechnological Advances*, **9**, 197–215.

VON ARNOLD, S., CLAPHAM, D., EGERTSDOTTER, U. and MO, L.H., 1996, Somatic embryogenesis
in conifers – A case study of induction and development of somatic embryos in *Picea abies*,
Plant Growth Regulation, **20**, 3–9.

VUYLSTEKE, D. and SWENNEN, R., 1990, Somaclonal variation in African plantains, *IITA Research*, **1**, 4–10.

WANG, W.C. and MARSHALL, D., 1996, Genomic rearrangement in long-term shoot competent
cell cultures of hexaploid wheat, *In Vitro Cellular and Developmental Biology – Plant*, **32**,
18–25.

WANG, X.H., LAZZERI, P.A. and LORZ, H., 1993, Regeneration of haploid dihaploid and diploid
plants from anther- and embryo-derived cell suspensions of wild barley (*Hordeum murinum*
L.), *Journal of Plant Physiology*, **141**, 726–732.

WHITE, P., 1963, *The Cultivation of Animal Cells and Plant Cells*, 2nd edition, New York: Roland
Press.

WICKREMESINHE, E.R.M., RODREGUEZ, E. and ARDITTI, J., 1990, Adventitious shoot regeneration and plant production from explants of *Apios americana*, *HortScience*, **25**, 1436–1439.

WILKINS, C.P., NEWBURY, H.J. and DODDS, J.H., 1988, Tissue culture conservation of fruit
trees, *FAO/IBPGR Plant Genetic Resources Newsletter*, **73/74**, 9–20.

WITHERS, L.A., 1991. Maintenance of plant tissue cultures, in KITSOP, B.E. and DOYLE, A.
(Eds), *Maintenance of Microorganisms: A Manual of Laboratory Methods*, 2nd edition, London: Academic Press.

WITHERS, L.A., 1995, Collecting *in vitro* for genetic resources conservation, in GUARINO, L., RAMANATHA RAO, V. and REID R. (Eds), *Collecting Plant Genetic Diversity: Technical Guidelines*, pp. 511–526, 162–185, Wallingford: CAB International.

YIDANA, J.A., WITHERS, L.A. and IVINS, J.D., 1987, Development of a simple method for collecting and propagating cocia germplasm *in vitro, Acta Horticulturae*, **212**, 95–98.

ZEL, J., DEBELJAK, N., UCMAN, R. and RAVNIKAR, M., 1997, The effect of jasmonic acid, sucrose and darkness on garlic (*Allium sativum* L. cv. Ptujski jesenski) bulb formation *in vitro*, *In Vitro Cellular and Developmental Biology – Plant*, **33**, 231–235.

ZIMMERMAN, R.H., BHARDWAJ, S.V. and FORDHAM, J.M., 1995, Use of starch gelled medium for tissue culture of some fruit crops, *Plant Cell Tissue and Organ Culture*, **43**, 207–213.

5

Phytosanitary Aspects of Plant Germplasm Conservation

ROBERT R. MARTIN AND JOSEPH D. POSTMAN

5.1 Introduction

The development of useful *ex situ* germplasm collections, whether for genetic resource conservation or in support of crop improvement programmes, requires that plants be collected from various sites around the world (see Hummer, Chapter 3, this volume) and introduced to new locations. A goal of many germplasm repositories is to represent the global genetic diversity of plant genera in their charge. One of the risks associated with collection of plant germplasm, especially from wild sources, is the inadvertent introduction of diseases or other pests along with plants or seeds. Some of the world's most destructive plant diseases have been the result of the accidental introduction of exotic pathogens during the importation of plant materials. Examples of such importations include downy mildew of corn and sorghum caused by *Peronosclerospora sorghi* (W. Weston & Uppal) C.G. Shaw, dutch elm disease (*Ophiostoma ulmi* (Buisman) Nannf.), white pine blister rust (*Cronartium ribicola* J.C. Fisch.), chestnut blight (*Cryphonectria parasitica* (Murril) Barr), and more recently karnal bunt of wheat (*Neovossia indica* (Mitra) Mundk.), sharka disease of *Prunus* sp. (Plum Pox Virus) and tomato spotted wilt virus (many hosts). Virus and virus-like diseases pose a special challenge as they are often symptomless in infected plants and require specialized tests to determine their presence. As plant breeders attempt to broaden the genetic base of our agricultural and horticultural crops there are increasing efforts to introduce new genetic material from areas of origin of the species involved. These introductions are often land races or wild species that are not closely examined for the presence of germplasm-borne pathogens.

5.2 Safe movement of germplasm

Quarantines have become the primary strategy for preventing the international movement of viruses and other pests along with plant material (Foster and Hadidi, 1998). Finding a balance between excluding pests through the use of quarantine measures and introducing germplasm for the improvement of agricultural production is often a challenge. The destruction of plant material found to be infected with a pathogen or the many years of testing often required before material can be released from quarantine, has frustrated farmers, horticulturists and breeders anxious to incorporate new genetic resources into their programmes (Plucknett and Smith, 1989). *In vitro* culture and molecular diagnostic techniques are helping to expedite the detection and elimination of pathogens from plant germplasm. The protection of virus tested clones in certification programmes helps to safeguard valuable plant material, and prevent the reintroduction of pathogens into clean stock. With increased international movement of germplasm the challenge to prevent introduction of exotic pathogens is also increased. The ease of air travel provides a multitude of opportunities for the introduction of a 'favourite' fruit, vegetable or flower; however, this introduction can lead to serious consequences if an exotic plant pathogen is inadvertently introduced at the same time. It is believed that this is how necrotic strain of potato virus Y (PVY^N) was introduced into eastern Canada, resulting in millions of dollars in losses to the seed potato industry in that region (MacDonald *et al.*, 1994).

5.2.1   *Quarantine*

Quarantines are the first line of defence against the movement of economically important plant pests between and within countries. Most countries have enacted quarantines to prevent the economic losses that result from the introduction of exotic insects or diseases. Introduced pests have been responsible for devastating losses of native plants and cultivated crops. Quarantines are not only used between countries, but are also an important administrative tool for preventing the spread of diseases within countries. In the United States, for example, certain states with pine based timber industries restrict the importation of *Ribes* species, which are an alternate host for *Cronartium ribicola* (white pine blister rust). The state of Oregon restricts the importation of *Corylus* species (hazelnut) from the eastern two-thirds of North America as well as from several Oregon counties to prevent the further spread of *Anisogramma anomala* Peck (e. Muller) (eastern filbert blight). Provinces in western Canada controlled the movement of potatoes from the Maritime Provinces during the early 1990s to protect their seed potato industries from the introduction of PVY^N.

5.2.2   *International cooperation*

The International Plant Protection Convention (IPPC) was adopted by a conference of the Food and Agriculture Organization (FAO) of the United Nations in 1952 with the purpose of 'securing common and effective action to prevent the spread and introduction of pests of plants and plant products and to promote measures for their control' (FAO, 1998). As of 1996, 105 countries had agreed to abide by the IPPC. The IPPC directs member countries to develop procedures for issuing phytosanitary certificates, and encourages international cooperation to prevent the spread of pests while minimizing interference with international trade. While most individual countries maintain their own plant quarantine regulations, several regional plant protection organizations have been formed by IPPC member countries in an attempt to provide regional consistency in regulations, and cooperation in the development of plant certification schemes (Table 5.1). Contact information for these regional organizations can be found at the FAO website (FAO, 1998). Three additional regional organizations administer plant protection programmes not associated with the IPPC (Table 5.2). The FAO has defined two categories of

Table 5.1 Regional plant protection organizations established under the international plant protection convention

APPPC Asia and Pacific Plant Protection Commission (established in 1956)	Australia, Bangladesh, Cambodia, China, Fiji, France (for French Polynesia), India, Indonesia, Laos, Malaysia, Myanmar, Nepal, New Zealand, Pakistan, Papua New Guinea, Philippines, Portugal (for Macau), Republic of Korea, Samoa, Solomon Islands, Sri Lanka, Thailand, Tonga, Vietnam
CPPC Caribbean Plant Protection Commission (established in 1967)	Barbados, Colombia, Costa Rica, Cuba, Dominica, Dominican Republic, France (for French Guiana, Guadeloupe, Martinique), Grenada, Guyana, Haiti, Jamaica, Mexico, Netherlands (for Aruba, Netherlands Antilles), Nicaragua, Panama, Saint Kitts and Nevis, Saint Lucia, Suriname, Trinidad and Tobago, United Kingdom (for British Virgin Islands), United States of America (for American Virgin Islands, Puerto Rico), Venezuela
COSAVE Comité Regional de Sanidad Vegetal para el Cono Sur (established in 1989)	Argentina, Brazil, Chile, Paraguay, Uruguay
EPPO European and Mediterranean Plant Protection Organization (established in 1950)	Albania, Austria, Belgium, Bulgaria, Croatia, Cyprus, Czech Republic, Denmark, Estonia, Finland, France, Germany, Greece, Guernsey, Hungary, Ireland, Israel, Italy, Jersey, Latvia, Luxembourg, Malta, Morocco, Netherlands, Norway, Poland, Portugal, Romania, Russian Federation, Slovakia, Slovenia, Spain, Sweden, Switzerland, Tunisia, Turkey, Ukraine, United Kingdom
IAPSC Inter-African Phytosanitary Council (established in 1956)	Algeria, Angola, Benin, Botswana, Burkina Faso, Burundi, Cameroon, Cape Verde, Central African Republic, Chad, Comoros, Congo, Côte d'Ivoire, Djibouti, Egypt, Equatorial Guinea, Ethiopia, Gabon, Gambia, Ghana, Guinea, Guinea-Bissau, Kenya, Lesotho, Liberia, Libyan Arab Jamahiriya, Madagascar, Malawi, Mali, Mauritania, Mauritius, Morocco, Mozambique, Niger, Nigeria, Rwanda, São Tomé and Principe, Senegal, Seychelles, Sierra Leone, Somalia, Sudan, Swaziland, Togo, Tunisia, Uganda, United Republic of Tanzania, Zaire, Zambia, Zimbabwe
JUNAC Junta del Acuerdo de Cartagena (established in 1969)	Bolivia, Colombia, Ecuador, Peru, Venezuela
NAPPO North American Plant Protection Organization (established in 1976)	Canada, Mexico, United States of America
OIRSA Organismo Internacional Regional de Sanidad Agropecuaria (established in 1955)	Costa Rica, Dominican Republic, El Salvador, Guatemala, Honduras, Mexico, Nicaragua, Panama
PPPO Pacific Plant Protection Organization (established in 1995)	Australia (including Norfolk Island), Cook Islands, Fiji, France (for French Polynesia, New Caledonia, Wallis and Futuna Islands), Kiribati, Marshall Islands, Micronesia (Federated States of), Nauru, New Zealand, Niue, Northern Mariana Islands (Commonwealth of), Palau, Papua New Guinea, Samoa, Solomon Islands, Tokelau, Tonga, Tuvalu, United Kingdom (for Pitcairn), United States of America (for American Samoa and Guam), Vanuatu

Table 5.2   Other regional organizations not established under the international plant protection convention with plant protection programmes

EU European Union	Austria, Belgium, Denmark, France, Germany, Greece, Ireland, Italy, Luxembourg, Netherlands, Portugal, Spain, Sweden, United Kingdom
IICA Instituto Interamericano de Cooperación para la Agricultura (established in 1980)	Antigua and Barbuda, Argentina, Bahamas (Commonwealth of the), Barbados, Belize, Bolivia, Brazil, Canada, Chile, Colombia, Costa Rica, Dominica, Dominican Republic, Ecuador, El Salvador, Grenada, Guatemala, Guyana, Haiti, Honduras, Jamaica, Mexico, Nicaragua, Panama, Paraguay, Peru, St Kitts and Nevis, St Lucia, St Vincent and the Grenadines, Suriname, Trinidad and Tobago, the United States of America, Uruguay and Venezuela
SPC South Pacific Commission (established in 1947)	Australia (including Norfolk Island), Cook Islands, Fiji, France (for French Polynesia, New Caledonia, Wallis and Futuna Islands), Kiribati, Marshall Islands, Micronesia (Federated States of), Nauru, New Zealand, Niue, Northern Mariana Islands (Commonwealth of), Palau, Papua New Guinea, Samoa, Solomon Islands, Tokelau, Tonga, Tuvalu, United Kingdom (for Pitcairn), United States of America (for American Samoa and Guam), Vanuatu

quarantinable pests. An 'A-1' pest is not yet present in an area and is the most significant from a quarantine standpoint. An 'A-2' pest may be present in an area but is not widely distributed. Quarantine regulations are generally more rigorous for A-1 than for A-2 pests.

Europe

The European Plant Protection Organization (EPPO) makes recommendations to its 39 member countries in Europe and the Mediterranean basin (Table 5.1) regarding plant quarantine issues related to the movement of commodities.

   The 15 nations of the European Union (EU) (Table 5.2) abide by EU phytosanitary laws, many of which are consistent with EPPO recommendations. EPPO has published a number of certification schemes, for example for strawberries (EPPO, 1994) and for fruit trees (EPPO, 1991, 1992), with specific suggestions for selection, production and maintenance of nuclear stock, and guidelines for pathogen testing and sanitation.

United States

In the United States, importation of nursery stock, plants, roots, bulbs, seeds and other plant products is controlled by foreign quarantine regulations (Title 7, Chapter III, Part 319 of the US Code of Federal Regulations). The Animal and Plant Health Inspection Service (APHIS), an agency of the US Department of Agriculture, is charged with enforcing plant quarantine regulations. For most plant genera, seeds have fewer restrictions than vegetative materials. Plant material imported for propagation will fall under one of the following categories, depending on the plant species and the country of origin:

1   *Restricted* – These plant propagules can be imported by any individual in any quantity but are subject to inspections for quarantine pests.
   (a)   Many plant species are not mentioned in the US quarantine regulations. Such plants or seeds may be inspected for visible evidence of diseases or insects at the

port of entry, and if the propagules appear healthy, they are released to the importing individual.

(b) Individuals must obtain a permit to import other plants. These plants or seeds must be inspected at the port of entry and if there is no sign of insects or disease, they are released to the permit holder.

(c) Plant materials which are at risk of harbouring more hazardous pests must be imported with a permit, inspected at an inspection station, and then grown according to certain 'post-entry' conditions. Post-entry restrictions generally require that plants be grown at an approved site, kept a certain distance away from other plants of the same genus, and inspected by an agricultural official several times during two growing seasons.

2 *Prohibited* – Certain vegetatively propagated plants, including many of the fruit and nut crops, cannot be imported directly by private individuals or organizations in commercial quantities. These plants can only be introduced in small quantities through a government approved quarantine facility, where they must be tested for economically important pathogens (especially viruses) before being released for general propagation and dissemination.

5.3 International guidelines

The International Plant Genetic Resources Institute (IPGRI) is another organization associated with FAO which promotes the conservation and use of genetic resources (http://www.cgiar.org/ipgri). In recognition of the phytosanitary hazards involved in the international movement of plant germplasm, IPGRI has funded a number of research programmes and conferences to develop appropriate technologies for reducing the risks of moving pathogens with plants (Frison and Diekmann, 1998). A series of crop-specific handbooks have been published by IPGRI to provide technical guidelines for the safe movement of a number of economically important crop plants (Frison and Diekmann, 1998; Diekmann, personal Communication) (Table 5.3).

These guides have especially targeted vegetatively propagated crops which carry a higher risk of carrying virus diseases and they suggest appropriate methods for excluding these diseases during germplasm exchange. Some general recommendations to promote the safe movement of germplasm include:

Table 5.3 Technical guidelines for the safe movement of germplasm published by FAO/IPGRI

Allium spp. (1997)	Pines (in press 1998)
Edible aroids (1989)	Potatoes (in press 1998)
Cassava (1991)	Small fruits (*Fragaria, Ribes, Rubus, Vaccinium*) (1994)
Citrus (1991)	Small cereal grains (1995)
Cocoa (1989)	Stone fruits (1996)
Coconut (1993)	Sugarcane (1993)
Eucalyptus (1996)	Sweet potato (1989)
Grapevine (1991)	Vanilla (1991)
Legumes (1990)	Yam (1989)
Musa (2nd edition 1996)	

- Germplasm should be obtained from a safe (pathogen tested) source.
- Movement of seed or pollen poses less risk than does movement of plants.
- *In vitro* cultures pose less risk than does vegetative plant materials potted in soil.
- *In vitro* material should be derived from pathogen tested sources or tested for viruses known to occur in the country of origin.
- Transfer of germplasm should be planned in consultation with quarantine authorities and a relevant indexing laboratory.
- Vegetative material should be subjected to full quarantine measures.

Reviews of successful certification programmes have recently been published for potatoes, grapes, ornamental plants, deciduous fruit trees, citrus, and strawberries (Hadidi *et al.*, 1998). Common features of certified stock schemes involve:

- Listing the pathogens of concern.
- Identifying appropriate methods for detecting the pathogens.
- Implementing strategies for eliminating pathogens from infected plants.
- Protecting foundation or nuclear stock from reinfection.
- Verifying that plants are correctly labelled or are 'true-to-type'.
- Preventing reinfection during subsequent propagation and distribution to commercial producers.

IPGRI recommends the development and use of 'broad-spectrum' virus detection techniques for quarantine indexing (Frison and Diekmann, 1998). Such tests may not identify specific pathogens, but rather will detect members of a larger group such as all plant phytoplasmas, or all members of a virus family.

5.4   Virus detection

Accurate disease diagnosis combined with sensitive, rapid and early detection of plant viruses is critical for effective management of most crop systems. However, in plant germplasm collections detection of a wide range of viruses, some of which may not be known, is required generally for a small number of samples. There have been several major improvements in virus detection over the past two decades. Serological detection was greatly improved with the introduction of ELISA to plant virus detection in the mid-1970s. This was enhanced further with the development of monoclonal antibody technology and its application to a large number of plant viruses in the 1980s. Similarly, nucleic acid hybridization has been used successfully for detecting many plant viruses and is the preferred method for detection of viroids. Cloning of plant viral nucleic acids and the development of non-radioactive detection methods have increased the utility of nucleic acid hybridization for virus detection. Most recently, the development of polymerase chain reaction (PCR) has greatly improved the sensitivity and utility of hybridization and other nucleic acid based assays. Immunocapture PCR combines the advantages of serology and PCR into a very sensitive method of detection (Chevalier *et al.*, 1995; Wetzel *et al.*, 1992).

   In germplasm repositories it is often necessary to test for viruses in accessions that are from diverse geographical locations and that may be infected with unknown viruses. Tests for common viruses known to occur in a genus should be carried out. However, repositories have the added need to determine whether these materials may be infected with yet undescribed viruses. This necessitates the use of broad spectrum techniques such as mechanical transmissions to herbaceous indicators, electron microscopy, double-stranded RNA analysis, grafting onto indicator hosts or looking for viral inclusions.

Monoclonal antibodies, nucleic acid hybridization and PCR provide the potential for the development of diagnostic reagents with desired specificities. Diagnostic reagents that detect all or most members of a virus group would be very useful in germplasm repositories. There are monoclonal antibodies that recognize most members of the potyviridae (Hammond and Jordan, 1991) or multiple luteoviruses (D'Arcy *et al.*, 1989) and oligonucleotides that can be used in PCR to amplify sequences from most luteoviruses (Robertson *et al.*, 1991) or geminiviruses (Rojas *et al.*, 1993).

5.4.1 *Requirements for detection and diagnosis*

The requirements for specificity and sensitivity of detection will vary in different situations. Germplasm repositories and clean plant programmes want to ensure that their 'nuclear' material is free of known viruses. In this situation, where much depends on the virus status of relatively few individual plants, several tests should be employed and the virus status of every plant may determined. While serological or nucleic acid-based tests often can be employed for detection of well characterized viruses known to occur in the crop, tests that detect a wide range of viruses are especially desirable. Mechanical transmission to selected herbaceous indicator plants, electron microscopic examination of leaf dips, double-stranded RNA (dsRNA) analysis and/or grafting may be desirable to ensure that the germplasm or 'nuclear' material is virus-free. 'Nuclear' material refers to the few plants that are the basis of all plant material in a clean plant propagation scheme and may also be referred to as mother block, foundation, or elite material (Martin, 1998). New material being added to virus-free collections may be put through a virus eradication programme prior to testing. This enhances the likelihood that the material will be free of viruses not reported previously in the crop and free of new strains of a virus which may not be detected with available antibodies or nucleic acid probes.

Germplasm repositories often have permits to bring in material to meet their mandate without having the plants go through a plant quarantine station. Therefore, their virus detection programmes should be similar to those of plant quarantine. In plant quarantine it is necessary to ensure that plant material is free of restricted viruses prior to its release to industry or breeding programmes. Quarantine programmes often work with only a few plants of a given genotype but the level of indexing on these few plants is as complete as possible. Maximum sensitivity is the goal of virus detection in quarantine programmes. These programmes often use an agreed upon test for each specific virus. The test may be grafting or mechanical transmission onto specific indicator plants, an ELISA, hybridization assay or PCR for specific viruses (Diekmann *et al.*, 1994). Detection of viruses of woody plants will often require graft transmission tests to indicator plants. When grafting is required for a single virus in a crop it might be more efficient to employ graft transmission tests for all viruses of quarantine significance since the work is already being done for the one virus. Thus, quarantine facilities may not opt for the newer laboratory techniques until they are available for all quarantinable viruses capable of infecting that crop.

Quarantine officials may also require extreme specificity in cases where a severe or resistance breaking strain of a virus occurs in some countries but not others. When there are several sources of plant material it might be easier to destroy all material that is infected with a particular virus rather than trying to determine which strain of a virus is present. There is a resistance breaking strain of raspberry bushy dwarf virus (RBDV) that occurs in Europe and is not known to occur in North America. The only way to differentiate the two strains is to graft onto resistant varieties of raspberry (Barbara *et al.*, 1984). Both strains are readily detected, but cannot be differentiated serologically and in practice any material being imported to North America from Europe that tests positive for RBDV is destroyed rather than risk introducing the resistance breaking strain of the virus. Recently, reverse transcription combined with PCR (RT–PCR) has been developed to differentiate these two strains of RBDV (Barbara *et al.*, 1995) and as this test is improved it may be possible to reliably detect resistance breaking strains of RBDV.

5.4.2 *Serological detection*

Enzyme-linked immunosorbent assay (ELISA) and dot immunobinding assay (DIBA) are currently the most widely used methods of serological detection for plant viruses. There have been relatively few changes in ELISA or DIBA for virus detection since the application of monoclonal antibodies (McAbs) (Halk and DeBoer, 1985; Jordan, 1990; Martin, 1998; Miller and Martin, 1988). ELISA has been reviewed elsewhere (Chu *et al.*, 1989; Clark and Bar-Joseph, 1984; Converse and Martin, 1990) and the general outline will not be covered here. Rather, some of the principles of the assays will be discussed. These assays are carried out on a solid phase (usually plastic multi-well plates, nitrocellulose membranes (Banttari and Goodwin, 1985) or filter paper (Haber and Knapen, 1989)) where each component of the test is applied successively and the reaction between virus and antibody is detected by enzymatic hydrolysis of a substrate that results in a colour change or light emission.

Assays carried out on nitrocellulose or filter paper are referred to as DIBA. The DIBA is about as sensitive as ELISA (Makkouk *et al.*, 1993) and offers the advantage that sample preparation can be very simple. A few microlitres of sap can be spotted onto the membrane or a freshly cut edge of a leaf, petiole or stem can be pressed gently against the absorbent membrane. The latter approach is referred to as tissue blotting and also provides some information on the location of the virus in the tissue, e.g. phloem (Holt, 1992). The other advantage of DIBA is that it is readily adapted to field situations and applications in areas with minimal laboratory facilities (Haber and Knapen, 1989; Makkouk *et al.*, 1993). The membranes can be taken to the field and plant tissue or insects blotted directly onto the membrane. Both of these assays can be performed without any specialized equipment.

The ELISA procedure is quite adaptable and many variations on the original method (Clark and Adams, 1977) have been described (Koenig and Paul, 1982). A standardized test protocol should be used in quarantine and certification schemes to ensure consistent results between laboratories or from year-to-year in the same facility. Standardization of an ELISA protocol may include factors such as the manufacturer of a microtitre plate used in the assay, identification of a specific monoclonal or polyclonal antiserum, part of plant to be sampled and at what stage of plant development sampling should take place, list of buffers to be used at each step of the procedure, the length and temperature of each incubation, how long the substrate should develop before absorbance values are taken, and what should be used as positive and negative controls.

ELISA in plant virology usually has two or three steps that use an antibody. The antibody applied directly to the microtitre plate is usually referred to as the trapping or coating antibody, since its purpose is to selectively bind the antigen of interest to the plate. The coating antibody is applied as purified immunoglobulin G (IgG) diluted in an appropriate buffer (usually carbonate or phosphate buffered saline). The second antibody (often the same source of IgG as the coating antibody) is conjugated with an enzyme in the standard double antibody sandwich (DAS) ELISA and is referred to as the conjugate or detecting antibody. In triple antibody sandwich (TAS) ELISA, the coating is done in the same manner as in DAS-ELISA but the second antibody (primary or detecting antibody) is specific for the antigen of interest and produced in a different animal than the trapping or coating antibody. The primary antibody is then followed by a conjugated antibody (secondary antibody) that is specific for antibodies produced by the animal that was the source of the primary antibody. For example, if the primary antibody was a McAb produced in a mouse, then the secondary antibody might be a rabbit–anti-mouse conjugate. We prefer a conjugate that is made in the same host species as the coating antibody to prevent cross reaction between conjugate and coating antibody.

When using monoclonal antibodies (McAbs) for virus detection or diagnosis, the substrate can usually be incubated overnight to increase the sensitivity of the assay. In this way, McAbs can be used to detect virus in individual aphids using a standard ELISA protocol (Martin and Ellis, 1986; Torrance, 1987). In our laboratory, we routinely take absorbance readings 1–2 hours after adding substrate and again after overnight incubation at room temperature. The most important aspect of developing an assay is to maximize the absorbance of infected/healthy samples. A good polyclonal or monoclonal antiserum can be used to develop an assay that does not require statistical analysis to differentiate between known positive and negative samples. If such a test is available, samples that give borderline results should be retested. It must be remembered however, that even with the best of detection methods, recently infected samples may well give a borderline or negative result. The application of statistics to determine when an absorbance value in ELISA represents a positive result has been presented elsewhere (Sutula *et al.*, 1986).

5.4.3   *Nucleic acid based assays*

Detection of pathogens by nucleic acid hybridization is based on the specific pairing between the target nucleic acid sequence (denatured DNA or RNA) and a complementary nucleic acid probe to form double-stranded nucleic acids. Thus, either RNA or DNA sequences may be used as probes. For detection of plant viruses, hybridizations are usually carried out on solid filter supports where the target nucleic acids are immobilized and the labelled nucleic acid probe is allowed to hybridize to them (Anderson and Young, 1985). Nitrocellulose and charged nylon are the most commonly used filters for hybridization. Nylon based filters are easier to handle and can be rehybridized several times (Gatti *et al.*, 1984). The use of nucleic acid hybridization assays for virus detection has been reviewed recently (Chu *et al.*, 1989; Hull, 1988). The greatest improvements in hybridization assays in recent years have been the advances in non-radioactive detection systems. Several methods of labelling nucleic acids are available. Incorporation of modified nucleotides such as biotin-11-UTP, or digoxigenin tagged UTP (Boehringer Mannheim, technical bulletin), can then be detected by streptavidin or an anti-digoxigenin antibody, respectively. Digoxigenin-labelled dUTP appears to be the most widely used non-radioactive tag for labelling probes for detection of plant viruses. Digoxigenin is linked via a spacer arm to dUTP and then incorporated into probe with the same enzymes used to make ^{32}P-labelled probes (Feinberg and Vogelstein, 1983). Antibodies specific for the digoxigenin and conjugated to an enzyme (usually alkaline phosphatase or horseradish peroxidase) are then used to complex with the bound probe. Either a precipitating or a light emitting substrate is then used to visualize the presence of the probe. Many biotechnology supply companies now market kits for tagging probes with non-radioactive labels.

The sensitivity of detection of plant viruses by nucleic acid hybridization is roughly similar to that of ELISA (Barbara *et al.*, 1987; Chu *et al.*, 1989; Fouly *et al.*, 1992; Mas *et al.*, 1993). Hybridization assays were more sensitive than ELISA with subterranean clover stunt virus (SCSV) when purified virus was used (Chu *et al.*, 1989); however, when infected tissues were used, ELISA was found to be more sensitive: the pathogen was detectable at a sap dilution of 1/625 while hybridization assays had a dilution limit of 1/125. The sensitivity of nucleic acid hybridization and ELISA were similar with barley yellow dwarf virus (BYDV) (Fouly *et al.*, 1992) and cherry leaf roll (Mas *et al.*, 1993) viruses.

Another consideration in the choice of detection procedure, is the ease of carrying out an assay. If two methods of detection are sensitive enough to meet the needs of an experiment, then the method that is simpler to carry out will probably be used. The expertise of the worker will determine which test is simpler. Someone who has extensive experience working with nucleic acids will be more inclined to use PCR or hybridization assays rather than ELISA; the converse is true for someone who is more familiar with serology than nucleic acid based tests.

There are many viruses of woody plants that have only been described in terms of symptomatology (Converse, 1987; Gilmer *et al.*, 1976). It is likely that nucleic acid based detection will be available before serological methods are developed for these viruses. Developments in cloning viral specific dsRNAs of plant viruses (Jelkmann *et al.*, 1989; Winter *et al.*, 1992) make it possible to readily make cDNA from dsRNA templates. Thus, for viruses where dsRNA can be extracted from diseased tissues, probes for nucleic acid hybridization or sequence information required to develop a PCR test are realistic short-term goals. Once the sequences are known, it will be possible to prepare antibodies to the coat proteins of these viruses for use in serological assays.

In the case of phytoplasmas, several groups have described universal oligonucleotides for amplification of phytoplasma specific DNA (Ahrens and Seemuller, 1992; Lee *et al.*, 1993; Smart *et al.*, 1996). Since this group of pathogens cannot be cultured, are not easily transmitted by grafting or mechanically, and can be difficult to detect by microscopy, the PCR based test is the preferred method for detecting phytoplasmas.

5.4.4   *Detection based on more traditional methods*

Serological or nucleic acid based diagnostic tools are not available to detect many of the viruses of woody plants because it has not been possible to purify the viruses. It is still necessary to carry out graft transmissions or mechanical inoculations onto herbaceous hosts for these viruses. Attempts to improve quarantine and certification schemes for tree fruit or small fruit crops is limited because these tests are labour intensive, and especially with grafting, require substantial amounts of space to grow test plants. It may be several years after grafting before the results of the test are obtained.

The introduction of new germplasm from collections made in the wild presents the possibility that undescribed viruses will be encountered (Spiegel *et al.*, 1993b). New viruses encountered will not be on any quarantine list but may still pose a biological risk. This material should be assayed with a broad spectrum test such as mechanical transmission, grafting or dsRNA analysis. Quite often we tend to be concerned about the viruses with which we are familiar and ignore those that have not yet been described.

5.4.5 *The significance of a test result*

In situations where a test result has significant biological or economic impact, one should use more than one type of test to confirm the presence or absence of a virus. A recent incidence of PVY^N in Canada and its implications for shipping seed potatoes to the USA is an excellent example of where test results had considerable economic impact and a confirmatory test could potentially have saved a lot of money and possibly prevented lawsuits. In this case the effect of mixed virus infections confounded the bioassay results on tobacco leading to a large percentage of false positives. As the movement of agricultural products between countries increases, quarantine restrictions based on plant pathogens will continue to be important. For example, several *Prunus* spp. were found to be infected with virus isolates which cross-reacted with plum pox potyvirus (PPV) antisera in ELISA (James *et al.*, 1994). In this instance, the impact of the test result was very significant since plum pox is an A-1 quarantine status virus not known to occur in North America. Plum pox is a member of the potyviridae that produce diagnostic inclusions in infected hosts. The prunus virus isolates did not induce these inclusions; their coat proteins were atypical of members of the potyviridae and RT–PCR tests with oligonucleotides specific to the 3′ or 5′ non-coding regions of plum pox virus did not amplify any fragments (James *et al.*, 1996). It is now thought that these prunus virus isolates are not plum pox virus, nor are they members of the potyviridae despite their reaction with antisera specific to plum pox virus. Reliance solely on the initial test results would have had very serious implications.

Another important consideration when using diagnostic tests that are not based on biological activity is the significance of the result. For example, many people are aware that PCR has been used to amplify DNA from insects embedded in amber for millions of years. This positive 'test' for insect DNA does not show that the insect is living. A similar situation could arrive when using laboratory tests to index plants for viruses. It has been shown that biologically active prunus necrotic ringspot virus is slowly lost from *Prunus pennsylvanica* seed (Fulton, 1964). Would ELISA, PCR or hybridization give positive results when biologically the virus was no longer seed transmitted? In the case of biological tests, it is also important to know that the symptoms observed are due to the virus in question rather than a mixed infection as was the case with some of the PVY^N testing mentioned above.

The legal implications of a decision to refuse a shipment of plant product based on a positive test that does not consider biological activity is an issue that needs to be considered. If a shipment of grain that has been fumigated with methyl bromide is assayed for a specific fungal pathogen by PCR, it would be possible to get a positive result based on non-living fungal tissue. Similarly, the results of a positive ELISA test for plum pox potyvirus could be used to turn back a shipment of *Prunus* planting stock, when in fact the material was free of plum pox potyvirus. We must remember to consider the biology of the pathogen and host rather than rely on a band in a gel or yellow colour development in a well of a microtitre plate.

5.5 Production of pathogen-free plants

Most plants that are vegetatively propagated from a virus-infected plant will be infected with the same viruses. These clonal plant populations can become infected with additional viruses if exposed to virus-carrying vectors such as nematodes or aphids. Many heirloom fruit cultivars have been clonally propagated for centuries and are universally virus infected, although the viruses may be latent or symptomless during much of the growing season. Pathogen-free plants can sometimes be identified by careful indexing, but often the only way to obtain healthy foundation stock for important clonally propagated plant cultivars is to subject them to virus-elimination therapy.

Exposure of growing plants or plant parts to elevated temperatures (heat therapy or thermotherapy), to apical meristem culture, or to antiviral chemicals (chemotherapy) has been used to eliminate viruses from infected plants. Improved virus elimination can often be achieved by combining these various techniques. The treated plants generally are not cured during therapy, but rather new plants are propagated from shoot tips or apical meristems following treatment. Plant tissue culture offers a convenient system for exposing plants to controlled temperatures or to controlled concentrations of antiviral chemicals. *In vitro* therapy also provides already-sterile plant tissue from which meristems can be dissected with no need for additional surface sterilization.

5.5.1   *Heat therapy*

While the exact effect of heat therapy on plant viruses is not well understood, it is known that replication of many viruses is significantly reduced at elevated temperatures (Spiegel *et al.*, 1993a; Walkey, 1980; Wang and Hu, 1980). The production of virus-encoded movement proteins and coat proteins may also be temperature sensitive (Mink *et al.*, 1998). These proteins are involved in cell-to-cell movement of viruses through plasmodesmata and long distance movement through the plant vascular system. Disruption in the production or activity of these proteins may also play a role in the effectiveness of heat therapy.

Nyland and Goheen (1969) reviewed the early history of using heat to eliminate viruses and other pathogens from infected plant material. Brief exposure of plants or plant parts to hot water at temperatures ranging from 30° to 70°C has been used successfully to eliminate many pathogens from infected plants, but nearly all of the pathogens eliminated by hot water baths were later discovered to be phytoplasmas and not viruses. Hot water may be a useful technique for sanitizing plant parts to eliminate arthropods, fungi, bacteria and phytoplasmas, but it has not been particularly useful for eliminating viruses (Fridlund, 1971, 1989; Nyland and Goheen, 1969).

Most modern virus therapy programmes involve growing whole plants or *in vitro* cultures at temperatures close to the threshold of normal plant growth. For most plants this is between 38° and 40°C. Mink *et al.* (1998) reviewed the effect of elevated temperatures on viruses and on plant physiology. Reduced synthesis of RNA at 40°C has been shown for several viruses, including tobacco mosaic virus (TMV) and cowpea chlorotic mottle virus (CCMV). At elevated temperatures synthesis of ssRNA stopped immediately for both of these viruses and synthesis of dsRNA stopped more gradually. When plants were returned to normal temperatures of 25°C, resumption of viral RNA synthesis lagged behind plant RNA synthesis by 4–8 hours for CCMV and by 16–20 hours for TMV.

Infected plants may be subjected to constant therapy temperatures, however alternating temperatures are commonly used to improve host survival. The lag in resumption of viral RNA synthesis behind that of the host plant may explain why alternating-temperature heat therapy is perhaps more successful than constant-temperature therapy. Plant survival is improved during alternating-temperature therapy. We have successfully eliminated many viruses from numerous fruit, nut, and oil crop genera (*Corylus, Fragaria, Humulus, Mentha, Mespilus, Pyrus, Rubus, Ribes, Sorbus, Vaccinium*) by growing either potted plants or *in vitro* plantlets at temperatures that alternate every four hours between 30° and 38°C (unpublished data). After two to four weeks of heat therapy, apical meristems are dissected and established *in vitro*. Once these meristem derived plants are rooted and established in soil, they are allowed to go through a natural dormant period and retested for viruses during the following growing season. The dormant period allows viruses, that may have been reduced to undetectable levels by therapy, to build up in the plant prior to indexing.

While typical virus therapy involves growing an infected plant at elevated temperatures prior to meristem culture, cucumber mosaic and alfalfa mosaic viruses have been eliminated by subjecting dissected meristems to heat therapy after they were removed from infected plants (Walkey, 1980). Cold therapy rather than heat therapy, followed by apical meristem culture, has successfully eliminated several viruses from infected plants (Walkey, 1980) and cold therapy has been particularly effective in the elimination of viroids, some of which are quite resistant to elevated temperatures (Lizarraga *et al.*, 1980; Paduch-Cichal and Kryczynski, 1987; Postman and Hadidi, 1995). Plants or *in vitro* cultures are typically grown at temperatures between 4° and 7°C for one to six months prior to removal of meristems to eliminate viroids including potato spindle tuber, chrysanthemum stunt, and apple scar skin viroids.

While there seems to be little agreement on the importance of humidity or light levels during therapy, it has been shown that elevated CO_2 enhances survival of some plant species at elevated temperatures. Blueberry (*Vaccinium* spp.) plants, which normally are difficult to keep alive at 38°C, were able to survive for extended periods at 40°C when the CO_2 level was increased to 1200 ppm, 3–4 times the normal concentration (Converse and George, 1987). We have observed improved survival of the infected host during *in vitro* heat therapy if the plantlet is cultured in a heat-sealed gas-permeable plastic pouch (Reed, 1991) rather than a test tube or other culture container. These plastic containers allow some gas exchange but are impervious to the exchange of moisture. Improved plant condition may be due to a decrease in fungal or bacterial contaminants, better water retention in the medium, or elevated CO_2 levels in these containers.

Viruses vary in their susceptibility to heat therapy. Apple mosaic ilarvirus, blueberry scorch carlavirus, and several of the mosaic viruses of raspberry are readily eliminated following therapy times of only 10–20 days and in the case of apple mosaic, following propagation of relatively large shoot tips over a centimetre in length (Postman and Mehlenbacher, 1994). Other viruses such as raspberry bushy dwarf idaeovirus, tobacco streak ilarvirus and apple stem grooving capillovirus persist in some meristems smaller than 0.5 mm dissected after 3–4 weeks of heat therapy.

5.5.2 Meristem tip culture

Virus elimination has been documented for a number of virus–host combinations following prolonged *in vitro* culture of the infected host plant, and in many additional instances following excision and culture of apical meristems. Faccioli and Marani (1998) recently reviewed the various roles that *in vitro* culture has played in elimination of viruses from infected plant material. The mechanisms for virus elimination during routine tissue culture are uncertain, however the presence of plant growth hormones, the physiological response of explants to repeated injuries, the induction of inhibitors or the disruption of enzymes needed for virus replication are several possibilities. Cherry leaf roll virus and arabis mosaic virus for example, have disappeared from infected plants after several months of growth *in vitro*, yet tobacco ringspot, cucumber mosaic and potato virus X are persistent.

It has been suggested that many viruses are unable to infect the apical meristem of a growing plant and that a virus-free plant can be produced if a small enough piece of apical tissue is propagated. Facciolo and Marani (1998), however, cite a number of studies where electron microscopy and fluorescence-linked antibodies have documented the presence of more than a dozen different viruses in apical dome tissue of assorted plant species. Yet plants regenerated from these infected meristems are often free of the viruses. The larger the size of an excised meristem the better the chance that it will survive *in vitro* culture, and the smaller the meristem size, the more likely it will be virus-free (Facciolo and Marani, 1998; Wang and Hu, 1980). The goal of many virus elimination programmes is to grow a meristem consisting of the apical dome and a pair of leaf primordia. Depending on the plant genotype, this shoot tip should be between 0.2 and 0.8 mm in length. Distribution of a virus within a plant may be uneven, especially towards the shoot tips (Fridlund, 1973; Gilmer and Brase, 1963), and a virus-free plant may be produced by random propagation of enough buds or shoot tips. Experience with a particular crop and the viruses that infect it will determine the size of the meristems that can be established *in vitro*, and how many meristems must be grown to have a reasonable probability that one of them will be virus-free. A single, virus-free, true-to-type plant is all that is necessary to produce a population of healthy plants.

Some woody plants are difficult to establish in *in vitro* culture from meristems, or difficult to root. These difficulties have been overcome in *Citrus* (Nauer *et al.*, 1983; Navarro *et al.*, 1975), *Malus* (Huang and Millikan, 1980), and *Prunus* (Deogratias and Lutz, 1986) by 'micrografting' shoot tips or apical meristems onto *in vitro* grown seedling rootstocks, which are later transplanted to soil after the grafts have become established. Pears which grew easily from *in vitro* meristems, but which were difficult to root, could be removed from tissue culture when they had elongated to about 1 cm and successfully cleft-grafted onto potted seedlings (Figure 5.1) (Postman and Hadidi, 1995). Selection of rootstock seedlings with distinct leaf colour or morphology aids in differentiating micrografts from rootstock sprouts as they grow out.

Although it is possible to eliminate viruses from plants following meristem tip culture alone, this procedure is almost always combined with heat therapy or chemotherapy to increase the likelihood of success. The combination of heat therapy followed by apical meristem culture has become the foundation for many virus elimination programmes at germplasm repositories, research institutes and commercial plant nurseries around the world.

5.5.3   Chemotherapy

Unlike fungi and bacteria, viruses cannot be eliminated from infected plants by protective chemical sprays. Several chemicals, however, can be used in the same manner as heat to inhibit the replication or movement of viruses thus producing a region of virus-free tissue. The chemicals can either be sprayed on growing plants or incorporated into tissue culture media. Following a period of chemotherapy, shoot-tips or apical meristems are excised and propagated as in heat therapy. While it may be possible in some instances to cure an infected plant, the cost of these chemicals precludes the application of chemotherapy to field trees (Hansen, 1988).

Many of the antiviral chemicals that have been used for plant virus chemotherapy are synthetic nucleotide analogues that had previously shown some effectiveness against animal

Figure 5.1 *In vitro* pear shoot from an apical meristem grafted onto a small pear rootstock. A bottle (inset) maintains high humidity, keeping the young shoot alive until the graft union forms

viruses (Dawson, 1984; Hansen, 1988; Kartha, 1986). The concentrations that are required during chemotherapy to inhibit virus multiplication are very close to the concentrations which are toxic to the host plant. The guanosine analogue ribavirin (Virazole (ICN, CA, USA); 1-D-ribofuranosyl-1,2,4-triazole-3-carboxamide), and the uracil analogue DHT (5-dihydroazauracil) are two substances which are particularly effective at inhibiting many different plant viruses (Hansen, 1988; Spiegel *et al.*, 1993a). Hansen (1988) reviewed applications of chemotherapy to virus infected plants. In early studies, antiviral substances were injected into stems or applied to whole plants as foliar sprays or root drenches. More recent work involved the incorporation of antiviral materials into tissue culture media leading to the elimination of viruses from many plants including peanut (*Arachis*), orchid (*Cymbidium*), potato (*Solanum*), and strawberry (*Fragaria*) (Albouy *et al.*, 1988; Borissenko *et al.*, 1985; Dunbar *et al.*, 1993; Kondakova and Schuster, 1991; Long and Cassells, 1986; Spiegel *et al.*, 1993a). Tissue culture methods permit more control over the concentration of chemical agents, the period of exposure, and the ability to combine chemotherapy with heat therapy or apical meristem culture. Combinations of more than one chemical are also more easily evaluated using tissue cultures.

The modes of action of the various plant virus chemotherapies are not well understood. Ribavirin, perhaps the best studied, has been shown in animal viruses to interfere with capping at the 5′ end of viral mRNA and many of the ribavirin-sensitive plant viruses also have a 5′ capped mRNA (Hansen, 1988). While chemotherapy is effective for eliminating a diverse array of viruses from many different plant species, the chemicals used, the effective concentrations, and the phytotoxicity levels vary greatly depending on the plant genotype. The possibility of genetic mutations when plants are exposed to antiviral chemicals poses risks that have not been adequately studied. Heat therapy poses a much lower risk of genetic change to treated plants, and the procedure is consistent regardless of the virus or the plant host species. Chemotherapy may supplement the other methods of plant virus elimination, but is not likely to replace them, except in situations where viruses are not easily eliminated by heat and meristem culture.

Not all plants resulting from therapy will be pathogen-free. Each plant must be subjected to follow-up indexing to confirm the virus status, and plants should be grown out and examined for trueness to type before being offered for commercial production. There are abundant opportunities for plants to become mislabelled during the various stages of propagation, heat therapy, *in vitro* culture, and re-establishment. Concerns have also been raised about the possibility of genetic mutations or selection of somaclonal variants especially following *in vitro* culture procedures which involve certain plant growth hormones (Swartz *et al.*, 1981). Such changes are less likely if regeneration from undifferentiated tissues or callus is avoided during *in vitro* culture. Heat therapy and apical meristem culture not only can eliminate viruses that have been previously detected, but may also eliminate exotic or latent viruses whose presence is unknown, as well as contaminating organisms such as bacteria, fungi or phytoplasmas (Fridlund, 1989).

5.6 Conclusions

Germplasm collections present a unique set of problems with regard to prevention of movement of plant viruses. Due to the nature of the plant material collected, there is always the possibility of introducing completely unknown viruses. A more complete set of virus tests needs be carried out on these plants compared to other virus testing programmes such as certification programmes dealing with a known set of viruses. Since germplasm collections deal with a broad range of genetic materials, sensitivity to thermotherapy or chemotherapy, and the ability to culture the plants *in vitro* may be more variable than in certification programmes where the genetics of a crop is more uniform.

Once pathogen-free germplasm is identified, whether through indexing of collected accessions, or through a combination of therapy and indexing, the plant material should be protected from reinfection. In conventional plant gene banks and certification programmes this involves isolating growing plants from virus vectors. Potted plants maintained in insect-proof enclosures will be protected from spread of viruses by insects and nematodes. Removal of flowers and exclusion of pollinators such as honeybees will prevent the movement of pollen-borne viruses. Conservation of clonal plant germplasm *in vitro* should assure that plants are protected from becoming infected with pathogens, including viruses, provided the source of the *in vitro* plants is virus-free.

References

AHRENS, U. and SEEMULLER, E., 1992, Detection of DNA of plant pathogenic mycoplasma-like organisms by a polymerase chain reaction that amplifies a sequence of the 16S rRNA gene, *Phytopathology*, **82**, 828–832.
ALBOUY, J., FLOUZAT, C., KUSIAK, C. and TRONCET, M., 1988, Eradication of orchid viruses by chemotherapy from in vitro cultures of *Cymbidium*, *Acta Horticulturae*, **234**, 413–419.

ANDERSON, M.L.M. and YOUNG, B.D., 1985, Quantitative filter hybridization, in HAMES, B.D. and HIGGINS, S.J. (Eds), *Nucleic Acid Hybridization, A Practical Approach*, pp. 73–111. IRL Press: Oxford.

BANTTARI, E.E. and GOODWIN, P.H., 1985, Detection of potato viruses S, X, Y, by enzyme-linked immunosorbent assay on nitrocellulose membranes (dot-ELISA), *Plant Disease*, **69**, 202–205.

BARBARA, D.J., JONES, A.T., HENDERSON, S.J., WILSON, S.C. and KNIGHT, V.H., 1984, Isolates of raspberry bushy dwarf virus differing in *Rubus* host range, *Ann. Appl. Biol.*, **105**, 49–54.

BARBARA, D.J., DAWATA, E.E., UENG, P.P., LISTER, R.M. and LARKINS, B.A., 1987, Production of complementary DNA clones from the MAV isolate of barley yellow dwarf virus, *J. Gen. Virol.*, **68**, 2419–2428.

BARBARA, D.J., MORTON, A., SPENCE, N.J. and MILLER, A., 1995, Rapid differentiation of closely related isolates of two plant viruses by polymerase chain reaction and restriction fragment length polymorphism analysis, *J. Virol. Methods*, **55**, 121–131.

BORISSENKO, S., SCHUSTER, G. and SCHMYGLA, W., 1985, Obtaining a high percentage of explants with negative serological reactions against viruses by combining potato meristem culture with antiphytoviral chemotherapy, *Phytopath. Z*, **114**, 185–188.

CHEVALIER, S., GREIF, C., CLAUZEL, J.M., WALTER, B. and FRITSCH, C., 1995, Use of an immunocapture-polymerase chain reaction procedure for the detection of grapevine virus A in Kober stem grooving-infected grapevines, *J. Phytopathol.*, **143**, 369–373.

CHU, P.W.G., WATERHOUSE, P.M., MARTIN, R.R. and GERLACH, W.L., 1989, New approaches to the detection of microbial plant pathogens, *Biotechnology and Genetic Engineering Reviews*, **7**, 45–111.

CLARK, M.F. and ADAMS, A.N., 1977, Characteristics of the microplate method of enzyme-linked immunosorbent assay for the detection of plant viruses, *J. Gen. Virol.*, **34**, 45–483.

CLARK, M.F. and BAR-JOSEPH, M., 1984, Enzyme immunosorbent assays in plant virology, *Methods in Virology*, **7**, 51–85.

CONVERSE, R.H. (Ed.), 1987, *Virus Diseases of Small Fruits*, USDA–ARS Agriculture Handbook, No. 631.

CONVERSE, R.H. and GEORGE, R.A., 1987, Elimination of mycoplasma like organisms in Cabot highbush blueberry with high carbon dioxide thermotherapy, *Plant Disease*, **71**, 36–38.

CONVERSE, R.H. and MARTIN, R.R., 1990, ELISA for plant viruses, in R. HAMPTON, E. BALL and S. DEBOER (Eds), *Serological Methods for Detection and Identification of Viral and Bacterial Plant Pathogens*, pp. 179–196, APS Press, St Paul, MN.

D'ARCY, C.J., TORRANCE, L. and MARTIN, R.R., 1989, Discrimination among luteoviruses and their strains by monoclonal antibodies and identification of common epitopes, *Phytopathology*, **79**, 869–873.

DAWSON, W.O., 1984, Effects of animal antiviral chemicals on plant viruses, *Phytopathology*, **74**(2), 211–213.

DEOGRATIAS, J.M. and LUTZ, A., 1986, *In vitro* micrografting of shoot tips from juvenile and adult *Prunus avium* L. and *Prunus persica* (L.) Batsch to produce virus-free plants, *Acta Horticulturae*, **193**, 139–145.

DIEKMANN, M., FRISON, E.A. and PUTTER, T. (Eds), 1994, *FAO/IPGRI Technical Guidelines for the Safe Movement of Small Fruit Germplasm*, Food and Agriculture Organization of the United Nations, Rome and International Plant Genetic Resources Institute, Rome.

DUNBAR, K.B., PINNOW, D.L., MORRIS, J.B. and PITTMAN, R.N., 1993, Virus elimination from interspecific *Arachis* hybrids, *Plant Disease*, **77**, 515–520.

EPPO, 1991, Certification scheme No. 1, Virus-free or virus-tested fruit trees and rootstocks Part I, Basic Scheme, *EPPO Bulletin*, **21**, 267–277.

EPPO, 1992, Certification scheme No. 1, Virus-free or virus-tested fruit trees and rootstocks: Part IV, Technical appendices, *EPPO Bulletin*, **22**, 277–283.

EPPO, 1994, Certification scheme No. 11, Pathogen tested strawberry, *EPPO Bulletin*, **24**, 875–889.

FACCIOLI, G. and MARANI, F., 1998, Virus elimination by meristem tip culture and tip micrografting, in A. HADIDI, R.K. KHETARPAL and H. KOGANEZAWA (Eds), *Plant Virus Disease Control*, pp. 346–380, St Paul, Minnesota: APS Press.

FAO, 1998, http://www.fao.org/waicent/faoinfo/agricult/agp/agpp/pq

FEINBERG, A.P. and VOGELSTEIN, B., 1983, A technique for radiolabelling DNA restriction endonuclease fragments to high specific activity, *Anal. Biochem.*, **132**, 6.13.

FOSTER, J.A. and HADIDI, A., 1998, Exclusion of plant viruses, in A. HADIDI, R.K. KHETARPAL and H. KOGANEZAWA (Eds), *Plant Virus Disease Control*, pp. 208–229, St Paul, Minnesota: APS Press.

FOULY, B.M., DOMIER, L.L. and D'ARCY, C.J., 1992, A rapid chemiluminescent detection method for barley yellow dwarf virus, *J. Virol. Methods*, **39**, 291–298.

FRIDLUND, P.R., 1989, Thermotherapy, in P.R. FRIDLUND (Ed.), *Virus and Virus-like Diseases of Pome Fruits*, pp. 284–295, Pullman: Washington State University.

FRIDLUND, P.R., 1971, Failure of hot-water treatment to eliminate Prunus viruses, *Plant Disease Reporter*, **55**(8), 738–740.

FRIDLUND, P.R., 1973, Distribution of chlorotic leaf spot virus in apple budsticks, *Plant Disease Reporter*, **57**(10), 865–869.

FRISON, E.A. and DIEKMANN, M., 1998, IPGRI's role in controlling virus diseases in plant germplasm, in A. HADIDI, R.K. KHETARPAL and H. KOGANEZAWA (Eds), *Plant Virus Disease Control*, pp. 230–236, St Paul, Minnesota: APS Press.

FULTON, R.W., 1964, Transmission of plant viruses by grafting, dodder, seed and mechanical inoculation, in M.K. CORBETT and H.D. SISLER (Eds), *Plant Virology*, pp. 39–67, Gainesville: University of Florida Press.

GATTI, R.A., CONCANON, P. and SALSER, W., 1984, Multiple use of Southern blots, *BioTechniques*, **2**, 148–155.

GILMER, R.M. and BRASE, K.D., 1963, Nonuniform distribution of prune dwarf virus in sweet and sour cherry trees, *Phytopathology*, **53**, 819–821.

GILMER, R.M., MOORE, J.D., NYLAND, G., WELSH, M.F. and PINE, T.S. (Eds), 1976, *Virus Diseases and Noninfectious Disorders of Stone Fruits in North America*, USDA–ARS Agriculture Handbook, No. 437.

HABER, S. and KNAPEN, H., 1989, Filter paper sero-assay (FiPSA): a rapid, sensitive technique for sero-diagnosis of plant viruses, *Can. J. Plant Pathol.*, **11**, 109–113.

HADIDI, A., KHETARPAL, R.K. and KOGANEZAWA, H. (Eds), 1998, *Plant Virus Disease Control*, St Paul, Minnesota: APS Press.

HALK, E.L. and DEBOER, S.H., 1985, Monoclonal antibodies in plant disease research, *Annu. Rev. Phytopathol.*, **23**, 321–350.

HAMMOND, J. and JORDAN, R.L., 1991, Monoclonal antibodies against potyvirus-associated antigens, hybrid cell lines producing these antibodies, and use therefore, Patent application, US Patents and Trademarks Office, Washington, DC 20231.

HANSEN, A.J., 1988, Chemotherapy of plant virus infections, in E. KURSTAK, R.G. MARUSYK, F.A. MURPHY and M.H.V. VAN REGENMORTEL (Eds), *Applied Virology Research*, Vol. 1, pp. 285–299, New York: Plenum Medical Book Co.

HOLT, C.A., 1992, Detection and localization of plant pathogens, in REID, P.D., PONT-LEZICA, R.F., CAMPILLO, E.D. and TAYLOR, R. (Eds), *Tissue Printing*, pp. 127–137, New York: Academic Press.

HUANG, S. and MILLIKAN, D.F., 1980, *In vitro* micrografting of apple shoot tips, *HortScience*, **15**(6), 741–743.

HULL, R., 1988, Rapid-diagnosis of plant virus infections by spot hybridization, *Trends Biotechnol.*, **2**, 88–91.

JAMES, D., THOMPSON, D.A. and GODKIN, S.E., 1994, Cross reactions of an antiserum to plum pox virus, *EPPO Bulletin*, **24**, 605–614.

JAMES, D., GODKIN, S.E., EASTWELL, K.C. and MACKENZIE, D.J., 1996, Identification and differentiation of *Prunus* virus isolates that cross react with plum pox virus and apple stem pitting virus antisera, *Plant Dis.*, **80**, 536–643.

JELKMANN, W., MARTIN, R.R. and MAISS, E., 1989, cDNA cloning of four plant viruses from dsRNA templates, *Phytopathology*, **79**, 1250–1253.

JORDAN, R.L., 1990, Strategy and techniques for the production of monoclonal antibodies; monoclonal antibody applications for viruses, in HAMPTON, R., BALL, E. and DEBOER, S. (Eds), *Serological Methods for Detection and Identification of Viral and Bacterial Plant Pathogens*, pp. 55–85, St Paul, MN: APS Press.

KARTHA, K.K., 1986, Production and indexing of disease-free plants, in L.A. Withers and P.G. Alderson (Eds), *Plant Tissue Culture and its Agricultural Applications*, pp. 219–238, London: Butterworths.

KOENIG, R. and PAUL, H.L., 1982, Variants of ELISA in plant virus diagnosis, *J. Virol. Methods*, **5**, 113–125.

KONDAKOVA, V. and SCHUSTER, G., 1991, Elimination of strawberry mottle virus and strawberry crinkle virus from isolated apices of three strawberry varieties by the addition of 2,4-dioxohexahydro-1,3,5-triazine to the nutrient medium, *J. Phytopathology*, **132**, 84–86.

LEE, I.M., HAMMOND, R.W., DAVIS, R.E. and GUNDERSEN, D.E., 1993, Universal amplification and analysis of pathogen 16S rDNA for classification and identification of mycoplasma-like organisms, *Phytopathology*, **83**, 834–842.

LIZARRAGA, R.E., SALAZAR, L.F., ROCA, W.M. and SCHILDE-RENTSCHLER, L., 1980, Elimination of potato spindle tuber viroid by low temperature and meristem culture, *Phytopathology*, **70**, 754–755.

LONG, R.D. and CASSELLS, A.C., 1986, Elimination of viruses from tissue cultures in the presence of antiviral chemicals, in WITHERS, L.A. and ALDERSON, P.G. (Eds), *Plant Tissue Culture and its Agricultural Applications*, pp. 239–248, London: Butterworths.

MACDONALD, J.G., KRISTJANSSON, G.T., SINGH, R.P., ELLIS, P.J. and McNAB, W.B., 1994, Consecutive ELISA screening with monoclonal antibodies to detect potato virus Y^N, *Amer. Potato J.*, **71**, 175–183.

MAKKOUK, K.M., HSU, H.T. and KUMARI, S.G., 1993, Detection of three plant viruses by dot-blot and tissue-blot immunoassays using chemiluminescent and chromogenic substrates, *J. Phytopathlogy*, **139**, 97–102.

MARTIN, R.R., 1998, Advanced diagnostic tools as an aid to controlling plant virus diseases, in A. HADIDI, R.K. KHETARPAL and H. KOGANEZAWA (Eds), *Plant Virus Disease Control*, pp. 381–392, St Paul, Minnesota: APS Press.

MARTIN, R.R. and ELLIS, P., 1986, Detection of potato leafroll and beet western yellows viruses in aphids, in *Proceedings of Workshop on Epidemiology of Plant Virus Diseases*, Orlando, Florida, August 6–8, 1986.

MAS, P. SANCHEZ-NAVARRO, J.A., SANCHEZ-PINA, M.A. and PALLAS, V., 1993, Chemiluminescent and colorigenic detection of cherry leaf roll virus with digoxigenin-labeled RNA probes, *J. Virol. Methods*, **45**, 93–102.

MILLER, S.A. and MARTIN, R.R., 1988, Molecular diagnosis of plant disease, *Ann. Rev. Phytopathol.*, **26**, 409–432.

MINK, G.I., WAMPLE, R. and HOWELL, W.E., 1998, Heat treatment of perennial plants to eliminate phytoplasmas, viruses, and viroids while maintaining plant survival, in A. HADIDI, R.K. KHETARPAL and H. KOGANEZAWA (Eds), *Plant Virus Disease Control*, pp. 332–345, St Paul, Minnesota: APS Press.

NAUER, E.M., ROISTACHER, C.N., CARSON, T.L. and MURASHIGE, T., 1983, *In vitro* shoot-tip grafting to eliminate citrus viruses and virus-like pathogens produces uniform budlines, *HortScience*, **18**(3), 308–309.

NAVARRO, L., ROISTACHER, C.N. and MURASHIGE, T., 1975, Improvement of shoot-tip grafting in vitro for virus-free citrus, *J. Amer. Soc. Hort. Sci.*, **100**(5), 471–479.

NYLAND, G. and GOHEEN, A.C. 1969, Heat therapy of virus diseases of perennial plants, *Ann. Rev. Phytopathology*, **7**, 331–354.

PADUCH-CICHAL, E. and KRYCZYNSKI, S., 1987, A low temperature therapy and meristem-tip culture for eliminating four viroids from infected plants, *J. Phytopathology*, **118**, 341–346.

PLUCKNETT, D.L. and SMITH, N.J.H., 1989, Quarantine and the exchange of crop genetic resources, *BioScience*, **39**(1), 16–23.

POSTMAN, J.D. and HADIDI, A., 1995, Elimination of apple scar skin viroid from pears by *in vitro* thermo-therapy and apical meristem culture, *Acta Horticulturae*, **386**, 536–543.

POSTMAN, J.D. and MEHLENBACHER, S.A., 1994, Apple mosaic virus in hazelnut germplasm, *Acta Horticulturae*, **351**, 601–609.

REED, B.M., 1991, Application of gas-permeable bags for *in vitro* cold storage of strawberry germplasm, *Plant Cell Reports*, **10**, 431–434.

ROBERTSON, N.L., FRENCH, R. and GRAY, S.M., 1991, Use of group-specific primers and the polymerase chain reaction for the detection and identification of luteoviruses, *J. Gen. Virol.*, **72**, 1473–1477.

ROJAS, M.R., GILBERTSON, R.L., RUSSELL, D.R. and MAXWELL, D.P., 1993, Use of degenerate primers in the polymerase chain reaction to detect whitefly-transmitted geminiviruses, *Plant Dis.*, **77**, 340–347.

SMART, C.D., SCHNEIDER, B., BLOMQUIST, C.L., GUERRA, L.J., HARRISON, N.A., AHRENS, U., *et al.*, 1996, Phytoplasm-specific PCR primers based on sequences of the 16S-23S rRNA spacer region, *Applied and Environmental Microbiology*, **62**, 2988–2993.

SPIEGEL, S., FRISON, E.A. and CONVERSE, R.H., 1993a, Recent developments in therapy and virus-detection procedures for international movement of clonal plant germplasm, *Plant Disease*, **77**(12), 1176–1180.

SPIEGEL, S., MARTIN, R.R., LEGGET, F., TER BORG, M. and POSTMAN, J., 1993b, Characterization and geographical distribution of a new ilarvirus from *Fragaria chiloensis*, *Phytopathology*, **83**, 991–995.

SUTULA, C.L., GILLET, J.M., MORRISSEY, S.M. and RAMSDELL, D.C., 1986, Interpreting ELISA data and establishing the positive-negative threshold, *Plant Disease*, **70**, 722–726.

SWARTZ, H.J., GALLETTA, G.J. and ZIMMERMAN, R.H., 1981, Field performance and phenotypic stability of tissue culture propagated strawberries, *J. Amer. Soc. Hort. Sci.*, **106**(5), 667–673.

TORRANCE, L., 1987, Use of enzyme amplification in an ELISA to increase sensitivity of detection of barley yellow dwarf virus in oats and in individual vector aphids, *J. Virol. Methods*, **15**, 131–138.

US Code, Title 7, Chapter III, Part 319 of the US Code

WALKEY, D.G.A., 1980, Production of virus-free plants by tissue culture, in D.S. INGRAM and J.P. HELGESON (Eds), *Tissue Culture Methods for Plant Pathologists*, pp. 109–117, Oxford: Blackwell.

WANG, P.J. and HU, C.Y., 1980, Regeneration of virus-free plants through *in vitro* culture, in A. FIECHTER (Ed.), *Advances in Biochemical Engineering*, Vol. 18, pp. 61–99, Berlin: Springer Verlag.

WETZEL, T., CANDRESSE, T., MACQUAIRE, G., RAVELONANDRO, M. and DUNEZ, J., 1992, A highly sensitive immunocapture polymerase chain reaction method for plum pox potyvirus detection, *J. Virol. Methods*, **39**, 27–37.

WINTER, S., PURAC, A., LEGGETT, F., FRISON, E.A., ROSSEL, H.W. and HAMILTON, R.I., 1992, Partial characterization and molecular cloning of a closterovirus from sweet potato infected with the sweet potato virus disease complex from Nigeria, *Phytopathology*, **82**, 869–875.

6

Cryopreservation

ERICA E. BENSON

6.1 Introduction

Cryopreservation is the preservation of viable cells, tissues and organs in liquid nitrogen, at −196°C. This storage procedure can be successfully applied to a wide range of organisms and biological tissues (Benson and Lynch, 1998; Benson *et al.*, 1998; Harding *et al.*, 1997) and it is being increasingly used to conserve crop plant germplasm (Ashmore, 1997). Thus, cryopreservation provides a long-term storage method for the conservation of plant genetic resources which cannot be maintained using conventional preservation methods, such as by seed banking. The application of cryopreservation must be considered in the context of other conservation options; it is best used as a complementary method, for example, when other storage protocols are not appropriate. Many examples of the appropriate application of plant cryopreservation are presented by the authors of chapters in this volume. Thus, although initial progress in plant cryopreservation was made using arable crop plant germplasm, more recently, it is being increasingly used to conserve endangered species (see Pence, Chapter 15 and González-Benito *et al.*, Chapter 16, this volume) and tropical rain forest trees (see Marzalina and Krishnapillay, Chapter 17, this volume). This increase in applications is largely due to the development of improved cryoprotection strategies which have made the technique more accessible to end users.

6.2 Principles of cryopreservation and germplasm preparation

Cryopreservation protocols comprise several component parts, many of which interface with *in vitro* manipulations such as tissue culture, shoot micropropagation, embryo rescue and somatic embryogenesis (see Lynch, Chapter 4, this volume, and Benson, 1994, 1995; Benson and Lynch, 1998; Benson *et al.*, 1999). However, cryoprotection is the key step, together with the application of pre-treatment strategies. Post-storage genetic stability is also an important consideration (see Harding, Chapter 7, this volume) and it is important that the cryoprotective and recovery strategies do not predispose germplasm to the risks of genetic instability (Harding, 1996). The different component steps of cryopreservation protocols are shown in Figure 6.1.

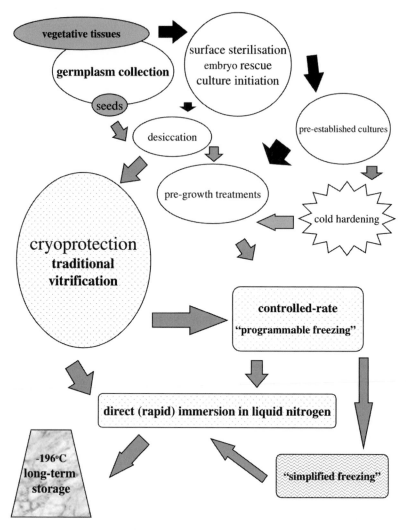

Figure 6.1   Component steps of cryopreservation protocols

6.2.1   *Preparing germplasm for cryopreservation*

Initially, it is important to consider the role of germplasm acquisition (see Hummer, Chapter 3 and Lynch, Chapter 4, this volume) in cryogenic storage. Cryopreservation can be applied to freshly collected seeds or vegetative germplasm, such as shoot-tips or buds which have been sampled from the field, using *in vitro* collection methods (see Pence, Chapter 15, this volume). However, cryopreservation is usually dependent on *in vitro* recovery methods and it is thus first important to surface sterilize the germplasm before it is placed in liquid nitrogen. In the case of recalcitrant seeds, embryo rescue is frequently performed as embryos are more amenable to cryogenic storage than whole seeds (see Marzalina and Krishnapillay, Chapter 17, this volume). In many cases, and especially for vegetatively propagated crops, cryopreservation is applied to already established micropropagated cultures (Figure 6.1). In this case it is important to note that tissue culture factors (Lynch *et al.*, 1994) can influence post-cryopreservation survival, as can time in culture (Benson *et al.*, 1991), morphogentic status (Lynch *et al.*, 1994) and light (Benson *et al.*, 1989).

6.2.2 Pre-treatments

Pre-treatments are usually applied to germplasm before cryoprotection; these manipulations do not usually support post-cryopreservation recovery, but they do enhance survival when used in combination with other cryoprotective strategies (Reed, 1996). Typical pre-treatments include: exposing temperate plant tissues to cold acclimation/hardening regimes; applying osmotic agents which reduce tissue water content prior to freezing; and pre-culturing tissues in media which contain 'anti-stress' agents such as proline, abscissic acid or trehalose (see Reed, 1996 for a review). The application of simple dehydrating pre-treatments in combination with sucrose and alginate bead encapsulation is an effective strategy for many different species (Dumet *et al.*, 1993a, 1993b; Gonzalez-Arnao *et al.*, 1998; Malaurie *et al.*, 1998; Mandal *et al.*, 1996; Thierry *et al.*, 1997).

6.3 Cryoprotection

There are two main approaches to cryoprotection; one is now termed 'traditional' and is based on the application of cryoprotective agents which were mainly developed by mammalian cryobiologists (Polge *et al.*, 1949). More recently, plant cryobiologists have placed greater emphasis on using procedures which circumvent the formation of ice during the cryopreservation process and these are based on vitrification techniques (see Figures 6.1, 6.2 and 6.3 and Section 6.3.2).

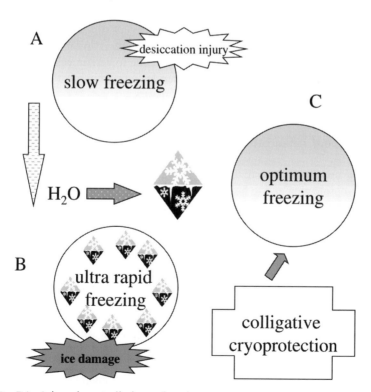

Figure 6.2 Principles of controlled rate freezing

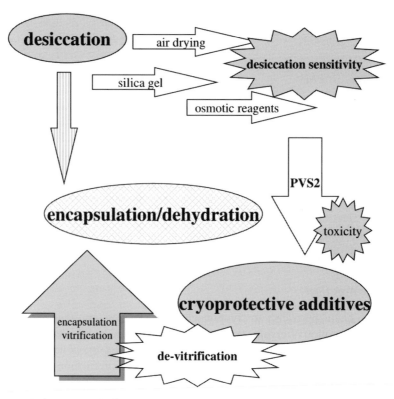

Figure 6.3   Pathways to vitrification

   Whilst cryoprotection is essential to successful cryopreservation, it is important to be aware that cryoprotectants can themselves be cytotoxic to certain cell types and species. Therefore, before embarking on a new cryopreservation programme it is first necessary to evaluate the effects of cryoprotectants on plant tissues.

6.3.1   Traditional cryoprotection and controlled rate cooling

In developing cryoprotective strategies it is important to understand the principles of traditional cryoprotection and to be aware of the central role of water in low temperature biology (Meryman and Williams, 1984). Water is thought to freeze at 0°C, and indeed it does, if a nucleus is present (e.g. dust particles or the side of a container) which can act as a template for ice crystallization. However, if nuclei are absent, water can super-cool far below zero and in the absence of an external nucleating agent, water can be super-cooled to –40°C (Figure 6.4). This is termed the 'temperature of homogeneous ice nucleation' and it is the point at which aggregates of water molecules can actually form ice (Meryman and Williams, 1984). Ice nucleation (or 'seeding' as it is sometimes called by cryobiologists) is accompanied by an exo-thermal event which releases energy as heat; this is termed the 'latent heat of freezing'. When water is super-cooled a heat deficit occurs and by cooling the water below its freezing point, some of the heat of fusion of ice is absorbed by the water. As a result, once the ice crystals are 'seeded', an ice front is rapidly propagated throughout the sample. This is accompanied by a rise in the freezing

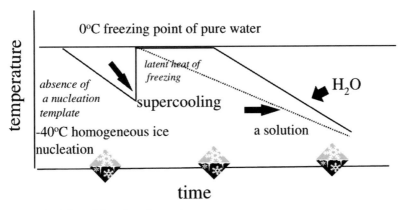

Figure 6.4   Principles of freezing dynamics

temperature to 0°C, the temperature of ice formation in pure water. In the case of pure water, the temperature will remain at zero until all the water is frozen out. However, the situation is different in the case of a solution which contains solutes and this is depicted by the dotted line shown in Figure 6.4. Thus, the presence of solutes in water will cause a freezing point depression. Solutes have a dynamic influence on the freezing point of the remainder of the water within the solution; as water freezes out, the solutes within the sample become more and more concentrated and the freezing point becomes depressed even further.

In the case of natural biological systems the freezing process occurs in the presence of cell solutes and the inter-play between solution and solvent effects is an important survival factor. Furthermore, cell freezing, under natural conditions usually occurs slowly and as a result ice nucleation occurs extra-cellularly (see Figure 6.2, schematic A). This creates a water vapour deficit between the inside and outside of the cell and as a consequence the unfrozen water from inside the cell moves to the extra-cellular compartment; this has the concomitant effect of causing cell dehydration and the intracellular solute concentration increases. By comparison, if biological tissues are frozen rapidly (e.g. by directly plunging them into liquid nitrogen) there is not enough time for the extra-cellular water to nucleate and intra- and extra-cellular freezing takes place at the same time (Figure 6.2, schematic B). This can have damaging consequences as intra-cellular ice formation is lethal.

The dynamic effects of 'slow-freezing' have been utilized advantageously by plant cryobiologists. By regulating the rate at which plant germplasm is frozen (and thus dehydrated) prior to immersion in liquid nitrogen it is possible to greatly enhance post-thaw survival after cryopreservation. At an optimum, controlled rate of freezing (see Figure 6.2, schematic C) the cells lose just enough intra-cellular water to ensure that when they are plunged into liquid nitrogen, ice damage is sufficiently limited to prevent cell death. However, it is also essential to consider the role of traditional cryoprotectants in the context of freezing dynamics (Figure 6.2); this is because cell dehydration can also be injurious during the extra-cellular freezing process. Cells suffer osmotic injury during water loss and this may be attributed to lethal changes in cell volume as well as the toxic concentration of cell solutes (Meryman and Williams, 1984).

Plants have naturally developed cryoprotective mechanisms which prevent environmental freezing damage and many species avoid freezing injury by simply failing to freeze and supercooling their cells to high sub-zero temperatures (Meryman and Williams, 1984), that is above the critical point of $-40\,^{\circ}C$ (the temperature of homogeneous ice nucleation). This is achieved by the natural accumulation of cell solutes which promote freezing point depression (Meryman and Williams, 1984). Similarly, in the case of cryopreservation, protective additives have been developed to overcome the problems of ice formation and dehydration injury, the two main factors involved in freezing injury. However, cryopreservation requires the storage of germplasm at $-196\,^{\circ}C$ (the temperature of liquid nitrogen), which is far below the natural sub-zero tolerance temperature of plants.

Traditional cryoprotectants exert their effects through colligative action and they may be considered as stabilizing 'solvents' for the solute component of frozen cells as they prevent the damaging effects of cell dehydration and volume change. Colligative cryoprotectants (e.g. glycerol or dimethyl sulphoxide (DMSO)) must be able to penetrate the cell and as such, they are equally distributed in the extra- and intra-cellular compartments. When extra-cellular ice forms in the presence of colligative cryoprotectants, water will be equally lost from both cell compartments and this prevents the toxic concentration of the cell's solutes through dehydration. Moreover, damaging cell volumes changes will be circumvented and the high cryoprotectant content of the cell will depress the freezing point to a very low sub-zero temperature at which damage, if it does occur, can be tolerated (Meryman and Williams, 1984). Cooling rate is also a major consideration when applying traditional cryopreservation protocols (see Figure 6.2). Plant cells contain vacuoles with relatively high water contents and it is important that sufficient time is allowed for the water to move from the intra-cellular compartment before immersion in liquid nitrogen. Failure to optimize cooling rates can lead to cryoinjury caused by ice formation. In this respect, non-penetrating cryoprotectants are also important in plant cryopreservation as their mode of action as osmotic agents assists the removal of potentially freezable water from the cell. This approach is especially important for plant cells which contain large vacuoles. Whilst osmotic and colligative cryoprotectants form the basis of most traditional cryopreservation protocols, it is important not to oversimplify the basis of cryoprotection in plants. Many cryoprotective additives function in other ways, for example, as antioxidants (Benson, 1990) and membrane and protein stabilizers (Finkle *et al.*, 1984).

6.3.2 *Vitrification*

Vitrification is the process by which water undergoes a phase transition from a liquid to an amorphous 'glassy state'; in this form, water does not possess a crystalline structure. Vitrification occurs when the solute concentration of a biological system becomes so high that ice nucleation is prevented, thus ice crystal formation and growth is inhibited by the highly viscous cell milieu and the water molecules form a glass. Vitrification has now become one of the main methods of cryoprotecting plant germplasm for cryopreservation and in effect, by using this approach, it is possible to conserve plant tissues in liquid nitrogen in the absence of ice. There are a number of different methods which can be applied to plant germplasm to promote the vitrification process and these are shown in Figure 6.3. The simplest involves the desiccation of tissues to a point at which the critical moisture content is so low that there is no water available for ice formation and the viscosity of the cell becomes so high that a glass is formed. Desiccation can be achieved through the treatment of germplasm with a sterile air flow (e.g. in a laminar air flow stream) or by drying over silica gel (Figure 6.3). These procedures are very simple and they only require the application of a simple desiccation process; however they can only be applied to germplasm which is not desiccation sensitive (Normah and Marzalina,

1996). A number of protocols have been developed which are dependent upon first dehydrating the germplasm with an osmotic agent such as sucrose before the desiccation treatment (Dumet *et al.*, 1993a, 1993b). Encapsulation of tissues in a calcium alginate matrix followed by osmotic dehydration and air or silica gel drying has also been applied as a cryoprotective strategy (Fabre and Dereuddre, 1990; Phunchindawan *et al.*, 1997). Vitrification can also be achieved through the treatment of tissues with high concentrations of cryoprotective additives; one such mixture is Plant Vitrification Solution Number 2 or 'PVS2' which was developed by Sakai and colleagues (Reinhoud *et al.*, 1995; Sakai *et al.*, 1990) and has been applied to many different plant species. However, PVS2 can be toxic and care must be taken in the application of the solution which comprises a highly concentrated cocktail of ethylene glycol, DMSO and glycerol. The duration of exposure to PVS2 and the temperature of its application are critical factors. PVS2 exerts its effects through osmotic dehydration (in the case of the non-penetrating cryoprotectants); however it is highly likely that DMSO penetrates the cells and thus exerts its effects by enhancing cell viscosity. The removal of vitrification solutions must be performed in such a way that prevents osmotic stress and PVS2 is usually 'unloaded' from the tissues by treatment with highly concentrated sucrose solutions (Reinhoud *et al.*, 1995).

Vitrified tissues may be directly plunged into liquid nitrogen, without the need of controlled rate cooling and indeed this is one of the main advantages of this technique as it circumvents the need to purchase expensive, controlled rate programmable freezers. However, the vitrified state is metastable and there does exist the potential for de-vitrification to occur on re-warming (Benson *et al.*, 1996a) and for this reason the control of re-warming rates after removal from cryogenic storage is critical. The usual procedure is to rapidly re-warm the vitrified tissues in a 45°C water bath; however, it is important to consider glass relaxation events as these can also damage specimens by fracturing and two-step re-warming procedures may be advisable. These incorporate a slow warming step to a temperature below the glass transition followed by a rapid re-warming step which ensures that the material does not de-vitrify as it passes through the glass transition point. Interestingly, the stability of glasses formed in cryopreserved tissues can be dependent upon the approach used to obtain the vitrified state and this can be investigated using differential scanning calorimetry (Benson *et al.*, 1996a). PVS2 solutions can be unstable and re-warming is critical; however, glasses obtained using encapsulation/dehydration may be more stable and encapsulated tissues can be re-warmed slowly at ambient temperatures without loss of survival (Benson *et al.*, 1996a).

6.4 Freezing and long-term cryogenic storage

Cryoprotected tissues and cells may be placed in cryogenic storage using two main routes (Figure 6.1), involving either direct immersion into liquid nitrogen or controlled rate (programmable) cooling. The former is usually applied for vitrified tissues and the latter for samples which are cryopreserved using traditional methodologies.

6.4.1 *Controlled rate freezing*

This usually involves a commercial instrument termed a 'programmable freezer' which comprises a freezing chamber into which liquid nitrogen is supplied from a pressurized dewar controlled by a solenoid valve. Samples can be placed into the chamber which contains special furniture which holds either cryovials or straws. The system is 'pro-grammed' with the aid of computer software and a range of freezing 'ramps' may be applied in which the initial temperature, freezing rate, terminal temperature (before transfer to liquid nitrogen) and holding time at the terminal transfer temperature can be precisely controlled. These parameters are critical to the success of traditional cryopreservation (see Section 6.3.1 and Figure 6.2). For some systems it may be essential to control the time and temperature of ice nucleation and the automatic or manual 'seeding' of ice in the cryovials may be an important step in the programming protocol.

As an alternative to expensive controlled rate freezers, 'low tech' simplified freezing systems are also commercially available. These comprise a small plastic vessel with inserts to take cryovials which are then fitted into a reservoir of solvent (usually propan-1-ol) of a specific volume and known thermal properties. The whole system is placed into a −20 or −80°C freezer and under the specifications of the design a cooling rate of −1°C/minute can be consistently achieved. The tank is maintained in the freezer until an optimum terminal transfer temperature is achieved and then the cryovials are transferred to a long-term liquid nitrogen dewar. These low tech approaches to cryopreservation have proved most effective for certain types of germplasm (Engelmann *et al.*, 1994) and are ideally suited for operators who cannot purchase expensive freezing units.

6.4.2 *Rapid freezing and long-term storage*

Rapid cooling usually involves the direct transfer of germplasm, in cryovials, to liquid nitrogen. It is important that the transfer takes place expediently, particularly if the tissues have been previously exposed to a terminal transfer temperature. The most practical means of ensuring rapid transfers to long-term storage dewars is to first place the samples in a small bench top liquid nitrogen dewar and then transfer this to the long-term reposit-ory. The samples can then be transferred efficiently to the larger dewar without risking warming, melting or devitrification.

There are a range of long-term storage dewars available on the commercial market, and choice will be cost and user dependent. However, it is important to evaluate the maximum holding time of the facility in order to ensure that the tanks are maintained at the required level. Weekly 'topping up' of the liquid nitrogen reservoir is a normal practice. Inventory systems are an important part of long-term cryopreservation manage-ment protocols. It is quite possible that cryovials will be maintained in storage for long periods of time and it is essential that retrieval from the dewars is undertaken without the disturbance of other vials or the need to expose the vials to ambient temperatures for long periods. Inefficient inventory systems can also lead to the misidentification of samples and this must be avoided.

6.5 Post-cryopreservation recovery

Assessments of post-storage viabilities are important and these can be performed, in the short-term, by viability testing. Vital stains such as fluorescein diacetate (FDA) and triphenyl tetrazolium acetate (TTC) are frequently applied (see Benson, 1994, for details of methodologies); however, viability tests must be confirmed by re-growth assessments. It is also imperative that recovering tissues are assessed for normal and abnormal growth and regeneration. In the case of organized structures such as shoot-tip or embryos it is important that growth after cryogenic storage does not involve a callus phase or adventitious shoot regeneration. These events may predispose plants recovered from cryogenic storage to somaclonal variation (see Harding, Chapter 7, this volume). In the case of secondary product producing cell lines it is important that biosynthetic capacity is maintained after cryopreservation (see Schumacher, Chapter, 9 this volume; Benson and Hamill, 1991). Post-storage tissue culture manipulations can greatly influence the survival and regeneration capacity of cryopreserved plant germplasm. The interface between *in vitro* and cryogenic factors is an important factor in determining the long-term performance (e.g. in terms of development, genetic stability and reproductive behaviour) of plants regenerated from cryopreserved germplasm (Benson and Hamill, 1991; Benson *et al.*, 1996b; Harding and Benson, 1994).

6.6 Cryopreservation protocols: techniques and practical considerations

Figures 6.5 and 6.6 illustrate examples of cryopreservation protocols which have been applied, on a routine basis, to a broad range of germplasm (for full methodology see

A. The Withers and King controlled rate freezing method for cell suspension cultures

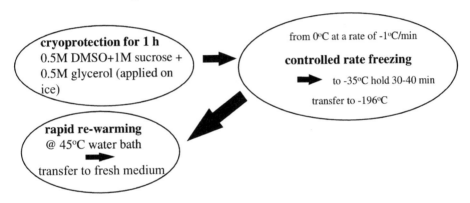

B. A PVS2 vitrification method for shoot-tips (adapted from Sakai *et al.*, 1990)

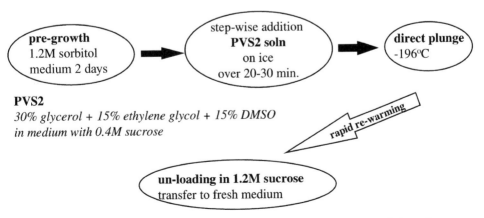

Figure 6.5 A summary of some frequently used cryopreservation protocols based on controlled rate freezing and vitrification

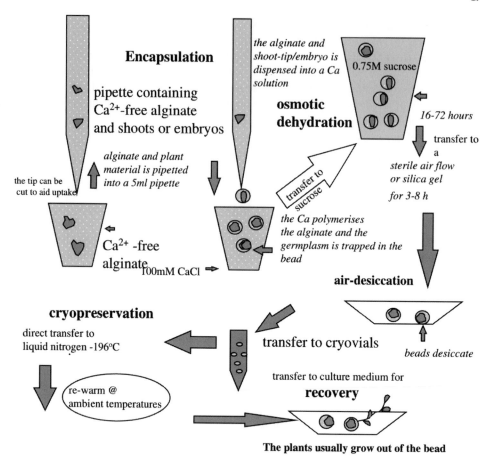

Figure 6.6 Summary of cryopreservation protocols for shoot-tips and embryos based on encapsulation–dehydration and desiccation techniques

Benson, 1994, 1995; Day and McClellan, 1995; Benson and Lynch, 1998). The Withers and King method established in 1980 (Figure 6.5, schematic A) has been successfully applied to a wide range of cell cultures and it provides an illustrative example of a typical programmable freezing method. In contrast, the PVS2 vitrification technique does not require a controlled rate cooling step and the samples may be directly plunged into liquid nitrogen. This protocol is based on the methods of Sakai and colleagues (Reinhoud *et al.*, 1995; Sakai *et al.*, 1990).

Figure 6.6 summarizes the essential steps of the encapsulation/dehydration technique which was first developed for *Solanum phureja* by Fabre and Dereuddre (1990), but has since been applied to a number of plant systems, many of which have previously proved recalcitrant to cryopreservation (Normah and Marzalina, 1996).

The three methods described in Figures 6.5 and 6.6 provide a spectrum of approaches which may be useful in the evaluation of cryopreservation responses for germplasm which has not been cryopreserved previously. However, when developing storage protocols it is also important to consider their practical aspects for routine use. In large scale genebanks and culture collections the controlled rate freezing approach may be the most cost effective and efficient method of conservation (B.M. Reed, personal communication) as it permits the large scale, simultaneous freezing of large numbers of samples. In contrast, to achieve the same end using vitrification and encapsulation methods would require time consuming operator handling. However, for laboratories which do not have access to specialized programmable freezers there may be no other option than to use the vitrification methods detailed in Figures 6.3 and 6.6. A further consideration is that some types of plant germplasm are differentially more responsive to one type of protocol as compared to another (Benson *et al.*, 1996a); thus choice of protocol will, in these cases, be dictated by the methods which produce the highest level of recovery.

A number of researchers (Blakesley *et al.*, 1997; Engelmann *et al.*, 1994; Lecouteux *et al.*, 1991; Yamada and Sakai, 1996) have placed emphasis on the application of simplified cryopreservation methods. These can be defined as simple because they are either: (a) protocols which are 'technologically simple', that is, they do not require expensive and specialized programmable freezing equipment; or (b) not demanding in terms of operator handling skills. Those organizations working towards the routine and large scale cryo-conservation of plant germplasm must take into consideration the various practical and technical attributes of the many different conservation methods now available.

6.7 Conclusions

Cryopreservation protocol development has reached a stage at which the storage method can be applied on a routine basis to a wide range of plant genetic resources. Whilst some plants still remain recalcitrant to cryopreservation, advances in plant cryobiology research offer the potential for overcoming the conservation difficulties associated with problematic species. The next phase of cryo-conservation will be to establish large scale cryopreserved genebanks and in this respect, the development of robust management systems for cryopreserved germplasm will be a necessary priority. The last decade has resulted in many outstanding developments in plant cryoprotection research. Most significantly, vitrification-based protocols and simplified procedures have made cryopreservation an accessible and cost-effective storage option for most laboratories who have a requirement for long-term *ex situ* conservation.

References

ASHMORE, S.E., 1997, *Status Report on the Development and Application of In Vitro Techniques for the Conservation and Use of Plant Genetic Resources*, Rome: International Plant Genetic Resources Institute.

BENSON, E.E., 1990, *Free Radical Damage in Stored Plant Germplasm*, Rome: International Board For Plant Genetic Resources.

BENSON, E.E., 1994, Cryopreservation, in R.A. DIXON and R.A. GONZALES (Eds), *Plant Cell Culture: A Practical Approach*, 2nd edn, pp. 147–168, Oxford: IRL Press, Oxford University.

BENSON, E.E., 1995, Cryopreservation of shoot-tips and meristems, in J.G. DAY and M.R. MCCLELLAN (Eds), *Methods in Molecular Biology*, Vol. 38, *Cryopreservation and Freeze-Drying Protocols*, pp. 121–132, Totowa, NJ: Humana Press Inc.

BENSON, E.E. and HAMILL, J.D., 1991, Cryopreservation and post-freeze molecular and biosynthetic stability in transformed roots of *Beta vulgaris* and *Nicotiana rustica*, *Plant Cell Tissue and Organ Culture*, **24**, 163–172.

BENSON, E.E. and LYNCH, P.T., 1998, Cryopreservation of rice tissue cultures, in R. HALL (Ed.), *Plant Cell Culture, Protocols, Methods in Molecular Biology*, Totowa, NJ: Humana Press Inc.

BENSON, E.E., HARDING, K. and SMITH, H., 1989, The effects of pre- and post-freeze light on the recovery of cryopreserved shoot-tips of *Solanum tuberosum, Cryo-Letters*, **10**, 323–344.

BENSON, E.E., HARDING, K. and SMITH, H., 1991, The effects of tissue culture on the post-freeze survival of cryopreserved shoot-tips of *Solanum tuberosum, Cryo-Letters*, **12**, 17–22.

BENSON, E.E., REED, B.M., BRENNAN, R.M., CLACHER, K.A. and ROSS, D.A., 1996a, Use of thermal analysis in the evaluation of cryopreservation protocols for *Ribes nigrum* L. germplasm, *Cryo-Letters*, **17**, 347–362.

BENSON, E.E., WILKINSON, M., TODD, A., EKUERE, U. and LYON, J., 1996b, Developmental competence and ploidy stability in plants regenerated from cryopreserved potato shoot-tips, *Cryo-Letters*, **17**, 119–128.

BENSON, E.E., LYNCH, P.T. and STACEY, G.N., 1998, Advances in plant cryopreservation technology: current applications in crop plant biotechnology, *AgBiotechNews and Information*, **10**, 133N–141N.

BENSON, E.E., HARDING, K. and DUMET, D.J., 1999, in SCRAGG, A. (Ed.), *Cryopreservation of Plant Cells Tissues and Organs: Theory and Practice, Encyclopaedia for Plant Cell Technology*, New York: John Wiley and Sons.

BLAKESELY, D., PERCIVAL, T., BHATTI, M.M. and HENSHAW, G.G., 1997, A simplified protocol for cryopreservation of embryogenic tissue of sweet potato (*Ipomea batatas*) using sucrose pre-culture only, *Cryo-Letters*, **18**, 77–80.

DAY, J.G. and McCLELLAN, M.R., 1995, Cryopreservation and freeze-drying protocols, *Methods in Molecular Biology*, Vol. 38, Totowa, NJ: Humana Press Inc.

DUMET, D., ENGELMANN, F., CHABRILLANGE, N., DUVAL, Y. and DEREUDDRE, J., 1993a, Importance of sucrose for the acquisition of tolerance to desiccation sensitivity and cryopreservation of oil palm somatic embryos, *Cryo-Letters*, **14**, 243–250.

DUMET, D.J., ENGELMANN, F., CHABRILLANGE, N. and DUVAL, Y., 1993b, Cryopreservation of oil palm (*Elaeis guineensis* Jacq.) somatic embryos involving a desiccation step, *Cryo-Letters*, **12**, 352–355.

ENGELMANN, F., DAMBIER, D. and OLLITRAULT, P., 1994, Cryopreservation of cell suspensions and embryogenic calluses of *Citrus* using a simplified freezing process, *Cryo-Letters*, **15**, 53–58.

FABRE, J. and DEREUDDRE, J., 1990, Encapsulation-dehydration: a new approach to cryopreservation of potato shoot-tips, *Cryo-Letters*, **11**, 413–426.

FINKLE, B.J., ZAVALA, M.E. and ULRICH, J.M., 1984, Cryoprotective compounds in the viable freezing of plant tissues, in KARTHA, K.K. (Ed.), *Cryopreservation of Plant Cells and Organs*, pp. 75–114, Florida: CRC Press.

GONZALEZ-ARNAO, M.T., ENGELMANN, F., URRA, C., MORENZA, M. and RIOS, A., 1998, Cryopreservation of citrus apices using the encapsulation-dehydration technique, *Cryo-Letters*, **19**, 177–182.

HARDING, K., 1996, Approaches to assess the genetic stability of plants recovered from in vitro culture techniques, in NORMAH, M.N., NARIMAH, M.K. and CLYDE, M.M. (Eds), *In Vitro Conservation. Proceedings of the International Workshop on In Vitro Conservation of Plant Genetic Resources*, July 4–6 1995, Kuala Lumpur, pp. 135–168, Plant Biotechnology Laboratory, Universiti Kebangsaan Malaysia Publication, Kuala Lumpur, Malaysia.

HARDING, K. and BENSON, E.E., 1994, A study of growth, flowering and tuberisation in plants derived from cryopreserved potato shoot-tips, *Cryo-Letters*, **15**, 59–66.

HARDING, K., BENSON, E.E. and CLACHER, K., 1997, Plant conservation biotechnology: an overview, *Agro-Food-Industry Hi-Tech*, **8**, 24–29.

LECOUTEUX, B., FLORIN, B., TESSEREAU, H., BOLLON, H. and PETIARD, V., 1991, Cryopreservation of carrot somatic embryos using a simplified freezing process, *Cryo-Letters*, **12**, 319–328.

LYNCH, P.T., BENSON, E.E., JONES, J., COCKING, E.C., POWER, J.B. and DAVEY, M.R., 1994, Rice cell cryopreservation; the influence of culture methods and the embryogenic potential of cell suspensions on post-thaw recovery, *Plant Science*, **98**, 185–192.

MALAURIE, B., TROUSLOT, M.-F., ENGELMANN, F. and CHABRILLANGE, N., 1998, Effect of pre-treatment conditions on the cryopreservation of in vitro-cultured yam (*Dioscorea alata* and *D. bulbifera*) shoot apices by encapsulation-dehydration, *Cryo-Letters*, **19**, 15–26.

MANDAL, B.B., CHANDEL, K.P.S. and DWIVEDI, S., 1996, Cryopreservation of Yam (*Dioscorea spp.*) shoot apices by encapsulation-dehydration, *Cryo-Letters*, **17**, 165–174.

MERYMAN, H.T. and WILLIAMS, R.J., 1984, Basic principles of freezing injury in plant cells: natural tolerance and approaches to cryopreservation, in KARTHA, K.K. (Ed.), *Cryopreservation of Plant Cells and Organs*, pp. 13–48, Florida: CRC Press.

NORMAH, M.N. and MARZALINA, M., 1996, Achievements and prospects of in vitro conservation for tree germplasm, in NORMAH, M.N., NARIMAH, M.K. and CLYDE, M.M. (Eds), *Techniques in In Vitro Conservation. Proceedings of the International Workshop on In Vitro Conservation of Plant Genetic Resources*, July 4–6 1995, Kuala Lumpur, pp. 253–261, Universiti Kebangsaan Malaysia Publishers, Kuala Lumpur.

PHUNCHINDAWAN, M., HITATA, K., SAKAI, A. and MIYAMOTO, K., 1997, Cryopreservation of encapsulated shoot primordia induced in horseradish (*Armoracia rusticana*) hairy root cultures, *Plant Cell Reports*, **16**, 469–473.

POLGE, C., SMITH, A.U. and PARKES, A.S., 1949, Revival of spermatazoa after vitrification and dehydration at low temperatures, *Nature*, **164**, 666.

REED, B.M., 1996, Pre-treatment strategies for cryopreservation of plant tissues, in NORMAH, M.N., NARIMAH, M.K. and CLYDE, M.M. (Eds), *Techniques in In Vitro Conservation. Proceedings of the International Workshop on In Vitro Conservation of Plant Genetic Resources*, July 4–6 1995, Kuala Lumpur, Malaysia, pp. 73–87, Universiti Kebangsaan Malaysia Publishers, Kuala Lumpur.

REINHOUD, P.J., URAGAMI, A., SAKAI, A. and VAN IREN, F., 1995, Vitrification of plant cell suspensions, in J.G. DAY and M.R. MCCLELLAN (Eds), *Cryopreservation and Freeze-drying Protocols. Methods in Molecular Biology*, Vol. 38, pp. 113–132, Totowa, NJ: Humana Press Inc.

SAKAI, A., KOBAYASHI, S. and OIYAMA, I., 1990, Cryopreservation of nucellar cells of navel orange (*Citrus sinensis* Osb. Var. Brasiliensis. Tanaka) by vitrification, *Plant Cell Reports*, **9**, 30–33.

THIERRY, C., TESSEREAU, H., FLORIN, B., MESCHINE, M.-C. and PETIARD, V., 1997, Role of sucrose for the acquisition of tolerance to cryopreservation of carrot somatic embryos, *Cryo-Letters*, **18**, 283–292.

WITHERS, L.A. and KING, P.J., 1980, A simple freezing unit and cryopreservation method for plant cell suspensions, *Cryo-Letters*, **1**, 213–220.

YAMADA, T. and SAKAI, 1996, Cryopreservation of cells and tissues using simplified methods, in NORMAH, M.N., NARIMAH, M.K. and CLYDE, M.M. (Eds), *In Vitro Conservation: Proceedings of the International Workshop on In Vitro Conservation of Plant Genetic Resources*, July 4–6 1995, Kuala Lumpur, Malaysia, pp. 89–103, Universiti Kebangsaan Malaysia Publishers, Kuala Lumpur.

7

Stability Assessments of Conserved Plant Germplasm

KEITH HARDING

7.1 Introduction

In establishing techniques for *in vitro* conservation (Ashmore, 1997), there are several important features to consider (Harding *et al.*, 1997). Plant cells exhibit the phenomenon of totipotency, that is the ability of a single cell to regenerate to a whole plant (Haberlandt, 1902). It is precisely this spontaneous regeneration process, where differentiated, organized tissues are taken from a plant and are introduced into culture and stimulated to divide which can result in cytological and chromosomal abnormalities. This manifestation is collectively known as 'somaclonal variation' (Larkin and Scowcroft, 1981) and is evidenced by heritable changes in plant phenotypes (Abdullah *et al.*, 1989; Stadelmann *et al.*, 1998), variation in chromosome numbers (Scowcroft, 1984), accumulation of gene mutations (Scowcroft *et al.*, 1984; Semal, 1986), alterations in levels of gene expression in ribonucleic acid (RNA), protein profiles and molecular changes in DNA sequences. Obviously, this has significance for *in vitro* conservation and other tissue culture-related technologies; consequently the genetic stability of plant tissues cultures has been reviewed in detail (Karp, 1993; Peschke and Phillips, 1992; Phillips *et al.*, 1994; Potter and Jones, 1990). Scowcroft (1984) indicated that some of this variation may well pre-exist in natural populations of plants taken from field collections or genebanks; moreover, variation itself may be generated *de novo* as a result of culture techniques. An important application of tissue culture techniques is the use of differentiated explants comprising organized structures like shoot-tips, roots and embryos. These are genetically programmed to develop into 'true to type' plants. If precautions are taken to avoid the dedifferentiated 'callus' phase, it is recognized that the induction of variation in regenerating plants is minimal (D'Amato, 1985).

7.2 Natural variation in populations

In natural populations of species, samples taken from field collections, by their very nature, represent a small fraction of the available diversity (see Harris, Chapter 2, this volume). This is true, especially for out-breeding species with long generation times, for instance tropical tree species. In relatively undisturbed habitats, species biodiversity is likely to be largely generated via normal evolutionary processes, i.e. genetic recombination and to some degree the products of mutagenesis attributed to failure in the DNA sequence replication error-proof reading systems. The extent of natural variation would mostly be dependent on the rates of genetic recombination and mutagenesis relative to the size of the plant genome. For example, the natural mutation rate for some (micro)organisms is recognized to be 1×10^{-6} to 1×10^{-9} bp/generation (Maynard-Smith, 1998), whereas it is known that field grown potato plants have to be rogued for aberrant forms, particularly as mutations (bolters) can occur at a relatively high frequency of 10^{-3} (Heiken, 1960). This has significance for stability assessments as these mutation rates specify the sample size, i.e. 1000 plants for potato, but this is dependent on the precision of the analytical techniques which are selected to detect DNA sequence changes (Powell *et al.*, 1997).

The development of procedures to determine natural variation in populations and the relationship between genotype and phenotype (morphological plasticity) should be considered within species. In view of the central dogma – DNA makes RNA makes protein as species evolve – DNA sequence information is a more reliable indicator of 'true to type' profiles compared to evaluations of gene products (i.e. proteins/enzymes and biochemical markers) as these are likely to be affected by environmental and climatic changes. The products of gene expression mask potential cryptic variation found in non-coding DNA sequences; moreover, profile variation can be induced by differential gene expression and by methylation imprinting DNA sequences.

7.3 Techniques to assess genetic stability

The analysis of plants regenerated from *in vitro* conservation procedures can be performed at the phenotypic, cytological, biochemical and molecular level with a range of techniques (Harding, 1996; see also González-Benito *et al.*, Chapter 16, this volume and Ng *et al.*, Chapter 13, this volume). The approaches taken to examine genetic stability in germplasm repositories are likely to be dependent on several practical factors such as the size of the germplasm collection, expertise, costs and labour.

7.4 Morphological variation

There are examples which show variability in mean recovery times, plant heights and modes of regeneration in plants regenerated from cryopreserved germplasm (Harding, 1996; Harding and Benson, 1994). In these assessments, it is important to note that the unstable phenotypic characters of the initial (R0 generation) plantlet can be strongly influenced by the source of nutrient and environmental conditions; however such variation is rarely inherited in seed progeny and plants of the R1 generation can lack the original aberrant characters. There are reports of phenotypic changes during the regeneration of plants from *in vitro* culture (Abdullah *et al.*, 1989; Scowcroft, 1984; Stadelmann *et al.*, 1998); however, this phenotypic plasticity can be misleading in stability assessments. In vegetatively propagated species, these features may be inherited somatically and persist in subsequent generations. The analysis of phenotypic variation is best performed with an approved list of plant descriptors (IBPGR, 1977). Phenotypic characters can be analysed by principal component analysis (Denton *et al.*, 1977) using a range of multivariate statistical techniques commercially available in statistical packages.

7.5 Cytological analysis

Genetic instability of tissue cultures resulting in chromosomal abnormalities is a well established phenomenon (Bayliss, 1980). The shoot-tips introduced into *in vitro* culture may comprise a heterogeneous collection of cell types. These may have undergone several rounds of DNA replication, and in the absence of cell division, give rise to polyploid cells (Halperin, 1986). The induction of cell division is likely to give rise to plants with extensive chromosomal rearrangements (Scowcroft, 1985). Chromosomal instability is also influenced by the genotype and tissue culture conditions (Gould, 1986). Gross chromosomal changes may include: polyploidy, aneuploidy, and mitotic abnormalities, for example multi-polar spindles, lagging chromosomes, fragments and asymmetric chromatid separation, large/small deletions, loss of satellites, translocations, chromosome fusions and bridges (Kovacs, 1985).

Cytological procedures are technically simple and stability assessments can be performed by analysis of the root-tips (or apical meristem) of regenerating plants but it does require skill, experience and patience. In most plant species, the condensation of chromatin during the cell cycle and the subsequent formation of chromosomes is affected by seasonal variation and daylight biorhythms. The techniques vary according to the laboratory (Kovacs, 1985; Pijnacker *et al.*, 1986).

There are a wide range of cytological techniques available, enabling these assessments to be performed on most plant species. Classical staining procedures, for instance Giemsa staining, are well proven and technical advances enable the examination of specific chromosomal regions (Sharma and Sharma, 1980). Chromosome structure can be examined by *in situ* hybridization (Wilkinson, 1992); other techniques include: FISH, the use of fluorescence *in situ* hybridization (McKeown *et al.*, 1992); genomic DNA in GISH, genomic *in situ* hybridization (Wilkinson *et al.*, 1995); and polymerase chain reaction (PCR) technology in PRINS, primed *in situ* labelling (Godsen *et al.*, 1991).

7.6 Biochemical analysis

Biochemical markers for stability assessments, for example plant metabolic/protein profiles, can be assayed by a range of spectrophotometric and chromatographic separation techniques. There is a wide choice of established analytical procedures to measure, for example, chlorophyll fluorescence and other secondary products/pigments (Charlwood and Rhodes, 1990). Analysis of these markers assumes that detectable changes are due to genetic alterations; however secondary metabolite synthesis is influenced by differences in cell physiological states. The expression of secondary product pathways is altered by external factors such as environmental stress factors, light, heat, humidity and nutrient levels. Isozymes have been used in plant breeding and population genetics as biochemical markers for classification and identification of plant species (Tanksley and Orton, 1983). The electrophoretic analysis of proteins or enzymes has been applied widely to breeding new varieties (Cooke, 1989).

7.7 Molecular analysis

There are a wide range of molecular techniques available to analyse DNA (Ayad *et al.*, 1997; Base paper, 1998; Karp *et al.*, 1997; see also Harris, Chapter 2, this volume);

however it is important to consider the concept of genomic complexity prior to the selection of an analytical technique.

7.7.1   *Genome structure*

A plant genome is highly complex; its molecular architecture comprises several levels of DNA sequence organization illustrated by variation in the size of the genome in a range of plant species (Arumuganathan and Earle, 1991). This variability in genome size is mostly due to repetitive DNA, high to moderately reiterated and single/low copy DNA sequences (Dean and Schmidt, 1995). For example, the 'power house' genes encoding ribosomal RNAs and proteins essential for functional ribosomes, which drive cellular reactions are repetitive DNA sequences present in several hundreds to thousands in plant species. The 'housekeeping' genes, i.e. protein products for key metabolic pathways are often found as single/low copy DNA sequences.

There are several classes of reiterative DNA sequences, which exist as long contiguous domains, i.e. tandem repeats or as those sequences interspersed between functional genes or sequences whose function is unknown. Some sequences do have a defined structure and according to their characteristics, they can be classified into groups, for example the long (LINE) or short (SINE) interspersed elements, transposable elements (McClintock, 1984; Ozeki *et al.*, 1997), retrotransposons (Flavell *et al.*, 1992; Hirochika *et al.*, 1996) or simple nucleotide repeats (Morgante and Olivieri, 1993).

7.7.2   *Techniques*

There are two general approaches to study genetic stability; from a practical viewpoint, these should be simple, easy, rapid, non-hazardous and cost-effective.

Polymerase chain reaction

This is an invaluable molecular biological tool for DNA analysis and it has numerous applications. It characteristically involves the use of Taq DNA polymerase, a thermostable enzyme with the cyclic amplification of genomic DNA. A typical reaction entails the heat denaturation of genomic DNA to melt the double-stranded molecules, therefore allowing the oligonucleotide primers to anneal to their complementary sequences. The DNA single strands with their primers act as templates for DNA amplification, where regions (up to 10 kilobases) defined by the distance between adjacent primer binding sites undergo replication. The controlled repetition of these reactions in sequence will selectively amplify the target sequence.

Polymerase chain reaction oligonucleotide primers of arbitrary nucleotide sequence or arbitrarily primed oligonucleotides produce randomly amplified polymorphic DNA (RAPDs) fragments (Welsh *et al.*, 1991; Williams *et al.*, 1991). These are useful markers for the detection of genetic change in plants regenerated from single protoplasts (Brown *et al.*, 1993; Wang *et al.*, 1993) and *in vitro* cultures (Rival *et al.*, 1998). Microsatellites are simple sequences composed of tandemly repeat motifs of four, three, two and one nucleotides found in most eukaryotic genomes. These regions are highly polymorphic and provide excellent genetic markers (Morgante and Olivieri, 1993). Other PCR applications include microsatellite and anchored-simple sequence repeats (Zietkiewicz *et al.*, 1994) to reliably fingerprint varieties of oil seed rape (Charters *et al.*, 1996). The resolution of PCR-based techniques for DNA and mRNA fingerprinting has been refined in AFLPs, amplified fragment length polymorphisms (Money *et al.*, 1996; Vos *et al.*, 1995).

PCR technology is both desirable and relevant to genetic stability studies; the techniques are reliable and require as little as 5–10ng of genomic DNA with the caveat that it is not uncommon to find DNA fragment profile changes between: (1) DNA preparations of the same sample, (2) thermo-cycling machines, (3) operators and laboratories and (4) preparations of Taq DNA polymerase (Lowe *et al.*, 1996; Hallden *et al.*, 1996).

DNA–DNA hybridization

This is one of the most powerful and established methods for genome analysis. It is a multi-step procedure (for full methodology as applied to germplasm conservation see Harding and Benson, 1995). In this type of analysis, it is important to select the correct restriction enzyme and hybridization probe combination to produce an informative DNA fragment profile. Prior to hybridization, the DNA probe is labelled either radioactively or preferably with a non-radioactive chemical tag, for example digoxigenin or biotin (Harding, 1992). The detection of homologous genomic DNA sequences generates a characteristic restriction fragment length polymorphism (RFLP) DNA profile. This is a useful diagnostic procedure for the identification of plant species and cultivars (Ainsworth and Sharp, 1989; Harding, 1991a; see also Harris, Chapter 2, this volume) and is proven for genetic stability assessments of cryopreserved germplasm (Harding, 1991b).

There are several examples illustrating relatively high frequency changes in DNA fragment profiles, thus a reduction in the number of ribosomal RNA gene (rDNA) copies was shown in regenerating tissue cultures (Landsmann and Uhrig, 1985; Potter and Jones, 1990), whereas unknown repetitive DNA sequence probes showed variation in the intensity of the hybridization signal in two somaclonal variants (Ball and Seilleur, 1986). Similar variation was observed in cryopreserved genomic DNA samples analysed for rDNA changes (Harding, 1997). Qualitative changes showing the appearance of a 'new' rDNA size class repeat unit was detected in protoplast-derived plants (Petyuch *et al.*, 1990). There are several useful multi-locus probes available to produce DNA profiles, for example, in DNA fingerprinting, the hypervariable 'minisatellite' sequences (Dallas, 1988), bacteriophage M13 (Nybom *et al.*, 1990) and PCR amplified microsatellite sequences (Morgante and Olivieri, 1993).

DNA methylation studies

Mechanisms have been proposed for the role of DNA methylation in the breakdown of normal developmental pathways which can lead to the genetic instability of plant tissue cultures (Karp, 1993; Phillips *et al.*, 1994). Evidence suggests specific DNA methylation patterns are stable during meiosis (Monk, 1990; Sano *et al.*, 1990) and are inherited in the F1 generation resulting in the phenomenon of DNA imprinting (Matzke and Matzke, 1993; Shemer *et al.*, 1996). The inheritance of genes normally obey the laws of Mendelian genetics, where genetic information is transmitted from the parents to their offspring. However, it appears that there are examples, where specific DNA methylation patterns are inherited as epigenetic mechanisms for gene expression (Kakutani *et al.*, 1996; Phillips *et al.*, 1994). The significance of inheriting new traits from imprinted methylation patterns has yet to be evaluated in plants derived from cryopreserved germplasm.

There are two approaches to analyse the methylation of DNA sequences: (1) DNA–DNA hybridization (Harding *et al.*, 1996) and (2) methylation specific PCR (Herman *et al.*, 1996). The finding that tissue cultures can contain methylated DNA (Harding *et al.*, 1996) is important, especially as *in vitro* techniques continue to play a role in the conservation of genetic resources for the international exchange of germplasm. Methylation may be induced in stressed, vitrified tissue cultures (Leonhardt and Kandeler, 1987) and cultures maintained under slow growth conditions can contain hypermethylated DNA (Harding, 1994; Smulders *et al.*, 1995). This semi-dormant 'slow growth' state of plantlets may activate specific DNA methylases, resulting in highly methylated domains within the genome which are adaptive responses to conditions of high osmotic stress (Britt, 1996). The biological significance of DNA methylation may be to conserve cellular resources during conditions of low metabolic activity (i.e. growth in mannitol supplemented medium). This may have implications for long-term storage procedures in which osmotically active solutions are employed as slow growth additives or cryoprotective agents. DNA methylation sequence changes may occur under these conditions leading to the induction of several genetic changes (Harding, 1994). These findings have implications for vegetatively propagated plants (e.g. tuber species), as changes due to methylated DNA may be inherited in the somatic progeny and altered phenotypes expressed in subsequent generations.

Genetic modification

This is a complex process and the release of genetically modified organisms (GMOs) into the environment must be considered in the context of regulatory issues. Controversy does exist in this area, but progress continues towards the increased use of GMO-related biotechnology in agriculture (Harding, 1995; Harding and Harris, 1997). The conservation of germplasm from GMOs is important and it is imperative that the stability of the transgenes is confirmed after *in vitro* storage. Several molecular biological studies have shown no detectable genetic variation after cryopreservation. The growth rates, secondary metabolite production and T-DNA structure were found to be unchanged after cryopreservation of root-tips from hairy root cultures of *Beta vulgaris* and *Nicotiana rustica* (Benson and Hamill, 1991). Moreover, DNA analysis of the integrated nptII selectable marker gene in transgenic *Citrus sinensis* showed no difference in size or number of nptII genes in both non-cryopreserved cells and cryopreserved cells (Kobayashi and Sakai, 1997; Sakai, 1995). This transgenic technology produces 'novel' germplasm suitable for the status of a new plant variety and where this 'novel germplasm' is conserved in a genebank it is likely to be subject to stability assessments. Where genetically modified plants contain single or multiple copies of a transgene, evidence indicates that DNA methylation plays a role in the transcription and expression of transgenes and the methylation status of these genes can be examined with the same pairs of isoschizomers. The restriction enzyme products show differential DNA methylation patterns between different transgenic lines (Harding, 1996), therefore establishing a genomic imprint as the basis of stability assessments for *in vitro* conservation.

Somatic hybridization studies

Plant protoplasts and tissue culture techniques have been instrumental in protoplast fusion programmes to produce somatic hybrids (Roest and Gilissen, 1993). There have been many successful attempts to produce somatic hybrids; these plants express traits including disease resistance to viruses, bacteria, fungi and insect pests (Cooper-Bland *et al.*, 1994). This technique has the potential for self-generating diversity in the numerous nuclear and cytoplasmic hybrid combinations (Kumar and Cocking, 1987). The analysis of these putative fusion products is important to confirm hybridity (Matthews *et al.*, 1997) and expression of the desirable traits, as it may well be useful germplasm for the production of new plant varieties. Thus in the future it will become necessary to assess the genetic stability of somatic hybrid germplasm which is conserved *in vitro*.

7.8 Conclusion

There are many repositories and established genebanks holding germplasm collections of vegetatively propagated crop species, where the use of tissue culture techniques continues to play a vital role in *ex situ* conservation. These procedures are satisfactory for the routine and genetically stable storage of cultures, however challenges do exist regarding stability assessments. The development of new and efficient analytical techniques which can be used to routinely evaluate the genetic stability of germplasm maintained *in vitro* continues to present a unique research opportunity for plant conservation biotechnologists.

References

ABDULLAH, R., THOMPSON, J.A., KHUSH, G.S. KAUSHIL, R.P. and COCKING, E.C., 1989, Protoclonal variation in the seed progeny of plants regenerated from rice protoplasts, *Plant Science*, **65**, 97–101.

AINSWORTH, C.C. and SHARP, P.J., 1989, The potential role of DNA probes in plant variety identification, *Plant Varieties and Seeds*, **2**, 27–34.

ARUMUGANATHAN, K. and EARLE, E.D., 1991, Nuclear DNA content of some important plant species, *Plant Molecular Biology Reporter*, **9**(3), 208–218.

ASHMORE, S.E., 1997, *Status Report on the Development and Application of In Vitro Techniques for the Conservation and Use of Plant Genetic Resources*, ISBN 92–9043–339–6, International Plant Genetic Resources Institute, Rome, Italy.

AYAD, W.G., HODGKIN, T., JARADAT, A. and RAO, V.R., 1997, *Molecular Genetic Techniques for Plant Genetic Resources*, Report of an IPGRI Workshop, ISBN 92–9043–315–9, International Plant Genetic Resources Institute, Rome, Italy.

BALL, S.G. and SEILLEUR, P., 1986, Characterisation of somaclonal variations, in SEMAL, J. (Ed.), *Potato: A Biochemical Approach, Somaclonal Variation and Crop Improvement*, pp. 229–235, Dordrecht: Martinus Nijhoff Publishers.

Base paper, 1998, Genome mapping and marker-assisted selection in plant breeding, in V.L. CHOPRA, R.B. SINGH and A. VARMA (Eds), *Crop Productivity and Sustainability: Shaping the Future. Proceedings of the 2nd International Crop Science Congress*, Oxford and IBH Publishing Co. PVT Ltd, New Delhi, Calcutta, India. pp. 999–1018.

BAYLISS, M.W., 1980, Chromosomal variation in plant tissue culture, *International Review Cytology*, **11A**, 113–143.

BENSON, E.E. and HAMILL, J.D., 1991, Cryopreservation and post-freeze molecular and biosynthetic stability in transformed roots of *Beta vulgaris* and *Nicotiana rustica, Plant Cell, Tissue and Organ Culture*, **24**, 163–172.

BRITT, A.B., 1996, DNA damage and repair in plants, *Annual Review Plant Physiology Plant Molecular Biology*, **47**, 75–100.

BROWN, P.T.H., LANGE, F.D., KRANZ, E. and LORZ, H., 1993, Analysis of single protoplasts and regenerated plants by PCR and RAPD technology, *Molecular General Genetics*, **237**, 311–317.

CHARLWOOD, B.V. and RHODES, M.J.C., 1990, *Secondary Products from Plant Tissue Culture, Proceedings of the Phytochemical Society of Europe (30)*, Oxford: Clarendon Press.

CHARTERS, Y.M., ROBERTSON, A., WILKINSON, M.J. and RAMSAY, G., 1996, PCR analysis of oilseed rape cultivars (*Brassica napus* L. ssp oleifera) using 5'-anchored simple sequence repeat (SSR) primers, *Theoretical Applied Genetics*, **92**, 442–447.

COOKE, R.J., 1989, The use of electrophoresis for the distinctness testing of varieties of autogamous species, *Plant Varieties and Seeds*, **2**, 3–13.

COOPER-BLAND, S., DE, MAINE, M.J., FLEMING, M.L.M.H., PHILLIPS, M.S., POWELL, W. and KUMAR, A., 1994, Synthesis of intraspecific somatic hybrids of *Solanum tuberosum*: assessments of morphological, biochemical and nematode (*Globodera pallida*) resistance characteristics, *J. Experimental Botany*, **45**(278): 1319–1325.

DALLAS, J.F., 1988, Detection of DNA 'fingerprints' of cultivated rice by hybridization with a human minisatellite DNA probe, *Proceedings National Academy Science (USA)*, **85**, 6831–6835.

D'AMATO, F., 1985, Cytogenetics of plant cell and tissue cultures and their regenerates, in CONGER, B.V. (Ed.), *Critical Reviews in Plant Sciences*, pp. 73–112, Boca Raton, Florida: CRC Press.

DEAN, C. and SCHMIDT, R., 1995, Plant genomes: a current molecular description, *Annual Review Plant Physiology Plant Molecular Biology*, **46**, 395–418.

DENTON, I.R., WESTCOTT, R.J. and FORD-LLOYD, B.V., 1977, Phenotypic variation of *Solanum tuberosum* L. cv. Dr McIntosh regenerated directly from shoot-tip culture, *Potato Research*, **20**, 131–136.

FLAVELL, A.J., DUNBAR, E., ANDERSON, R., PEARCE, S.R., HARTLEY, R. and KUMAR, A., 1992, Ty1copia group retrotransposons are ubiquitous and heterogeneous in higher plants, *Nucleic Acids Research*, **20**(14), 3639–3644.

GODSEN, J., HANRATTY, D., STARLING, J., FANTES, J. and PORTEOUS, D., 1991, Oligonucleotide primed *in situ* DNA synthesis (PRINS): a method for chromosome mapping, banding and investigation of sequence organisation, *Cytogenetics and Cell Genetics*, **57**, 100–104.

GOULD, A.R., 1986, Factors controlling generation of variability *in vitro*, in VASIL, I.K. (Ed.), *Cell Culture and Somatic Cell Genetics of Plants, Vol. 3, Plant Regeneration and Genetic Variability*, pp. 549–567, New York: Academic Press Inc.

HABERLANDT, G., 1902, Kultinversuche mit isolierten Pflanzellen, *Sber. Akad. Wiss. Wien*, **111**, 69–92.

HALLDEN, C., HANSEN, M., NILSSON, N.O. and HJERDIN, A., 1996, Competition as a source of errors in RAPD analysis, *Theoretical Applied Genetics*, **93**, 1185–1192.

HALPERIN, W., 1986, Attainment and retention of morphogenetic capacity *in vitro*, in VASIL, I.K. (Ed.), *Cell Culture and Somatic Cell Genetics of Plants*, Vol. 3, pp. 3–47, New York: Academic Press.

HARDING, K., 1991a, Restriction enzyme mapping of the ribosomal RNA genes in *Solanum tuberosum*: potato rDNA restriction enzyme map, *Euphytica*, **54**, 245–250.

HARDING, K., 1991b, Molecular stability of the ribosomal RNA genes in *Solanum tuberosum* plants recovered from slow growth and cryopreservation, *Euphytica*, **55**, 141–146.

HARDING, K., 1992, Detection of ribosomal RNA genes by chemiluminescence in *Solanum tuberosum* L.: a rapid and non-radioactive technique for the characterisation of potato germplasm, *Potato Research*, **35**, 199–204.

HARDING, K., 1994, The methylation status of DNA derived from potato plants recovery from slow growth, *Plant Cell, Tissue and Organ Culture*, **37**, 31–38.

HARDING, K., 1995, Biosafety of selectable marker genes, *AgBiotech News and Information*, **7**(2), 47–52.

HARDING, K., 1996, Approaches to assess the genetic stability of plants recovered from *in vitro* culture, in NORMAH, M.N. (Ed.), *Proceedings of the International Workshop on In Vitro Conservation of Plant Genetic Resources*, pp. 137–170, ISBN 983–9647, Kuala Lumpur, Malaysia: Plant Biotechnology Laboratory, UKM.

HARDING, K., 1997, Stability of the ribosomal RNA genes in *Solanum tuberosum* L. plants recovered from cryopreservation, *Cryo-Letters*, **18**, 217–230.

HARDING, K. and BENSON, E.E., 1994, A study of growth, flowering, and tuberisation in plants derived from cryopreserved potato shoot-tips: implications for *in vitro* germplasm collections, *Cryo-letters*, **15**, 59–66.

HARDING, K. and BENSON, E.E., 1995, Biochemical and molecular methods for assessing damage, recovery and stability in cryopreserved plant germplasm, in GROUT, B.W.W. (Ed.), *Genetic Preservation of Plant Cells In Vitro*, pp. 103–151, Heidelberg, Springer-Verlag.

HARDING, K. and HARRIS, P.S., 1997, Risk assessment of the release of genetically modified plants: a review. *Agro-food-Industry, Hi-tech*, **8**(6), 8–13.

HARDING, K., BENSON, E.E. and ROUBELAKIS-ANGELAKIS, K.A., 1996, Changes in genomic DNA and rDNA associated with the tissue culture of *Vitis vinifera*: a possible molecular basis for recalcitrance, *Vitis*, **35**(2), 79–85.

HARDING, K., BENSON, E.E. and CLACHER, K., 1997, Plant conservation biotechnology: an overview, *Agro-food-Industry, Hi-tech*, **8**(3), 24–29.

HEIKEN, A., 1960, Spontaneous and X-ray-induced somatic aberrations, in *Solanum tuberosum* L., Almqvisy and Wiksell. Stockholm.

HERMAN, J.G., GRAFF, J.R., MYOHANEN, S., NELKIN, B.D. and BAYLIN, S.B., 1996, Methylation-specific PCR: a novel PCR assay for methylation status of CpG islands, *Proceedings National Academy Sciences* (USA), **93**, 9821–9826.

HIROCHIKA, H., SUGIMOTO, K., OTSUKI, Y., TSUGAWA. H. and KANDA, M., 1996, Retrotransposons of rice involved in mutations induced by tissue culture, *Proceedings National Academy Sciences (USA)*, **93**, 7783–7788.

IBPGR, 1977, *Descriptors for the Cultivated Potato*, Z. HUAMAN, J.T. WILLIAMS, W. SALHUANA and VINCENT, L. (Eds) Rome, Italy. AGPE:IBPGR/77/32.

KAKUTANI, T., JEDDELOH, J.A., FLOWERS, S.K., MUNAKATA, K. and RICHARDS, E.J., 1996, Developemntal abnormalities and epimutations associated with DNA hypomethylation mutations, *Proceedings National Academy Sciences (USA)*, **93**, 12406–12411.

KARP, A., 1993, Are your plants normal? – Genetic instability in regenerated and transgenic plants, *Agro-Food-Industry Hi-Tech*, May/June, 7–12.

KARP, A., KRESOVICH, S., BHAT, K.V., AYAD, W.G. and HODGKIN, T., 1997, *Molecular Tools in Plant Genetic Resources Conversation: A Guide to Technologies*, ISBN 92-9043-323-X, IPGRI Technical Bulletin 2, International Plant Genetic Resources Institute, Rome, Italy.

KOBAYASHI, S. and SAKAI, A., 1997, Cryopreservation of *Citrus sinensis* cultured cells, in M.K. RAZDAN and E.C. COCKING (Eds), *Conservation of Plant Genetic Resources In Vitro*, pp. 202–223, Science Publishers, Inc., USA.

KOVACS, E.I., 1985, Regulation of karyotype stability in tobacco tissue cultures of normal and tumorous genotypes, *Theoretical Applied Genetics*, **70**, 548–554.

KUMAR, A. and COCKING, E.C., 1987, Protoplast fusion, a novel approach to organelle genetics in higher plants, *American J. Botany*, **74**(8), 1289–1303.

LANDSMANN, J. and UHRIG, H., 1985, Somaclonal variation in *Solanum tuberosum* detected at the molecular level, *Theoretical Applied Genetics*, **71**, 500–505.

LARKIN, P.J. and SCOWCROFT, W.R., 1981, Somaclonal variation – a novel source of variability from cell cultures for plant improvement, *Theoretical Applied Genetics*, **60**, 197–214.

LEONHARDT, W. and KANDELER, R., 1987, Ethylene accumulation in culture vessels – a reason for vitrification, *Acta Horticulturae*, **212**, 223–229.

LOWE, A.J., HANOTTE, O. and GUARINO, L., 1996, Standardisation of molecular genetic techniques for the characterisation of germplasm collections: the case of random amplified polymorphic DNA (RAPD), *Plant Genetic Resources Newsletter*, **107**, 50–54.

MATTHEWS, D., McNICOLL, J., HARDING, K. and MILLAM, S., 1997, The detection of alien DNA sequences in somatic hybrids, *Plant Molecular Biology Reporter*, **15**(1), 62–70.

MATZKE, M. and MATZKE, A.J.M., 1993, Genomic imprinting in plants: parental effects and trans-inactivation phenomena, *Annual Review Plant Physiology Plant Molecular Biology*, **44**, 53–76.

MAYARD-SMITH, J., 1998, *Evolutionary Genetics*, Oxford, New York: Oxford University Press.

MCCLINTOCK, B., 1984, The significance of responses of the genome to challenge, *Science*, **226**, 792–801.

MCKEOWN, C.M.E., WATERS, J.J., STACEY, M., NEWMAN, B.F., CARDY, D.L.N. and HULTEN, M., 1992, Rapid interphase FISH diagnosis of trisomy 18 on blood smears, *Lancet*, **340**, 499.

MONEY, T., READER, S., QU, L.J., DUNFORD, R.P. and MOORE, G., 1996, AFLP-based mRNA fingerprinting, *Nucleic Acids Research*, **24**(13), 2616–2617.

MONK, M., 1990, Variation in epigenetic inheritance, *Trends in Genetics*, **6**(4), 110–114.

MORGANTE, M. and OLIVIERI, A.M., 1993, PCR-amplified microsatellites as markers in plant genetics, *The Plant Journal*, **3**(1), 175–182.

NYBOM, H., ROGSTAD, S.H. and SCHAAL, B.A., 1990, Genetic variation detected by use of the M13 'DNA fingerprint' probe in *Malus*, *Prunus* and *Rubus* (*Rosaceae*), *Theoretical Applied Genetics*, **79**(2), 153–156.

OZEKI, Y., DAVIES, E. and TAKEDA, J., 1997, Somatic variation during long term subculturing of plant cells caused by insertion of a transposable element in a phenylalanine ammonia-lysae (PAL) gene, *Molecular General Genetics*, **254**, 407–416.

PESCHKE, V.M. and PHILLIPS, R.L., 1992, Genetic implications of somaclonal variations in plants, *Advances in Genetics*, **30**, 41–75.

PETYUCH, G.P., BORISYUK, N.V., KUCHKO, A.A. and GLEBA, Y.Y., 1990, Variability of ribosomal RNA genes amongst potato protoplast derived plants, p. 163 (A4-78), VIIth International Congress on Plant Tissue and Cell Culture, Amsterdam.

PHILLIPS, R.L., KAEPPLER, S.M. and OLHOFT, P., 1994, Genetic instability of plant tissue cultures: breakdown of normal controls, *Proceedings National Academy Science (USA)*, **91**, 5222–5226.

PIJNACKER, L.P., HERMELIN, J.H.M. and FERWERDA, M.A., 1986, Variability of DNA content and karyotype in cell cultures of an interdihaploid *Solanum tuberosum*, *Plant Cell Reports*, **5**, 43–46.

POTTER, R.H. and JONES, M.G.K., 1990, Molecular analysis of genetic stability, in DODDS, J.H. (Ed.), *In vitro Methods for Conservation of Plant Genetic Resources*, pp. 71–89, London: Chapman and Hall.

POWELL, W., MORGANTE, M., ANDRE, C., HANAFEY, M., VOGEL, J., TINGEY, S. and RAFALSKI, A., 1997, The comparison of RFLP, RAPD, AFLP and SSR (microsatellite) markers for germplasm analysis, *Molecular Breeding*, **2**, 225–238.

RIVAL, A., BERTRAND, L., BEULE, T., COMBES, M.C., TROUSLOT, P. and LASHERMES, P., 1998, Suitability of RAPD analysis for the detection of somaclonal variants in oil palm (*Elaeis guineensis* Jacq), *Plant Breeding*, **117**, 73–76.

ROEST, S. and GILISSEN, L.J.W., 1993, Regeneration from protoplasts – a supplementary literature review, *Acta Botany Netherlands*, **42**(1), 1–23.

SAKAI, A. 1995, Cryopreservation for germplasm collection in woody plants, in JAIN, S.M., GUPTA, P.K. and NEWTON, R.J. (Eds), *Somatic Embryogenesis in Woody Plants, Vol. 1, History, Molecular and Biochemical Aspects and Application*, pp. 293–315, Netherlands: Kluwer Academic Publishers.

SANO, H., KAMADA, I., YOUSSEFIAN, S., KATSUMI, M. and WABIKO, H., 1990, A single treatment of rice seedlings with 5-azacytidine induces heritable dwarfism and undermethylation of genomic DNA, *Molecular General Genetics*, **220**, 441–447.

SCOWCROFT, W.R., 1984, *Genetic Variability in Tissue Culture: Impact on Germplasm Conservation and Utilisation*, International Board for Plant Genetic Resources, Report (AGPG:IBPGR/84/152), Rome.

SCOWCROFT, W.R., 1985, Somaclonal variation: the myth of clonal uniformity, in HOHN, B. and DENNIS, E.S. (Eds) *Genetic Flux in Plants*, pp. 217–243, New York: Springer-Verlag.

SCOWCROFT, W.R., RYAN, S.A., BRETTEL, R.I.S. and LARKIN, P.J., 1984, Somaclonal variation: a 'new' genetic resource, in HOLDEN, J.H.W. and WILLIAMS, J.T. (Eds), *Crop Genetic Resources: Conservation and Evaluation*, pp. 258–267, London: George Allen & Unwin Ltd.

SEMAL, J., 1986, *Somaclonal Variations and Crop Improvement*, Dordrecht: Martinus Nijhoff Publishers.

SHARMA, A.K. and SHARMA, A., 1980, *Chromosome Techniques: Theory and Practice*, London: Butterworths.

SHEMER, R., BIRGER, Y., DEAN, W.L., REIK, W., RIGGS, A.D. and RAZIN, A., 1996, Dynamic methylation adjustment and counting as part of imprinting mechanisms, *Proceedings National Academy Sciences (USA)*, **93**, 6371–6376.

SMULDERS, M.J.M., RUS-KORTEKAAS, W. and VOSMAN, B., 1995, Tissue culture-induced DNA methylation polymorphisms in repetitive DNA of tomato calli and regenerated plants, *Theoretical Applied Genetics*, **91**, 1257–1264.

STADELMANN, F.J., BOLLER, B., SPANGENBERG, G., KOLLIKER, R., WANG, Z.Y., POTRYKUS, I. and NOSBERGER, J. 1998, Field performance of cell-suspension-derived *Lolium* perenne L. regenerants and their progenies, *Theoretical and Applied Genetics*, **96**, 634–639.

TANKSLEY, S.D. and ORTON, T.L., 1983, *Isozymes in Plant Genetics and Breeding (Parts A and B)*, Oxford: Amsterdam, Elsevier.

VOS, P., HOGERS, R., BLEEKER, M., REIJANS, M., VAN DE LEE, T., HORNES, M., FRIJTERS, A., POT, J., PELEMAN, J., KUIPER, M. and ZABEAU, M., 1995, AFLP: a new technique for DNA fingerprinting, *Nucleic Acids Research*, **23**(21), 4407–4414.

WANG, Z.Y., NAGEL, J., POTRYKUS, I. and SPANGENBERG, G., 1993, Plants from cell suspension-derived protoplasts in *Lolium* species, *Plant Science*, **94**, 179–193.

WELSH, J., HONEYCUTT, R.J., MCCLELLAND, M. and SOBRAL, B.W.S., 1991, Parentage determination in maize using the arbitrarily primed polymerase chain reaction, *Theoretical Applied Genetics*, **82**, 473–476.

WILKINSON, D.G., 1992, *In situ Hybridisation, A Practical Approach*, Oxford: IRL Press.

WILKINSON, M.J., BENNETT, S.T., CLULOW, S.A., ALLAINGUILLAUME, J., HARDING, K. and BENNETT, M.D., 1995, Evidence for somatic translocation during potato dihaploid induction, *Heredity*, **74**, 146–151.

WILLIAMS, J.G.K., KUBELIK, A.R., LIVAK, K.J., RAFALSKI, J.A. and TINGEY, S.V., 1991, DNA polymorphisms amplified by arbitrary primers are useful as genetic markers, *Nucleic Acids Research*, **18**, 6531–6535.

ZIETKIEWICZ, E., RAFALSKI, A. and LABUDA, D., 1994, Genome fingerprinting by simple sequence repeat (SSR)-anchored polymerase chain reaction amplification, *Genomics*, **20**, 176–183.

PART II

Applications of Biotechnology in Plant Diversity Conservation

8

Conservation Strategies for Algae

JOHN G. DAY

8.1 Introduction

Algae are an ancient and extremely diverse group of plantlike organisms, with representatives of the blue–green algae (cyanobacteria) being present for the last 3550 million years (Schopf and Walter, 1982). They range in morphology and size from microscopic picoplanktonic cyanobacteria (<2 μm in diameter) which are prokaryotic and closely resemble other eubacteria, a variety of unicellular, multicellular, filamentous and thalloid forms, to giant kelps that may be up to 60 m long. Their taxonomy is problematic, but it is clear on the basis of both traditional taxonomy and modern molecular techniques that they are polyphyletic (Bold and Wynn, 1985; Cavalier-Smith, 1993). The 'amount' of algal biodiversity, as in other groups of organisms, is largely unknown, however, the advent of molecular biological techniques and improvements in electron microscopy have greatly increased our knowledge-base and assisted in elucidating inter-relationships. Approximately 37 000 species of algae have been recognized/described (Table 8.1) but estimates of the total number of algal species vary from a relatively conservative 40 000 to >10 000 000 (Hawksworth and Mound, 1991; John, 1994).

From an economic perspective, algae are often considered a nuisance. Those associated with the water industry may think of them as a source of irritation – blocking water filters, causing 'off flavours' etc. – or even a major economic problem when toxic cyanobacteria may result in a reservoir being unusable as a source of drinking water. However, others see the group as a rich source of valuable chemicals or novel pharmacologically active agents (Glombitza, 1979; Lincoln et al., 1990). Approximately 500 species of algae are used as human food or food products, and about 160 species are considered commercially valuable (Abbott, 1988). Products from algae may have significant financial value; the Japanese harvest of Porphyra (Nori) is worth US$1 billion annually (Mumford and Miura, 1988) and the annual value of algal polysaccharides, primarily agars and carageanans, is US$500 million (Jensen, 1993). Other commercially important products from algae include: health foods (Lee, 1997); pigments (Cannell, 1990); aquaculture feeds (Borowitzka, 1997) and lipids (Shifrin and Chisholm, 1980). They are also widely used in ecotoxicity testing (OECD, 1984) and may be used to treat waste water (Oswald, 1988). Moreover, algae as a group are responsible for fixing

Table 8.1   Biodiversity of algae

Algal group	Division	Class	Common name	No.[a]
Cyanobacteria	Cyanophyta	Cyanophyceae	Blue–green algae	2000
Green algae	Charophyta	Charophyceae	Stoneworts	11 000
	Chlorophyta	Chlorophyceae		3600
		Ulvophyceae		
	Prasinophyta	Prasinophyceae		120
Chromophyte algae	Bacillariophyta	Bacillariophyceae	Diatoms	10 000
		Fragilariophyceae		
	Chrysophyta	Chrysophyceae		1000
		Synurophyceae		
	Dictyochophyta	Dictyochophyceae		10
		Pelagophyceae		
	Eustigmatophyta	Eustigmatophyceae		12
	Phaeophyta	Phaeophyceae	Brown algae	900
	Prymnesiophyta	Prymnesiophyceae		300
	Raphidophyta	Raphidophyceae		15
	Xanthophyta	Xanthophyceae		600
Red algae	Rhodophyta	Rhodophyceae		4000
Cryptomonads	Cryptophyta	Cryptophyceae		200
Dinoflagellates	Pyrrophyta	Pyrrophyceae		2000
Euglenoids	Euglenophyta	Euglenophyceae		900
Glaucophytes	Glaucophyta	Glaucophyceae		13

[a] Approximate no. species described; data from Norton *et al.* (1996).

40 per cent of the earth's carbon (Bolin *et al.*, 1977) and as such are major carbon sinks and also oxygen producers.

8.2   Alternative strategies employed to conserve algae

As with other groups of organisms two basic options are available for the long-term conservation of algae: conservation *in situ* in managed or non-managed ecosystems, and *ex situ* conservation. The former has the advantage that the algae will continue to interact with the other biological and physico-chemical factors in their environment and will not vary from 'wild-type' strains. This type of algal conservation occurs in marine parks, or other areas protected from the excesses of man's activities (Phillips, 1998). However, in reality this approach is not appropriate for many organisms. Where access to an algal strain is needed quickly, or an axenic culture is required, *ex situ* maintenance is the only realistic option. Further stimuli to the *ex situ* conservation of living materials have been the Convention on Biodiversity (CBD), specifically Article 9: *ex situ* conservation (UNEP, 1992) and the parallel development of bioprospecting for products with commercial value (Day, 1993).

8.3 Roles of genetic resource centres and culture collections

The primary role of an algal culture collection is the same as any other collection of living material that is to be a repository for cultures. In microbial service collections, including algal collections, this role is often associated with other products and services including: provision of authentic specimens for research; material for education; material for bioassay use; aquaculture starter cultures; identification; training; acting as a depository for patent purposes; consultancy; and other commercial applications. All of these require the maintenance of viable, healthy, physiologically and genetically stable cultures.

There are more than 11 000 strains of algae including representatives of approximately 1600 different species retained in protistan collections around the world (Miyachi *et al.*, 1989), with more than 80 per cent of these being maintained in the six largest algal culture collections (Table 8.2). These collections provide the scientific community with cultures and their associated information, as well as a variety of other services (see above and Table 8.2).

8.4 Methodological strategies employed to conserve algae

Any conservation methodology adopted should guarantee long-term stability of the morphological, physiological and genetic characteristics of the preserved organism. A variety of methods have been applied to the long-term stabilization/preservation of algae (McLellan *et al.*, 1991; Warren *et al.*, 1997). However, the most commonly used techniques involve the routine serial subculture of the algae under controlled environmental conditions. The alternative approaches, which involve less routine maintenance of the conserved specimens/cultures, all depend on the removal of water and/or altering the cellular physicochemical environment with respect to water activity. These techniques fall into three main categories: drying; freeze-drying and cryopreservation.

8.4.1 *Maintenance by serial subculture*

Historically algae have been maintained *ex situ* by regular serial subculture (Leeson *et al.*, 1984; Pringsheim, 1946). This continues to be the method of choice for most phycologists, particularly when relatively small numbers of cultures are involved. There is an extensive literature on culture techniques and maintenance conditions (Day and McLellan, 1995; McLellan *et al.*, 1991; Stein, 1973; Warren *et al.*, 1997). Medium composition depends both on the requirements of the algae (e.g. diatoms require the inclusion of a silica source) and the preferences of the researcher. Information on medium suitability and full recipes are listed in all major collection catalogues (Andersen *et al.*, 1997; Nerad, 1993; Schlösser, 1994; Starr and Zeikus, 1993; Tompkins *et al.*, 1995; Watanabe and Hiroki, 1997). In general, cultures are maintained under sub-optimal temperature and light regimes ($<20°C$ and <50 µmol photon $m^{-2}s^{-1}$); this maximizes the interval between subcultures and thus minimizes handling/transfers of the strain. Alternative techniques include the use of medium containing organic carbon for maintaining axenic strains capable of heterotrophic or mixotrophic growth and solidified, rather than liquid medium may be employed to maximize the period between transfer to fresh medium.

Table 8.2 Activities of major[a] algal culture collections

Name of catalogue	Acronym	Country	No. strains	Latest printed	Available on WWW	Use of Cryopreservation	Patent Deposits	WFCC No.
Algensammlung am Institut fur Botanik	ASIB	Austria	1600	1985	−	−	−	505
American Type Culture Collection	ATCC	USA	200	1993	+	+	+	1
Culture Collection of Algae and Protozoa	CCAP	UK	1700	1995	+	+	+	140/522
Provasoli–Guillard Center for Culture Collection of Marine Phytoplankton	CCMP	USA	1450	1997	+	+/−	−	2
National Institute for Environmental Studies Collection	NIES	Japan	1000	1997	−	+	−	591
Sammlung von Algenkulturen	SAG	Germany	1630	1994	−	−	−	192
Culture Collection of Algae at the Univ. of Texas at Austin	UTEX	USA	2100	1993	+	+/−	−	606

[a] Service collections retaining > 1000 algal strains with the addition of the American Type Culture Collection (ATCC).
+   Available/currently used.
+/− Under development.
−   Not available/not used.
Data from Takishima et al. (1989) and Day (1996).

Although serial subculture has proven very successful, with some isolates being maintained for more than 80 years, it is widely recognized as being sub-optimal. It is a labour and consumables intensive process and the continuing increases in costs act as a stimulus to the development of long-term preservation techniques. Furthermore, this technique can potentially lead to selection of a population which may not be representative of the parent culture. In extreme cases this may include changes in important physiological and morphological characteristics, for example, irreversible shrinkage of diatoms (Jaworski *et al.*, 1988), loss of spines in *Micractinium pusillum* and alteration of pigment composition in a number of algae (Warren *et al.*, 1997).

8.4.2 Maintenance by storage in liquid medium

Some species of algae, particularly those that may form resistant structures, may be maintained long-term in biphasic medium (Tompkins *et al.*, 1995). This approach has been successfully employed to preserve algal zygotes and cysts for up to 20 years (Coleman, personal communication).

8.4.3 Drying techniques

Some algae are extremely resistant to desiccation and algal cysts/spores may survive prolonged exposure to dry conditions and high temperatures (Buzer *et al.*, 1985). Examples of this include air dried soil samples containing *Haematococcus pluvialis* aplanospores that can regenerate fresh cultures after 27 years storage (Leeson *et al.*, 1984) and the cyanobacterium *Nostoc commune* that has been revived from herbaria specimens after 107 years storage (Cameron, 1962).

Drying, generally air drying, may be used successfully for a wide range of cyst forming protists and some strains, e.g. the achlorophylous euglenoid *Polytoma*, are commonly transported as dried material on filter paper (Alexander *et al.*, 1980; Nerad, 1993). Furthermore, storage of cyanobacterial cultures in dried soil, or non-perfumed cat-litter, has been used by some researchers to maintain 'back-up' cultures for periods of several years (Parker, personal communication). However, drying has not been widely applied as a method of long-term conservation of algae in major service collections. This is primarily due to the low levels of recovery for some organisms and the relatively short shelf-life of stored material (Day *et al.*, 1987).

More recent research using a controlled drying protocol demonstrated that the method has some potential for a number of green algae (Malik, 1993). This study employed equipment which vacuum dried the algae in a suspending medium incorporating protective chemicals including skimmed milk, neutral activated charcoal, *meso*-inositol or raffinose. This approach may be utilizable for several algae, but it is unlikely to be satisfactory for the long-term preservation of more fragile organisms and, as yet, no long-term data on viability have been published.

Freeze-drying techniques

Lyophilization using conventional freeze-drying equipment and protocols employing 20 per cent (w/v) skimmed milk or 12 per cent (w/v) sucrose as protective agents in the suspending medium has been successfully applied to preserve a range of cyanobacteria and eukaryotic microalgae (Holm-Hanson, 1973; McGrath *et al.*, 1978). This technique has been regularly used at the American Type Culture Collection (ATCC) for a number of organisms (Daggett and Nerad, 1992). However, levels of post-lyophilization viability may be low, 10^{-2} to less than 10^{-7} per cent of the original level being recorded by McGrath *et al.* (1978). On using this approach at the Culture Collection of Algae and Protozoa (CCAP) low levels or no viability was observed post-rehydration of freeze-drying eukaryotic microalgae. The highest level of viability observed was 1 per cent for *Chlorella emersonii*; however, no viability was detected after storage for one year (Day *et al.*, 1997). It is worth noting that *C. emersonii* has previously been demonstrated to survive up to two years storage using this technique (Day, 1987). In this material viability levels were extremely low, in the range 10^{-4} to less than 10^{-7} per cent of the original level (Day, unpublished data). Freeze-drying cyanobacteria, using the method of Kolkowski and Smith (1995) has proven more successful, with survival of both unicellular and filamentous forms (Day *et al.*, unpublished data). However, where non-axenic isolates were examined, on suspension of the lyophilized samples in fresh medium, the protective agents stimulate bacterial 'blooms' that had a deleterious effect on the recovery of the preserved cyanobacterium. Although this technique may be employed, low levels of viability, problems associated with non-axenic cultures and the possibility of selecting for a tolerant sub-population have dissuaded the major collections from adopting it as a technique to conserve algae.

8.4.4   *Cryopreservation techniques*

The general theory and principles of cryopreservation are outlined by Benson (Chapter 6, this volume). Cryopreservation is the optimal method of long-term storage of algae, where high post-thaw viability can be guaranteed. At ultra-low temperatures (less than $-135°C$) no further deterioration of stored material can occur and viability levels should be independent of storage duration measured in decades (Grout, 1995). Therefore, assuming there are no perturbations in the storage regime, long-term stability of the frozen specimens can effectively be guaranteed. As yet there are no published reports on the genetic stability, or otherwise, of cryopreserved algae. However, a selection of the algae originating from different ecological niches, and from different algal Divisions and Classes, were demonstrated to have retained the same levels of post-thaw viability after up to 22 years storage in the CCAP Cryostore (Day *et al.*, 1997). These factors and the significant savings in costs associated with serial subculture have stimulated all the major collections to consider employing cryopreservation, with most of them currently using or developing the technique to preserve a proportion of their holdings (Table 8.2).

    In comparison to drying or freeze-drying, cryopreservation can result in high levels of post-thaw viability, with levels in excess of 95 per cent for some members of the Chlorococcales (Morris, 1978). Where levels of viability are high, the possibility of selecting a preservation tolerant sub-population is minimized and also the time required to re-grow a preserved culture to a suitable density for distribution is restricted. At the CCAP, and some of the other major collections, levels of viability in excess of 50 per cent, for non-clonal cultures, are required before cultures are retained only in the cryo-preserved state. However, other collections including the ATCC do not impose arbitrary minimum levels of post-thaw viability (Nerad, personal communication). Currently at the CCAP approximately 35 per cent of the algal strains lodged in the collections are maintained in a cryopreserved state (Table 8.3). At least a further 5 per cent, primarily those

Table 8.3 Biodiversity of algal strains maintained at CCAP

Algal group	Division	No. strains maintained	
		Cryopreserved	Total
Cyanobacteria	Cyanophyta	152	228
Green algae	Chlorophyta	433	871
	Prasinophyta	5	118
Chromophyte algae	Bacillariophyta	8	97
	Chrysophyta	0	20
	Eustigmatophyta	15	24
	Phaeophyta	1	3
	Prymnesiophyta	1	34
	Xanthophyta	43	55
Red algae	Rhodophyta	4	79
Cryptomonads	Cryptophyta	1	62
Dinoflagellates	Pyrrhophyta	1	30
Euglenoids	Euglenophyta	30	120
Total		694	1741

Data from Tompkins *et al.* (1995) and Day (1998).

strains with less than 50 per cent post-thaw viability, are retained both in a cryopreserved state and by serial subculture.

Two-step cooling

Most of the freezing protocols that have been developed utilize a two-step system with controlled/semi-controlled cooling from room temperature to an intermediate holding temperature (−30°C being commonly employed). This allows cryo-dehydration (see Benson, Chapter 6, this volume) of the cells to occur, prior to plunging into liquid nitrogen (−196°C). The frozen material is then stored in either liquid or vapour phase nitrogen in an appropriate liquid nitrogen storage system. Although some organisms can be successfully cryopreserved and stored at higher subzero temperatures (>−70°C), viability levels rapidly fall on storage (Brown and Day, 1993). Therefore it is necessary to maintain frozen cultures at extremely low temperatures, optimally in liquid nitrogen at −196°C.

The development of most algal cryopreservation protocols has been empirical (Bodas *et al.*, 1995; Morris, 1978). These techniques have been effective in preserving a relatively limited taxonomic range of algae and have largely been restricted to morphologically uncomplicated or small species (Table 8.3). A variety of approaches have been employed to improve post-thaw viability; these may be divided into cooling protocol development and freeze-tolerance improvement. Factors that can be varied to reduce, or prevent, damage on freezing and thawing include: type, concentration and duration of exposure to the cryoprotectant (Morris, 1976a); cooling regime employed (Day and Fenwick, 1993); storage temperature/thermal stability of the cryostore; rate of thawing and post-thaw manipulations (Day *et al.*, 1997; Morris, 1976a). Pre- and post-preservation growth conditions of the culture can be altered to increase tolerance to freezing and aid post-thaw recovery; these include: age of culture (Morris, 1978); light intensity (Beaty and Parker, 1992); incubation temperature (Morris, 1976b); osmotic potential of the medium (Canavate and Lubian, 1995); nutrient limitation (Ben-Amotz and Gilboa, 1980) and nutritional mode (Morris *et al.*, 1977). The majority of effective protocols use late log/early station-ary phase cultures; however, the key factor is the 'vigour' of the culture. Cryopreservation of senescent, stressed or damaged cells will invariably result in lower levels of post-thaw survival compared to the same protocol being applied to a healthy 'vigorous' culture of the same algal strain. Full, step-wise descriptions of protocols are available in the liter-ature (see Alexander *et al.*, 1980; Bodas *et al.*, 1995; Day and DeVille, 1995; Lee and Soldo, 1992; see also Benson, Chapter 6, this volume).

Encapsulation dehydration

Encapsulation in alginate gel followed by dehydration in conjunction with the non-penetrating cryoprotectant sucrose has been used to preserve gametophytes of *Laminaria digitalis* (Vigneron *et al.*, 1997). These had survival levels in the range 25–75 per cent depending on age, sex and stress (Vigneron *et al.*, 1997). This approach has also been employed to preserve a range of microalgae and cyanobacteria (Hirata *et al.*, 1996). The technique was found to be suitable for six of the seven marine algae examined; how-ever, only one freshwater alga, *Chlorella pyrenoidosa*, survived (Hirata *et al.*, 1996). This species is extremely robust and should survive most standard cryopreservation protocols. An alternative approach employed at the CCAP to preserve *Euglena gracilis* involved encapsulation in calcium alginate and cryopreservation using a standard two-step protocol. This resulted in high levels of post-thaw viability (Fleck, 1998). The mechanism(s) of the protection afforded by encapsulation are not fully understood; however, it appears to: assist in dehydration of the cells; provide support during the freezing process; protect against freeze-fracture; and possibly provide some protection against free-radical medi-ated injury by functioning as an exogenous antioxidant (Fleck, 1998).

Vitrification

Vitrification using the method of Sakai (Uragami *et al.*, 1989) has been successfully employed to preserve the multicellular alga *Enteromorpha intestinalis,* with 100 per cent of the thalli surviving exposure to the cryoprotectant solution (PVS2) and subsequent immersion in liquid nitrogen (Fleck, 1998). On applying this approach to other algae including *E. gracilis, Vaucheria sessilis* and *Microcystis aeruginosa* no viability follow-ing treatment with the vitrification solution was observed (Fleck, 1998). In addition to the toxicity of the vitrification solution, stability of the vitrified material may be a problem during storage or on thawing. Temperatures of up to $-110°C$ may be observed on removal of cryovials from the bottom of a Cryostore inventory system (Day *et al.*, 1997). At this temperature damage could theoretically occur to stored material, as it is above the limit of ice crystal growth $-139°C$ (Morris, 1981). In vitrified samples devitrification and/or ice crystal growth could occur at temperatures as low as $-130°C$ (Taylor, 1987); this could potentially result in cellular damage. In vitrified *E. intestinalis* freeze-fracturing of the thallus occurred during thawing; however, all filament sections were capable of regenerating new thalli. In other systems including mammalian tissues this event has been identified as lethal (Pegg *et al.*, 1997) and it is probable that for unicellular algae freeze-fracture would be lethal.

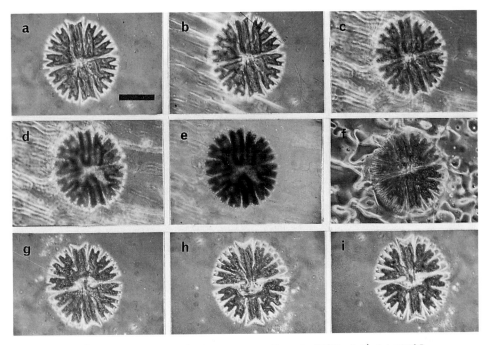

Figure 8.1 Changes in *Micrasterias rotata* on cooling at −30°C min⁻¹. (a) +20°C, (b) −1°C, (c) −2°C, (d) −3°C, (e) −30°C, (f–i) warming from −30°C to +20°C at +50°C min⁻¹. Bar = 50 μm (reproduced with permission of Drs G.J. Morris and M. Engels)

Mechanisms of cell damage associated with cryopreservation

Successful empirically developed cryopreservation protocols are effective on the basis that they reduce osmotic stress, cold shock and potential damage by intracellular and extracellular ice formation, before and during freezing, and on thawing. Improvements to existing methods and the preservation of a greater diversity of algae require a greater understanding of the mechanisms of the fundamental modes of cell damage during freezing and thawing (see Figure 8.1).

Conventional two-step cryopreservation has not proven effective for many of the larger, more morphologically complex, or multicellular algae examined. On employing super-optimal cooling rates, which allow insufficient time for dehydration of the cell, intracellular ice may be formed (Figure 8.1e). In all reported cases in algae, with the exception of *Chlorella prothecoides* (Morris *et al.*, 1977), intracellular ice formation (IIF) was lethal. By manipulation of the cooling and cryoprotectant regimes it may be possible to avoid IIF, however, in some algae, e.g. *E. gracilis*, even under optimal conditions a proportion of the cells undergo IIF (Figure 8.2). The filamentous Chlorophyte *Spirogyra grevilleana* was killed by intracellular ice and gas bubbles on being cooled and frozen using a standard protocol at −10°C min⁻¹ (Morris and McGrath, 1981). The filamentous diatom *Fragilaria cortonensis* was lethally injured at fast cooling rates by intracellular ice and at slow rates by freeze-induced hypertonic stress (McLellan, 1989). Furthermore, gametophytes of *Undaria pinnatifida* which were reported to be successfully cryopreserved, only survived four days post-thaw (Arbault *et al.*, 1990), indicating that a significant amount of cell damage occurred during the process. In the coenocytic alga *Vaucheria sessilis*, on cooling using a conventional cooling protocol: −1°C min⁻¹ ⇒ −35°C ⇒

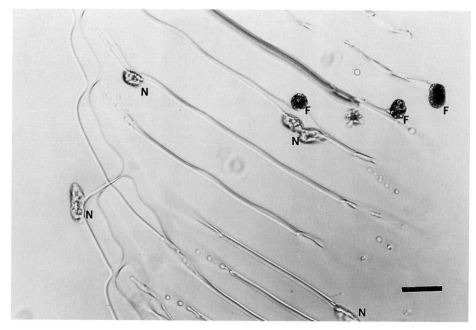

Figure 8.2 *Euglena gracilis* at −30°C cryopreserved under optimal conditions [methanol 10% (v/v), cooling at 0.5°C min^{-1} 0°C, −60°C liquid nitrogen]. Note cells which have 'flashed' (F) due to the presence of refractive intracellular ice crystals and other cells which do not appear to contain intracellular ice (N). Bar = 50 μm

Liquid N$_2$ [5 per cent (w/v) DMSO], lack of cellular compartmentalization allowed the propagation of intracellular ice throughout the thallus, resulting in death of the alga (Fleck *et al.*, 1997).

In addition to the above, significant damage has been observed at the ultrastructural level with physical disruption of cellular organelles and membranes (Fleck, 1998; McLellan, 1989). Other factors causing both lethal and sublethal injuries include: pre-cooling manipulations (e.g. centrifugation), cryoprotectant toxicity and chilling damage (Fleck, 1998). These effects can most readily be detected employing vital staining, measurement of oxygen evolution capacity or gross changes in morphology e.g. flagellar loss (Fleck, 1998). Furthermore, free-radical mediated damage and fluctuations in antioxidant levels have also been implicated in freeze-induced damage in both plant and animal systems (Benson, 1990). Recent studies indicate that this may be an important factor in the apparent freeze-recalcitrance of some algae (Fleck, 1998).

8.5   Concluding comments

There are limitations in our current understanding of the modes of cryopreservation-induced damage and specific sites of injury in algae. Future research will invariably involve the use of techniques including flowcytometry, cryomicroscopy and electron microscopy. In parallel, studies on oxidative stress/injury, other freeze-induced biochemical injuries and the responses of the alga's endogenous protective mechanisms (notably level and composition of antioxidants) to cryopreservation are required.

It is anticipated that elucidation of the key sites of injury will assist in the improvement of existing cryopreservation methodologies and the development of alternative approaches. Areas that present significant challenges include the cryopreservation of large/complex unicellular and multicellular algae. An additional challenge is the improvement of techniques to assess viability. In most published studies, survival has been determined on the basis of post-treatment growth, reaction to vital staining, fluorescence or loss of pigmentation. All of these approaches have shortfalls, regrowth is difficult to assess for non-unicellular algae and other techniques may significantly overestimate post-thaw viability. However, techniques including flowcytometry, measurement of oxygen evolution and response to specific stimuli, e.g. wound healing in *V. sessilis*, may form the basis of alternative rapid viability assays.

In conclusion, although large proportions of the holdings of all the major collections are nominally freeze-recalcitrant, it is probable that if resources were available the majority would be amenable to cryopreservation. The ultimate challenge is to develop approaches that are robust, reliable and result in high levels of post-thaw viability.

Acknowledgements

The author thanks all those who are currently, and have been previously, associated with cryopreservation research at the CCAP, particularly Drs G.J. Morris, R.A. Fleck and M.R. McLellan.

References

ABBOTT, I.A., 1988, Food and food products from seaweeds, in LEMBI, C.A. and WAALAND, J.R. (Eds), *Algae and Human Affairs*, pp. 135–147, Cambridge: Cambridge University Press.

ALEXANDER, M., DAGGETT, P.-M., GHERNA, R., JONG, S. and SIMIONE, F., 1980, *American Type Culture Collection Methods I. Laboratory Manual on Preservation, Freezing and Freeze-drying as Applied to Algae, Bacteria, Fungi and Protozoa*, Rockville: American Type Culture Collection.

ANDERSEN, R.A., MORTON, S.C. and SEXTON, J.P., 1997, Culture Collection of Marine Phytoplankton Catalogue of strains, *Journal of Phycology*, **33**, Supplement 1.

ARBAULT, S., RENARD, P., PEREZ, R. and KASS, R., 1990, Cryopreservation trials on gametophytes of the food alga *Undaria pinnatifida* Laminariales, *Aquatic Living Resources*, **3**, 207–216.

BEATY, M.H. and PARKER, B.C., 1992, Cryopreservation of eukaryotic algae, *Virginia Journal of Science*, **43**, 403–410.

BEN-AMOTZ, A. and GILBOA, A., 1980, Cryopreservation of marine unicellular algae. II. Induction of freezing tolerance, *Marine Ecology Progress Series*, **2**, 221–224.

BENSON, E.E., 1990, *Free Radical Damage in Stored Plant Germplasm*, Rome: International Board for Plant Genetic Resources.

BODAS, K., BRENNIG, C., DILLER, K.R. and BRAND, J.J., 1995, Cryopreservation of blue-green and eukaryotic algae in the culture collection at the University of Texas at Austin, *Cryo-Letters*, **16**, 267–274.

BOLD, H.C. and WYNNE, M.J., 1985, *Introduction to the Algae: Structure and Reproduction*, Englewood Cliffs, NJ: Prentice Hall Inc.

BOLIN, B., DEGENS, E.T., DUVIGNEAU, D.P. and KEMP, S., 1977, The global biogeochemical carbon cycle, in BOLIN, B., DEGENS, E.T., KEMP, S. and KETNER, P. (Eds), *The Global Carbon Cycle*, pp. 1–53, New York: Wiley & Son.

BOROWITZKA, M., 1997, Microalgae for aquaculture: opportunities and constraints, *Journal of Applied Phycology*, **9**, 393–401.

BROWN, S. and DAY, J.G., 1993, An improved method for the long-term preservation of *Naegleria gruberi*, *Cryo-Letters*, **14**, 347–352.

BUZER, J.S., DOHMEIER, R.A. and DU TOIT, D.R., 1985, The survival of algae in dry soils exposed to high temperatures for extended time periods, *Phycologia*, **24**, 249–251.

CAMERON, R.E., 1962, Species of *Nostoc* Vauch. Occurring in the Sonoran Desert in Arizona, *Transcripts of the American Microscopy Society*, **81**, 379–384.

CANNELL, R., 1990, Algal Biotechnology, *Applied Biochemistry and Biotechnology*, **21**, 85–105.

CAVALIER-SMITH, T., 1993, Kingdom Protozoa and its 18 phyta, *Microbiological Reviews*, **57**, 953–994.

CANAVATE, J.P. and LUBIAN, L.M., 1995, The relation of cooling rate, cryoprotectant concentration and salinity with the cryopreservation of marine microalgae, *Marine Biology*, **124**, 325–334.

DAGGETT, P.-M. and NERAD, T.A., 1992, Long-term maintenance of *Chlamydomonas* by cryopreservation and freeze-drying, in LEE, J.J. and SOLDO, A.T. (Eds), *Protocols in Proto-zoology*, A-65.1, Lawrence, Kansas: Society of Protozoologists.

DAY, J.G., 1987, Physiological and biotechnological aspects of immobilized photoautotrophs, unpublished PhD Thesis, University of Dundee.

DAY, J.G., 1993, *United Kingdom Federation for Culture Collections Newsletter*, **22**, 1.

DAY, J.G., 1996, Conservation of microbial biodiversity: the roles of algal and protozoan culture collections, in SAMSON, R.A., STALPERS, J.A., van der MEI, D. and STOUTHAMER, A.H. (Eds), *Culture Collections to Improve the Quality of Life. Proceedings of ICCC 8*, pp. 173–177, Baarn, The Netherlands: CBS.

DAY, J.G., 1998, Cryo-conservation of microalgae and cyanobacteria, *Cryo-Letters*, Supplement 1, 7–14

DAY, J.G. and DEVILLE, M.M., 1995, Cryopreservation of algae, *Methods in Molecular Biology*, **38**, 81–90.

DAY, J.G. and FENWICK, C., 1993, Cryopreservation of members of the genus *Tetraselmis* used in aquaculture, *Aquaculture*, **118**, 151–160.

DAY, J.G. and MCLELLAN, M.R., 1995, Conservation of algae, in GROUT, B. (Ed.), *Genetic Preservation of Plant Cells* in Vitro, pp. 75–98, Berlin: Springer.

DAY, J.G., PRIESTLEY, I.M. and CODD, G.A., 1987, Storage reconstitution and photosynthetic activities of immobilized algae, in WEBB, C. and MAVITUNA, F. (Eds), *Plant and Animal Cells, Process Possibilities*, pp. 257–261, Chichester: Ellis Harwood Ltd.

DAY, J.G., WATANABE, M.M., MORRIS, G.J., FLECK, R.A. and MCLELLAN, M.R., 1997, Long-term viability of preserved eukaryotic algae, *Journal of Applied Phycology*, **9**, 121–127.

FLECK, R.A., 1998, Mechanisms of cell damage and recovery in cryopreserved freshwater protists, unpublished PhD Thesis, University of Abertay Dundee.

FLECK, R.A., DAY, J.G., RANA, K.J. and BENSON, E.E., 1997, Visualisation of cryoinjury and freeze events in the coenocytic alga *Vaucheria sessilis* using cryomicroscopy, *Cryo-Letters*, **18**, 343–355.

GLOMBITZA, K.-W., 1979, Antibiotics from algae, in HOPPE, H.A., LEVRING, T. and TANAKA, Y. (Eds), *Marine Algae in Pharmaceutical Science*, pp. 303–342, Berlin: Walter de Gruyter.

GROUT, B.W.W., 1995, Introduction to the *in vitro* preservation of plant cells, tissues and organs, in GROUT, B. (Ed.), *Genetic Preservation of Plant Cells* in Vitro, pp. 1–20, Berlin: Springer.

HAWKSWORTH, D.I. and MOUND, L.A., 1991, Diversity data-bases: the crucial significance of collections, in HAWKSWORTH, D.L. (Ed.), *The Biodiversity of Microorganisms and Insects*, pp. 17–29, Wallingford: CAB Int.

HIRATA, K., PHUCHINDAWAN, M., TUKAMOTO, J., GODA, S. and MIYAMOTO, K., 1996, Cryopreservation of microalgae using encapsulation-dehydration, *Cryo-Letters*, **17**, 321–328.

HOLM-HANSEN, O., 1973, Preservation by freezing and freeze-drying, in STEIN, J. (Ed.), *Hand-book of Phycological Methods: Culture Methods and Growth Measurements*, pp.195–206, Cambridge: Cambridge University Press.

JAWORSKI, G.H.M., WISEMAN, S.W. and REYNOLDS, C.S., 1988, Variability in sinking rate of the freshwater diatom *Asterionella formosa*: the influence of colony morphology, *British Phycological Journal*, **23**, 167–176.

JENSEN, A., 1993, Present and future needs for algae and algal products, *Hydrobiologia*, **261**, 15–24.

JOHN, D.M., 1994, Biodiversity and conservation: an algal perspective, *The Phycologist*, **38**, 3–15.

KOLKOWSKI, J.A. and SMITH, D., 1995, Cryopreservation and freeze-drying fungi, *Methods in Molecular Biology*, **38**, 49–62.

LEE, Y.-K., 1997, Commercial production of microalgae in the Asia-Pacific rim, *Journal of Applied Phycology*, **9**, 403–411.

LEE, J.J. and SOLDO, A.T. (Eds), 1992, *Protocols in Protozoology*, Lawrence, Kansas: Society of Protozoologists.

LEESON, E.A., CANN, J.P. and MORRIS, G.J., 1984, Maintenance of algae and protozoa, in KIRSOP, B.E. and SNELL, J.J.S. (Eds), *Maintenance of microorganisms*, pp. 131–160, London: Academic Press.

LINCOLN, R.A., STRUPINSKI, K. and WALKER, J.M., 1990, Biologically active compounds from diatoms, *Diatom Research*, **5**, 337–349.

MALIK, K.A., 1993, Preservation of unicellular green-algae by liquid-drying, *Journal of Microbiological Methods*, **18**, 41–46.

MCGRATH, M.S., DAGGETT, P.-M. and DILWORTH, S., 1978, Freeze-drying of algae: Chlorophyta and Chrysophyta, *Journal of Phycology*, **14**, 521–525.

MCLELLAN, M.R., 1989, Cryopreservation of diatoms, *Diatom Research*, **4**, 301–318.

MCLELLAN, M.R., COWLING, A.J., TURNER, M.F. and DAY, J.G., 1991, Maintenance of algae and protozoa, in KIRSOP, B. and DOYLE, A. (Eds), *Maintenance of Microorganisms and Cultured Cells*, pp. 183–208, London: Academic Press Ltd.

MIYACHI, S., NAKAYAMA, O., YOKOHAMA, Y., HARA, Y., OHMORI, M., KOMOGATA, K., SUGAWARA, H. and UGAWA, Y. (Eds), 1989, *World Catalogue of Algae*, Tokyo: Japan Scientific Societies Press.

MORRIS, G.J., 1976a, The cryopreservation of *Chlorella* 1. Interactions of rate of cooling, protective additive and warming rate, *Archives of Microbiology*, **107**, 57–62.

MORRIS, G.J., 1976b, The cryopreservation of *Chlorella* 2. Effect of growth temperature on freezing tolerance, *Archives of Microbiology*, **107**, 309–312.

MORRIS, G.J., 1978, Cryopreservation of 250 strains of Chlorococcales by the method of two step cooling, *British Phycological Journal*, **13**, 15–24.

MORRIS, G.J., 1981, *Cryopreservation. An Introduction to Cryopreservation in Culture Collections*, Cambridge: Institute of Terrestrial Ecology.

MORRIS, G.J. and MCGRATH, J.J., 1981, Intracellular ice nucleation and gas bubble formation in *Spirogyra*, *Cryo-Letters*, **2**, 341–352.

MORRIS, G.J., CLARKE, K.J. and CLARKE, A., 1977, The cryopreservation of *Chlorella* 3. Effect of heterotrophic nutrition on freezing tolerance, *Archives of Microbiology*, **114**, 309–312.

MUMFORD, T.F. and MIURA, A., 1988, *Porphyra* as food, in LEMBI, C.A. and WAALAND, J.R. (Eds), *Algae and Human Affairs*, pp. 87–117, Cambridge: Cambridge University Press.

NERAD, T.A., 1993, *ATCC Catalogue of Protists*, Rockville: ATCC.

NORTON, T.A., MELKONIAN, M. and ANDERSEN, R.A., 1996, Algal biodiversity, *Phycologia*, **35**, 308–326.

OECD, 1984, *OECD Guidelines for Testing Chemicals, Section 2 – Effects on Biotic Systems, No. 201, 'Alga, Growth Inhibition Test'*, Paris: Organization for Economic Development.

OSWALD, W.J., 1988, Micro-algae and waste-water treatment, in BOROWITZKA, M. and BOROWITZKA, L.J. (Eds), *Micro-algal Biotechnology*, pp. 305–328, Cambridge: Cambridge University Press.

PEGG, D.E., WUSTEMAN, M.C. and BOYLAN, S., 1997, Fractures in cryopreserved elastic arteries, *Cryobiology*, **34**, 183–192.

PHILLIPS, J.A., 1998, Marine conservation initiatives in Australia: their relevance to the conservation of macroalgae, *Botanica Marina*, **41**, 95–104.

PRINGSHEIM, E.G., 1946, *Pure Cultures of Algae*, Cambridge: Cambridge University Press.

SCHLÜSSER, U.G., 1994, SAG Sammlung von Algenkulturen at University of Göttingen, *Botanica Acta*, **107**, 3–186.

SCHOPF, J.W. and WALTER, M.R., 1982, Origin and early evolution of the cyanobacteria: the geological evidence, in CARR, N.G. and WHITTON, B.A. (Eds), *The Biology of the Cyanobacteria*, pp. 543–564, Oxford: Blackwells Scientific Press.

SHIFRIN, N.S and CHISHOLM, S.W., 1980, Phytoplankton lipids: environmental influences on production and possible commercial applications, in SHELEF, G. and SOEDER, C.J. (Eds), *Algae Biomass*, pp. 627–645, Amsterdam: Elsevier/North Holland Biomedical Press.

STARR, R.C. and ZEIKUS, J.A., 1993, UTEX – The culture collection of algae at the University of Texas at Austin, *Journal of Phycology*, **29**, 1–106.

STEIN, J. (Ed.), 1973, *Handbook of Phycological Methods: Culture Methods and Growth Measurements*, Cambridge: Cambridge University Press.

TAKISHIMA, Y., SHIMURA, J., UGAWA, Y. and SUGAWARA, H. (Eds), 1989, *Guide to World Data Center on Microorganisms – A list of Culture Collections of the World*, Saitama, Japan: WFCC World Data Center on Microorganisms.

TAYLOR, M.J., 1987, Physico-chemical principles in low temperature biology, in GROUT, B.W.W. and MORRIS, G.J. (Eds), *The Effects of Low Temperatures on Biological Systems*, pp. 3–71, London: Edward Arnold.

TOMPKINS, J., DEVILLE, M.M., DAY, J.G. and TURNER, M.F. (Eds), 1995, *Culture Collection of Algae and Protozoa Catalogue of Strains*, Ambleside: Culture Collection of Algae and Protozoa.

UNEP, 1992, *Convention on Biological Diversity*, Nairobi: United Nations Environment Programme Environmental Law and Institutions Program Activity Centre.

URAGAMI, A., SAKAI, A., NAGAI, M. and TAKAHASHI, T., 1989, Survival of cultured cells and somatic embryos of *Asparagus officinalis* cryopreserved by vitrification, *Plant Cell Reports*, **8**, 418–421.

VIGNERON, T., ARBAULT S. and KAAS, R., 1997, Cryopreservation of gametophytes of *Laminaria digitata* (L) by encapsulation dehydration, *Cryo-Letters*, **18**, 117–126.

WARREN, A., DAY, J.G. and BROWN, S., 1997, Cultivation of Protozoa and Algae, in HURST, C.J., KNUDSEN, G.R., MCINERNEY, M.J., STEZENBACH, L.D. and WALTER, M.V. (Eds), *Manual of Environmental Microbiology*, pp. 61–71, Washington DC: ASM Press.

WATANABE, M.M. and HIROKI, M., 1997, *NIES – Collection List of strains*, Tsukuba: NIES.

Cryo-conservation of Industrially Important Plant Cell Cultures

HEINZ MARTIN SCHUMACHER

9.1 Introduction: the biotechnological use of dedifferentiated plant cell cultures

It is not more than 100 years ago that chemical synthesis started to replace plants as the major source of organic compounds for human use. Today, plants serve as an important source of secondary metabolites used in pharmacy, biotechnology and food technology. The first practical applications of *in vitro* techniques to plants occurred in the 1920s, with differentiated structures such as zygotic embryos. Efforts to grow dedifferentiated, isolated plant cells *in vitro* were unsuccessful until 1939 and it was the discovery of auxin in 1930 that finally led to success in achieving the production of continuously growing plant cell cultures.

Soon after the Second World War, many examples of interesting biosynthetic processes in dedifferentiated plant cells became known. In some plant cell cultures even higher concentrations of secondary metabolites were found as compared to the intact plants (for a review see Carew and Staba, 1965). At the same time, cell culture techniques for the cultivation of suspension cultures were established and improved. It may have been the commercial production of antibiotics by fungi that led plant physiologists to think that the production of secondary plant metabolites by dedifferentiated cell cultures was a possibility. Although, for a long time, the idea to produce commercially important secondary metabolites by dedifferentiated suspension cultures remained the major concern of plant biotechnology, up to now, there have been very few examples of commercial applications. This situation is due to a number of limitations. Unfortunately, many compounds of high commercial value (e.g. morphine, codeine or cardiac glycosides) are not formed by dedifferentiated cell lines. In other cases, compounds are only formed in small amounts, such that commercial exploitation seems to be impossible. Different strategies have been developed to overcome these problems such as the enhancement of production rate by the alternating use of growth and production media, the application of biotic and abiotic elicitors, and the selection of high yielding strains (for a review see Berlin, 1997).

The basis for the selection of high yielding strains is the heterogeneity that occurs in almost all cell lines. Chromosomal aberrations have been observed for single cells of many plant cell cultures (Bayliss 1980; see also Harding, Chapter 7, this volume). More important for biotechnology is the fact that cells can also differ largely in their production capacity for certain compounds. In most cases it has not been clearly demonstrated whether this heterogeneity has a genetic or an epigenetic basis (for review see Wilson, 1990; see also Harding, Chapter 7, this volume). Unfortunately, it is often the case that selected, high yielding strains show considerable instability when a selection pressure is no longer maintained (Deus-Neumann and Zenk, 1984). It is obvious that the stable formation of a desired compound is an essential prerequisite for any commercial biotechnological application, especially for the capital-intensive large scale production of plant metabolites. Therefore, the major concern of cryopreservation of cell cultures for biotechnological applications is not just survival (sometimes not even genetic stability), but mainly the stability of biosynthetic capacity. Genetic changes concerning genes that are not expressed in culture may not be of importance; on the other hand irreversible epigenetic changes, that is, changes in gene expression are crucial for the maintenance of biosynthetic capacity. Since the biosynthetic capacity may vary from cell to cell, even in selected high yielding strains, the major risk for conservation using cryopreservation is an undesired selection process, which might occur if the survival rates vary from cell type to cell type.

The following review will therefore concentrate on studies in which cryopreservation has been applied for the conservation of secondary product forming cells. Emphasis will be given to investigations which assess the stability of product formation after the recovery of cells from cryopreservation. Cryopreservation methods will be outlined; however, for full details of the principles of cryopreservation methodology see Benson (Chapter 6, this volume).

9.2   Stability of product formation after cryopreservation

9.2.1   *Anthocyanins*

The first study on the retention of product formation after cryopreservation was performed by Dougall and Whitten (1980). These authors investigated the formation of anthocyanins in 25 sub-lines of a wild carrot cell culture before and after storage at $-140°C$. Anthocyanins are normally formed spontaneously in *Daucus carota* cell lines and they are not of great pharmaceutical interest, but some are used in food technology. The low costs of existing production methods would not normally justify the production of these secondary metabolites by cell culture technology (Berlin, 1997).

Dougall and Whitten (1980) used a single, simple freezing method for all 25 different sub-lines. Thus, without a pre-culture phase, cell density was adjusted to a specific value; medium with dimethyl sulphoxide (DMSO) was added to a final concentration of 5 per cent. A freezing rate of $-1°C/min$ was used to a final temperature of $-70°C$ at which the samples were then immersed into liquid nitrogen. Unfortunately, no measurements of survival rates were reported by these workers, but they state that from two samples per cell line, at least one re-grew. For detecting the total content of anthocyanins, the absorbance of extracts at OD_{530} was measured and Dougall and Whitten (1980) compared product formation in a maintenance and a production medium before and after cryopreservation. Before freezing, the high accumulating cell lines showed less growth than in the production medium. Although growth rate after cryogenic storage was decreased, especially for the low yielding cell lines, the capacity for product formation remained the same for the different cell lines. Thus, 'high yielding' strains remained high yielding and low yielding strains remained 'low yielding'.

9.2.2 *Ginsenosides*

One of the early targets for the commercial application of cell culture technology was *Panax ginseng*, the root extracts of which are traditionally used in Asian medicine and in increasing amounts in the Western hemisphere. Conventional plant production is time consuming and costly. The plants need a cultivation period of five to six years and harvesting the roots normally destroys the plants. Fortunately, cell cultures spontaneously produce almost the same pattern of ginsenosides as the whole roots of plants and the contents of ginsenosides can quite often exceed that of roots. Work on *Panax ginseng* cell cultures has also led to cryopreservation studies.

The first paper on the subject was published by Butenko *et al.* (1984). They used a programmable freezing approach (see Benson, Chapter 6, this volume). For pre-culture, they combined osmotic treatments with cold hardening. A comparison of different methods revealed that the best result was achieved using a pre-culture procedure which decreased the cultivation temperature gradually from an ambient temperature to 2°C and, simultaneously increasing the sucrose content of the medium from 3 per cent to 20 per cent within 18 days. For cryoprotection, 20 per cent sucrose yielded the highest survival rates (which was 51 per cent, measured by phenosafranine staining). The rather complicated cooling programme included a seeding (ice nucleation) step (see Benson, Chapter 6 this volume) and the cells were transferred into liquid nitrogen from −70°C. In terms of post-cryopreservation stability assessments, only the growth curve of control and recovered cultures were compared one year after recovery; they appeared unchanged. No data on product formation were published (Butenko *et al.*, 1984).

Detailed data on product formation of recovered cultures were later produced by Seitz and Reinhard (1987). They used different approaches, all based on programmed freezing and they also used the slightly modified procedure of Butenko *et al.* (1984). Butenko's method based on cold hardening proved a superior approach to using mannitol or sorbitol as pre-culture treatments. Using a more simple freezing programme, Seitz and Reinhard (1987) even achieved survival rates of up to 40 per cent. With four different methods they were also able to recover actively growing cell cultures and all recovered cell lines showed the same growth characteristics as the controls. For assessing chemical properties, Seitz and Reinhard (1987) measured not only the contents of total saponins but compared the production patterns of eight ginsenosides. The interesting result was that those which yielded much lower survival rates, had recovered cell lines with an unchanged product pattern. It was also notable that one of the methods showing less than 10 per cent survival showed better recovery growth in the survivors, than two other methods showing much higher survival rates.

Mannonen *et al.* (1990) preserved a *Panax ginseng* cell line using the standard procedure of Withers and King (1980) (see also Benson, Chapter, 6, this volume). Unfortunately, no post-freeze survival rates were given; however, the total amount of ginsenosides, together with product pattern was unchanged after cryopreservation. Furthermore, these authors demonstrated that alternative conservation methods, such as preservation under mineral oil for six months as well as continuous subculturing for 14 months failed to preserve the biosynthetic capacity of the cells.

9.2.3 *Diosgenine*

Diosgenine has been a compound of extraordinary pharmaceutical importance for many years. It was used as raw material for the semi-synthetic production of steroidal pharmaceuticals and until 1975, the main product source was *Dioscorea* roots collected in Yucatan, that is, until the plant almost became extinct. Thus, parallel efforts were made to find chemical structures for synthesis, and alternative plant sources were explored. The use of cultured cells of *Dioscorea* to produce diosgenine was investigated and cryopreservation became an important means of storing high producing cell lines. Thus, Butenko *et al.* (1984) published a successful cryopreservation method for *Dioscorea deltoidea* cell cultures. They used a programmable freezing method (see also Benson, Chapter 6, this volume) and for pre-culture, amino acids in low concentrations (0.01–0.02 M) were added. Asparagine and alanine yielded the best results; proline was less effective. DMSO 7 per cent was used as cryoprotectant and programmed freezing down to −90°C before immersion in liquid nitrogen was applied. Extracts of recovered and control cultures were analysed for their saponin and phytosterol content by GC. Analysis was performed in the second and fifth passage after thawing and the authors quantified diosgenine, sitosterol and stigmasterol. The secondary product content of samples recovered from liquid nitrogen were equal to unfrozen controls. Even more impressive was the fact that the GC scans of recovered cryopreserved cells and control cultures had identical profiles, in terms of product pattern and the magnitude of their side peaks.

9.2.4 *Rosmarinic acid*

The caffeic acid derivative, rosmarinic acid is formed spontaneously and in high amounts by cell cultures of several plants. It gained commercial interest because of its antiphlogistic activity and rosmarinic acid was produced even under large scale conditions. For a long time the contents of rosmarininc acid in *Coleus blumei* cell cultures was the highest of all secondary metabolites measured in plant cell cultures (Berlin, 1997).

   Cryopreservation of dedifferentiated cell cultures of *Coleus blumei* was carried out in Tübingen with an extensive study on the formation of rosmarinic acid performed by Reuff (1987). The optimized freezing method is very simple: cells from the logarithmic growth phase are incubated for eight hours in medium containing 1 M sorbitol, cooled at −1°C/min to −40°C, held at this temperature for 40 minutes and then immersed into liquid nitrogen. Analyses of recovered cultures showed that the production rate of rosmarinic acid, as measured by HPLC, was not changed by cryogenic storage. Even more interesting, is that Reuff (1987) analysed the production rate of rosmarinic acid during several steps of the optimization procedure. Thus, changing the transfer temperature to liquid nitrogen from −100°C to −40°C, for example, increased the survival rate from almost 0 to 35 per cent. Even cultures recovered from experiments using a suboptimum transfer temperature had unchanged rosmarinic acid contents as compared to unfrozen controls. The same result was obtained when different cooling rates were used. Cultures cooled at a rate of −10°C/min produced survival rates of 20 per cent compared to the 40 per cent rate achieved under optimum conditions; they still retained their capacity for rosmarinic acid formation. Reuff (1987) also compared the growth and rosmarinic acid production of control cultures and cultures after successive freezing and recovery cycles. All recovered cultures showed the same growth and production characteristics as the controls. Stability was also demonstrated for recovery after different storage periods in liquid nitrogen (from one day to 15 months storage). This work clearly shows that cultures were stable even under sub-optimum freezing conditions, or after successive cryopreservaiton cycles. Unfortunately no data are available regarding subsequent freezing studies performed under sub-optimum conditions.

9.2.5 *Biotin*

Similar conclusions regarding the successful application of cryopreservation can be made from freezing experiments with biotin producing callus cultures. Watanabe *et al.* (1983) published a successful cryopreservation method for biotin-producing callus cultures of *Lavandula vera*. They immersed small pieces of callus taken from the logarithmic growth phase in liquid medium and added an equal volume of a solution of 20 per cent D-glucose and 10 per cent DMSO, which was gradually applied over one hour. Cells were frozen with a rate of $-1°C/min$ to $-40°C$ and then immersed into liquid nitrogen. Although the biotin content of recovered cells was sometimes higher and sometimes lower compared to that of the initial calli, no selection for producing or non-producing cells could be observed. Using the same freezing method Kuriyama *et al.* (1990) later improved cell recovery considerably in this system by adding activated charcoal to the regrowth medium. These results also indicate that in the early experiments of Watanabe *et al.* (1983) the cell cultures could be preserved under sub-optimum freezing conditions, without losing their biosynthetic capacity.

9.2.6 *Indole alkaloids*

The *Apocynaceae* are a plant family of great pharmaceutical interest and, apart from the cardenolide containing *Oleander* and *Strophanthus*, *Catharanthus* and *Rauvolfia* plants, they produce indole alkaloids which are of importance because of their pharmaceutical uses. Examples include: the heart antiarrhythmic ajmaline and the antihypertensive agent, reserpine. Because of their highly complex chemical structure, indole alkaloids are still not amenable to chemical synthesis. Nevertheless, the major reason for intensive studies on *Catharanthus* cell cultures is because of the anticancer drugs vincristine and vinblastine. These dimeric alkaloids are only present in very low concentrations in *Catharanthus* plants and they belong to some of the most expensive groups of pharmaceuticals produced. Unfortunately, these compounds could never be isolated from cell cultures; however, although no process of commercial interest could be finally realized, cell lines derived from *Catharanthus roseus* plants belong to some of the most intensively studied plant cell cultures.

Many *Catharanthus* cell cultures contain low levels of monoterpene indole alkaloids and major components include ajmalicine, serpentine and catharanthine. The formation of these compounds can be increased by cell selection. Deus-Neumann and Zenk (1984) were able to increase the levels of ajmalicine and serpentine from 14 mg/l to >350 mg/l using this technique. But the same authors also showed that the high production rates drop if the selection pressure is not continuously maintained. Thus, cryogenic storage is important for the conservation of high producing cell lines of *Catharanthus* and many 'classical studies' regarding the assessment of cryopreservation for maintaining biosynthetic capacities have been performed using this species. Correspondingly, many papers (see below) report the successful cyropreservation of *Catharanthus roesus* cell cultures and most have used the standard protocol of Withers and King (1980). Although these reports do not always present data on product formation after cryogenic storage, those that did present findings of extraordinary importance.

In their first attempt to cryopreserve cells of *Catharanthus roseus* Kartha *et al*. (1982) selected a non-producing cell line (no. 916), for the reason that this cell line consisted of very small and dense cells which contained many small vacuoles, but were lacking one large central one. The authors pre-cultured their cells for 24 hours in 5 per cent DMSO and then increased the concentration to 7.5 per cent DMSO for cryoprotection. After one hour of incubation, the cells were frozen using rates of 0.5 to 1°C/min to −40°C and immersed into liquid nitrogen. By this method they achieved a 50 per cent survival rate compared to untreated and unfrozen control cells. They then demonstrated that mitotic index as well as the frequency distribution of the DNA content of re-grown cultures was unchanged by cryogenic storage. Nevertheless, microscopic observations showed that the sub-cellular structure of the cells, re-grown after storage had changed as the small vacuoles had fused to form one large central vacuole.

In a later study, Chen *et al*. (1984a) found that a modified method (using only one hour preculture) was not successful for the preservation of a high alkaloid producing cell line. However, in contrast to the previous non-producing cell line, this one consisted of much larger cells which had less dense cytoplasm and a huge central vacuole. These authors carried out further investigations to study the mode of action of the cryoprotectants. By NMR techniques they measured the amount (percentage) of water that remains unfrozen in a solution at a given temperature in the state of equilibrium. By using these techniques Chen *et al*. (1984a) measured the percentage of unfrozen water in the medium and the cryoprotective solutions as well as in mixtures of these solutions which contained cells. They found that the addition of cryoprotective substances increased the level of unfrozen water in the mixtures and established a simple equation between cell survival and the percentage of unfrozen water in a suspension at a given temperature. Cell survival positively correlated with the percentage of unfrozen water. They demonstrated that, using the same cryoprotective solution, the percentage of unfrozen water was lower in the alkaloid producing as compared to the non-producing strain. That this result could not be mimicked by salt solutions showed that cryoprotection is not simply an osmotic effect. It was also shown that the integrity of the membranes had a considerable influence on the percentage of unfrozen water that was moderated by the application of cryoprotective solutions. Nevertheless, by improving their methods, even alkaloid producing strains could finally be manipulated to survive cryopreservation. For the preservation of alkaloid forming cultures, cells were precultured on sorbitol containing agar medium for four days and a mixture of sorbitol (1 M) and DMSO (5 per cent) was applied for pretreatment and cryoprotection. In a subsequent paper Chen *et al*. (1984b) reported the cryopreservation of three different alkaloid producing cell lines. For the non-producing cell line pretreatment and cryoprotection with DMSO only was sufficient; however, the alkaloid producing strains had to be pre-cultured for 20 hours in medium supplemented with 1 M sorbitol. A mixture of 1 M sorbitol and 5 per cent DMSO was used as cryoprotectant and conditions for programmed freezing and thawing were the same for the producing and non-producing strains. Washing after thawing was performed with the non-producing strain but this eventually turned out to be detrimental for the alkaloid producing strains.

Analysis of the alkaloid content of recovered samples revealed that the total amount of alkaloids formed and the patterns of their product distribution (although data on the alkaloid pattern are only expressed qualitatively as ++ or + for different substances) are not altered by the cryogenic pretreatments or freezing. These results show that although the cryopreservation of non-producing strains is more difficult, freezing of alkaloid-producing strains by improved methods does not specifically select for non-producing cells. Unfortunately, no efforts were made to describe the homogeneity of the investigated cell lines.

Mannonen *et al.* (1990) used exactly the same method as Chen *et al.* (1984b) for the cryopreservation of their alkaloid producing cell line of *Catharanthus roseus*. They analysed the formation of catharanthine and ajmalicine in recovered cells (as they had already done for *Panax ginseng*, see Section 9.2.2) and they demonstrated that freezing preserves the biosynthetic capacity of cells far better than continuous subculturing (for 12 months) or medium-term storage under mineral oil (for six months). The ajmalicine content of cells recovered from freezing dropped to 20 per cent of the initial value, and the catharanthine content slightly increased. Unfortunately, Mannonen *et al.* (1990) did not describe the shape of their cells. From the work of Chen *et al.* (1984a, 1984b) it cannot be ascertained as to whether it is levels of alkaloid production or simply the shape of cells which makes certain lines more difficult to cryopreserve. Later studies by Suk Weon Kim *et al.* (1994) demonstrate a clear relationship between alkaloid production and the shape of cells in *Catharanthus* cultures. They isolated several cell lines from the same initial culture and found that alkaloid production is dramatically increased when the cell aspect ratio (cell length:cell width) exceeds a certain threshold. From these results it can be expected that all high producing cells are of the elongated type in the case of *Catharanthus* and therefore these may be more difficult to cryopreserve.

9.2.7 *Berberine*

Berberine belongs to the protoberberine alkaloids and although several alkaloids of this type are of pharmaceutical importance most attention has been paid to berberine; it is widely used, especially in the Asian market. *Berberis*, *Coptis* and *Thalictrum* cultures are the main systems which have been investigated for protoberberine production. Although protoberberines are formed by cell cultures of several plants and sometimes in high quantities, spontaneously high yielding strains can also be established by cell selection from *Coptis japonica* cell lines (Berlin, 1997).

Cell cultures of *Berberis wilsoniae* have been investigated for berberine production by Reuff (1987) who made great efforts to cryopreserve these cell cultures. She tried almost all approaches for cryopreservation which were known at that time, including different pre-culture methods, cryoprotectants and freezing methods. Although she was able to increase survival rates to 25 per cent, she never achieved re-growth of thawed cultures. Reuff (1987) found that berberine leaking out of damaged cells was not toxic, but other low molecular weight and thermo-stable compounds from autoclaved cell sap of *Berberis* cultures were toxic to *Berberis* cells. Thus, these compounds produced by damaged cells may kill the few cells surviving cryopreservation after thawing.

9.3 Stability of a biosynthetic capacity: cardiac glycosides

Even in modern medicine, the drugs digoxin and digitoxin still belong to the most widely used and most expensive groups of plant pharmaceuticals. This has stimulated interest to explore *in vitro* techniques as a means of improving the production of the drugs, or even to produce them using cell culture bioreactors. Unfortunately, the corresponding glycosides are not formed in dedifferentiated cell cultures of *Digitalis*. Nevertheless, cell cultures of *Digitalis* are able to perform biotransformations of these compounds and the less important digitoxin can be transformed to the more valuable digoxin (Seitz *et al.*, 1983). Efforts have been made to use this biotransformation capacity commercially. Another transformation based on the the the glycosylation of digoxin, digitoxin and gitoxin has also been investigated (Diettrich *et al.*, 1982).

These transformation reactions have also been used by Diettrich *et al.* (1982) as 'markers' to prove the stability of cryopreserved cell cultures of *Digitalis lanata*. For programmed freezing of the cells mannitol was used as the pre-culture osmotic, for one week. A mixture of sucrose, glycerol and DMSO yielded the best cryoprotective results. The freezing rate was between −0.5 to −2°C/min and −60°C and the best survival rate obtained by this method was 51 per cent. The authors investigated the occurrence of membrane damage by microscopy and they also tested the capacity of the cells to glycosylate digitoxin, gitoxin and digoxin, three passages after recovery of the frozen samples. All products (measured only for digitoxin) were transformed and the transformation rate of the cultures recovered after cryopreservation was not changed. The high similarity of the initial and the re-cultivated strains was also demonstrated by measuring the frequency distribution of the DNA content of the cell nuclei by microdensitometry.

A similar freezing method was used by Seitz *et al.* (1983) for *Digitalis* cell lines. They used a higher mannitol concentration (6 per cent) for a shorter period of time (three days). For cryoprotection a slightly different mixture of glycerol, sucrose and DMSO was used and the freezing rate was −1°C/min and they achieved a survival rate of over 50 per cent. For testing stability, they measured the transformation of β-methyldigitoxin to β-methyldigoxin and the transformation rate was unchanged in cultures recovered from cryopreservation, compared to unfrozen controls. Small scale and large scale 20 l biorectors were used for these studies; this was the first report which tested the retention of a biosynthetic capacity of cryopreserved cultures under large scale conditions. It was thus remarkable that stability of *Digitalis* strains after cryopreservation was demonstrated by measuring two different biotransformation reactions, one of them even under large scale conditions. In addition the frequency distribution of DNA content in a cell line was measured and found stable.

9.4   Transformed root cultures for secondary metabolite production

The fact that valuable compounds (such as atropine) found in plant roots are sometimes not expressed in dedifferentiated cell cultures (for a review, see Berlin, 1997) led to the idea of producing these with root cultures obtained from transformation with *Agrobacterium rhizogenes*. These cultures normally exhibit the same metabolism as normal roots and they can grow without application of plant hormones. Although the problem of generating genetic instability in differentiated cultures is less risky than for dedifferentiated cell lines a preservation method is important to ensure the stabilization of the transgenes. Benson and Hamill (1991) showed that secondary product forming hairy root cultures are also amenable to cryopreservation. They preserved transformed root cultures of *Beta vulgaris* and *Nicotiana rustica*. The simple ultra-rapid freezing of small root tips of 2–4 mm on hypodermic needles was successful, however, for practicability of storage they also used programmed freezing. In contrast to apical meristems, a preculture period of excised root tips turned out to be disadvantageous. Although high survival rates were achieved (46 per cent for *Beta vulgaris* and 83 per cent for *Nicotiana rustica*) the capacity for root conversion was less and a high degree of variation among replicates occurred throughout the whole procedure. Culture age before freezing and the hormone treatment after recovery had a remarkable influence on root regeneration. Nevertheless, for regenerated roots the post-freeze stability of the T-DNA could be demonstrated as well as the stability of secondary metabolite formation. For *Nicotiana rustica* roots the ratio of nicotine/anatobine as well as that of nicotine/nornicotine was measured and the betaxanthin and betacyanin levels for *Beta vulgaris* roots showed stabilty in synthesis patterns after cryogenic storage.

9.5 Conclusions

Although the use of cryopreservation for certain secondary product producing plant cells, tissues and organs is still far from being applied on a routine basis, the first technique that could be addressed as a 'standard method' was the programmable freezing protocol developed by Withers and King (1980). From that time, many different approaches for the cryopreservation of plant cells have been devised (see Benson, Chapter, 6, this volume). Thus, apart from using controlled rate freezing, today's techniques include: ultra-rapid freezing, vitrification and encapsulation/dehydration. However, most research of dedifferentiated cell cultures which is considered to be of biotechnological importance has been carried out by using programmed, controlled rate freezing. The main reason is that most cryopreservation studies using secondary metabolite producing cell lines were performed during the period 1980 to 1990, at a time when programmed freezing was still the method of choice for plant cell conservation. Other approaches, such as vitrification, were still under development at this time. Another reason for the dominance of the programmed freezing methods is that they still offer the most practical way of handling large numbers of suspension cells.

In almost all cases tested, cryopreservation is able to conserve biochemical stability, and maintain production rates of secondary compounds in cells which have been cryo-conserved; similarly biotransformation rates are also stable (see Table 9.1).

In many cases the stability of other characters such as growth rate, mitotic index, and the frequency distribution of DNA also accompanied positive biochemical stability assessments of the cryopreserved cell lines. The danger of an undesired cell selection after cryogenic storage seems to be less important than expected. It is even more remarkable that on careful analysis of the reports on cryopreservation procedures, in some cases it has been shown that even the application of sub-optimum storage methods did not change the important characters of the preserved cell lines.

However, another important consideration has to be made. In most of the cases where cryopreservation was successful in maintaining biosynthetic capacity, the work was done on cell lines for which the production of the desired compounds normally occurred spontaneously and in high quantities in cell cultures. This is true for ginsenoside production of *Panax ginseng* cells, rosmarinic acid production in *Coleus blumei* cells, diosgenine production in *Dioscorea* cell lines and for the biotransformations of *Digitalis* cells (Berlin, 1997). Since all cultured cells seem to produce these compounds, cell selection would be of no consequence for product formation. In the case of *Catharanthus* cells it is, however, obvious that non-producing and producing cell lines react differently to cryopreservation. It is remarkable that in the case of these cell lines a specific cell shape seems to be linked to alkaloid production. The non-producing cell shape is clearly more amenable to cryopreservation than the shape of alkaloid producing cells. This may be

Table 9.1 Cell cultures that have been investigated for the retention of chemical traits after cryopreservation

Cell culture	Chemical trait measured	A	B	Reference
Daucus carota	Anthocyanin production	+	+	Dougall and Whitten (1980)
Panax ginseng	Ginsenoside production	+		Butenko *et al.* (1984)
		+	+	Seitz *et al.* (1983)
		+	+	Mannonen *et al.* (1990)
Dioscorea composita	Diosgenine production	+	+	Butenko *et al.* (1982)
Coleus blumei	Rosmarinic acid production	+	+	Reuff (1987)
Lavandula vera	Biotin production	+	+	Watanabe *et al.* (1983)
Digitalis purpurea	Glycosylation	+	+	Diettrich *et al.* (1982)
	β-Hydroxylation	+	+	Seitz *et al.* (1983)
Catharanthus roseus	Indole alkaloid production	+	+	Chen *et al.* (1984b)
		+	+	Mannonen *et al.* (1990)
Berberis wilsoniae	Berberine production	–		Reuff (1987)
*Nicotiana tabacum**	Nicotine/nornicotine ratio	+	+	Benson and Hamill (1991)
*Beta vulgaris**	Betaxanthine, betacyanin production	+	+	Benson and Hamill (1991)

A = cryopreservation carried out successfully; B = chemical trait retained after recovery.
* Root cultures obtained from transformation with *Agrobacterium rhizogenes*.

reflected by the molecular structure of the membranes. In the future, it may be important to demonstrate more precisely the role of sub-cellular structures, membranes and cell morphology in the formation of the indole alkaloids.

The key enzyme that combines the iridoglucoside secologanin and the amino acid derived tryptamine to form the basic structure of the indole alkaloid strictosidine, may be localized in the vacuole. Further steps of the biosynthesis seem to take place in the cytoplasm, whereas peroxidases oxidizing ajmalicine to the quaternary serpentine could also be localized in the vacuole. Quaternary alkaloids accumulate in the vacuole either by an ion-trap mechanism or even by active transport (Hashimoto and Yamada, 1994; Kutchan, 1995). These examples show that alkaloid formation requires highly complex membrane structures. It is easily possible that membranes of alkaloid producing cells are much more complex than those of non-producing cells and that they are also more vulnerable to freezing damage.

Even more complex, is the situation for protoberberines. Although of considerable commercial interest in the 1980s *Berberis wilsoniae* cell lines still cannot be cryopreserved. *Berberis* cultures, even non-selected ones, normally form high yields of protoberberine alkaloids. The quaternary forms of these alkaloids are highly cytotoxic because of their intercalation with DNA. Therefore, complex membrane systems have developed for their biosynthesis. Several of the biosynthetic enzymes including the berberine bridge enzyme (BBE), *S*-tetrahydroprotoberberine oxidase, *S*-canadine oxidase and columbamine *O*-methyltransferase are located in subcellular alkaloid-forming vesicles, which are characterized by their specific gravity. These vesicles also contain the same composition of alkaloids as the central vacuole. Alkaloids seem to be sythesized inside these vesicles and finally accumulate in the central vacuole by the fusion of these vesicles with the tonoplast. How these vesicles are formed and which parts of the biosynthetic pathway take place inside these vesicles is still under discussion (for review see Hashimoto and Yamada, 1994; Kutchan, 1995). They do seem to be specific for alkaloid biosynthesis and their role may be to prevent cytoplasmic and DNA damage by the toxic alkaloids.

Reuff (1987) demonstrated that berberine leaking out of damaged cells does not poison the surviving cells or cause failure of re-growth of cell lines thawed after freezing. Later, it was shown that *Coptis* and *Thalictrum* cells detoxify exogenously applied berberine and exhibit an uptake system following Michaelis–Menten kinetics. The complex membrane structures for the biosynthesis of these quaternary alkaloids protects the cells as long as they are functioning. But it seems to be possible that even partial membrane damage and loss of compartmentalization during the freezing process may lead to an intoxication of otherwise surviving cells by endogenous quaternary protoberberines produced during recovery growth.

It is a pity that no published data are available as yet regarding the cryopreservation of cell lines which have been used commercially (such as *Lithospermum erythrorhizon* cultures for the production of shikonin or the highly regarded *Taxus brevifolium* cultures which will probably be used in the future for taxol production). Finally, it may be concluded that when routine cryopreservation procedures are developed and applied on larger scales in the secondary products sector, it can be achieved with some confidence. The present data indicate that the biosynthetic potential of dedifferentiated cell lines can be best maintained by cryopreservation. This also takes into account the expected problems associated with the inherent heterogeneity of many secondary product forming cell lines.

References

BAYLISS, M.W., 1980, Chromosomal variation in plant tissues in culture, *Ann. Rev. Cytol.*, Suppl., 11A, 113–144.

BENSON, E.E. and HAMILL, J.D., 1991, Cryopreservation and post freeze molecular and biosynthetic stability in transformed roots of *Beta vulgaris* and *Nicotiana rustica*, *Plant Cell Tissue and Organ Culture*, **24**, 163–172.

BERLIN, J., 1997, Secondary products from plant cell cultures, in REHM H.J. and REED G. (Eds), *Biotechnology*, Vol. 7, 2nd edition, pp. 593–640, Weinheim, New York, Basel, Cambridge: VCH Verlagsgesellschaft.

BUTENKO, R.G., POPOV, A.S., VOLOKOVA, L.A., CHERNYAK, N.D. and NOSOV, M., 1984, Recovery of cell cultures and their biosynthetic capacity after storage of *Dioscorea deltoidea* and *Panax ginseng* in liquid nitrogen, *Plant Science Letters*, **33**, 285–292.

CAREW, D.P. and STABA E.J., 1965, Plant tissue culture: 1st fundamentals, application and relationship to medicinal plant studies, *Lloydia*, 28, 1–26.

CHEN, T.H.H., KARTHA, K.K., CONSTABLE, F. and GUSTA, L.V., 1984a, Freezing characteristics of cultured *Catharanthus roseus* (L.) G. DON cells treated with dimethylsufoxide and sorbitol in relation to cryopreservation, *Plant Physiol.*, 75, 720–725.

CHEN, T.H.H., KARTHA, K.K., LEUNG, N.L., KURZ, W.G.W., CHATSON, K.B. and CONSTABLE, F., 1984b, Cryopreservation of alkaloid-producing cell cultures of periwinkle (*Catharanthus roseus*), *Plant Physiol.*, 75, 726–731.

DELUCA, V., 1993, Enzymology of indole alkaloids, in DEY, P.M. and HARBORNE, M.P. (Eds), *Methods in Plant Biochemistry*, Vol. 9, pp. 345–368, London: Academic Press.

DEUS-NEUMANN, B. and ZENK, M.H., 1984, Instability of indole alkaloid production in *Catharanthus roseus* cell suspension cultures, *Planta Medica*, **50**(5), 427–431.

DIETTRICH, B., POPOV, A.S., PFEIFFER, B., NEUMANN, D., BUTENKO, R. and LUCKNER, M., 1982, Cryopreservation of *Digitalis lanata* cell cultures, *Planta Medica*, **46**, 82–87.

DOUGALL, D.K. and WHITTEN, G.H., 1980, The ability of wild carrot cell cultures to retain their capacity for anthocyanin synthesis after storage at −140°C, *Planta Medica*, Supplement, 129–135.

HASHIMOTO, T. and YAMADA, Y., 1994, Alkaloid biogenesis: molecular aspects, *Ann. Rev. Plant Physiol. Plant Mol. Biol.*, **45**, 257–285.

KARTHA, K.K., LEUNG, N.L., GAUDET-LAPRAIRIE P. and CONSTABLE, F., 1982, Cryopreservation of periwinkle, *Catharanthus roseus* cells cultured *in vitro*, *Plant Cell Reports*, **1**, 135–138.

KURIYAMA, A., WATANABE, K., UENO, S. and MITSUDA, H., 1990, Effect of post-thaw treatment on the viability of cryopreserved *Lavandula vera* cells, *Cryo-Letters*, **11**, 171–178.

KUTCHAN, T.M., 1995, Alkaloid biosynthesis – The basis for metabolic engineering of medicinal plants, *The Plant Cell*, **7**, 1059–1070.

MANNONEN, L., TOIVONEN, L. and KAUPPINEN, V., 1990, Effects of long-term preservation on growth and productivity of *Panax ginseng* and *Catharanthus roseus* cell cultures, *Plant Cell Reports*, **9**, 173–177.

REUFF, I., 1987, Untersuchungen zur Kryokonservierung pflanzlicher Zellkulturen am Beispiel von Coleus blumei und Berberis wilsoniae, PhD Thesis, Tübingen.

SATO, H., KOBAYASHI, Y., FUKUI, H. and TABATA, M., 1990, Specific differences in tolerance to exogenous berberine among plant cell cultures, *Plant Cell Reports*, **9**, 133–136.

SEITZ, U. and REINHARD, E., 1987, Growth and ginsenoside pattern of cryopreserved *Panax ginseng* cell cultures, *J. Plant Physiol.*, **131**, 215–223.

SEITZ, U., ALFERMANN, A.W. and REINHARD, E., 1983, Stability of biotransformation capacity in *Digitalis lanata* cell cultures after cryogenic storage, *Plant Cell Reports*, **2**, 273–276.

SUK WEON KIM, KYUNG HEE JUNG, SANG SOO KWAK and JANG RYOL LIU, 1994, Relationship between cell morphology and indole alkaloid production in suspension cells of *Catharanthus roseus*, *Plant Cell Reports*, **14**, 23–26.

WATANABE, K., MITSUDA, H. and YAMADA, Y., 1983, Retention of metabolic and differentiation potentials of green *Lavandula vera* callus after freeze-preservation, *Plant & Cell Physiol.*, **24**(1), 119–122.

WILSON, G., 1990, Screening and selection of cultured plant cells for increased yields of secondary metabolites, in Dix, Ph.J. (Ed.), *Plant Cell Line Selection*, pp. 187–217, Weinheim, New York, Basel, Cambridge: VCH Verlagsgesellschaft.

WITHERS, L.A. and KING, P., 1980, A simple freezing unit and cryopreservation method for plant cell suspensions, *Cryo-Letters*, **1**, 213–220.

YAMAMOTO, H., SUZUKI, M., KITAMURA, T., FUKUI, H. and TABATA, M., 1989, Energy-requiring uptake of protoberberine alkaloids by cultured cells of *Thalictrum flavum*, *Plant Cell Reports*, **8**, 361–364.

In Vitro Conservation of Temperate Tree Fruit and Nut Crops

BARBARA M. REED

10.1 Introduction

Many germplasm facilities for the preservation and distribution of fruit and nut germplasm are now instituting slow-growth and cryopreservation strategies (Ashmore, 1997; Brettencourt and Konopka, 1989). Some genera have well-defined methods, while the techniques for others are still under investigation. Primary collections of plant germplasm are often in field plantings that are vulnerable to disease, insect, and environmental stresses. Slow-growth techniques provide a secondary storage method for clonal field collections (see Lynch, Chapter 4, this volume). Alternative germplasm storage technologies also provide storage modes for experimental material, allow for staging of commercial tissue culture crops, and provide a reserve of germplasm for plant distribution. Cryopreservation in liquid nitrogen (LN) provides a low-input method for storing a base collection (long-term backup) of clonal materials. Recent improvements in cryopreservation methods make these long-term collections of clonal germplasm feasible. Both *in vitro* and cryopreserved collections provide insurance against the loss of valuable genetic resources and may provide alternative distribution methods.

Medium-term storage of clonal plants involves slow-growth strategies such as temperature reduction, environmental manipulation, or chemical additions in the culture medium. Storage techniques developed thus far provide several options so it is now possible to match improved techniques with a facility's needs and resources for the best possible plant preservation.

New techniques and improvements in cryopreservation research have greatly increased the number of cryopreserved species. Suspension or callus cultures, dormant buds, *in vitro* grown apical meristems, isolated embryonic axes, seeds, somatic embryos, and pollen are now stored in LN. Cryopreserved collections of temperate plants of economic importance are now established in several countries. The storage of most temperate horticultural crops as base collections in liquid nitrogen is now feasible.

10.2   Literature review of progress

10.2.1   *Medium-term storage at above freezing temperatures*

In vitro collections play an important role in storing and distributing germplasm throughout the world. Certification programmes often incorporate *in vitro* culture as a standard
technique for producing virus-negative plants from stock collections. *In vitro* culture
systems are available for most temperate fruit and nut crops, but information on medium-
term storage is limited for many genera. Most studies involve temperatures near freezing,
but some tests of room temperature storage and chemical inhibition are also available
(see Lynch, Chapter 4, this volume). Published research is available for *Malus, Morus,
Prunus, Punica* and *Pyrus*; however most studies are restricted to a few genotypes and
storage conditions. Published reports of *in vitro* storage systems for temperate nut trees
are very limited; however, a species by species report of conservation methods currently
applied to temperate tree fruit and nut crops is given below.

Corylus

More than 80 genotypes of hazelnut (*Corylus* sp.) *in vitro* cultures are stored at 4°C in
low light (5 µmol m^{-2} s^{-1}) at NCGR–Corvallis, Oregon, USA (Reed, unpublished). Storage was also successful in total darkness where the mean storage duration for accessions
held at 4°C in the dark is 1.26 years with a range of 8 months to 2.5 years (Reed and
Chang, 1997).

Malus

Lundergan and Janick (1979) first suggested the feasibility of *in vitro* germplasm storage
after successfully storing *Malus domestica* Borkh. cv. Golden Delicious *in vitro* for 12
months at 1°C and 4°C. Success with other *Malus* species and cultivars included storage
ranging from 9 months to 3.5 yr (Druart, 1985; Orlikowska, 1991, 1992; Wilkins *et al.*,
1988). Wilkins *et al.* (1988) studied cultures of five *Malus domestica* cultivars, *M. prunifolia*
(Willd.) Borkh., and *M. baccata* (L.) Borkh. and successfully stored them for 12 to 28
months at 4°C with a 16 hour photoperiod on an agar-based multiplication medium. Some
rootstock cultivars were successfully stored on liquid medium. Charcoal in the medium
(see Lynch, Chapter 4, this volume) increased storage time, but eliminated proliferation during storage. Two apple rootstock cultivars kept at 4°C in the dark stored better on
medium with 6-benzylaminopurine (BAP) than on medium lacking the growth regulator
(Orlikowska, 1991, 1992). Storing plants immediately after subculture was important for
the survival of multiple-shoot tufts, but shoot tips and nodal segments survived whether
stored at 0, 10, or 20 days after subculture. Eckhard (1989) stored apple-shoot cultures at
2 to 4°C under low light for 1.5 years on a reduced sucrose, low BAP medium. Druart
(1985) stored topped, partially submerged shoots of three *Malus* rootstocks and 'Golden
Delicious' on hormone-free medium in the dark at 2°C for 1.5 to 3.5 years with 100 per
cent survival. Apple species and cultivars (150 genotypes) are stored at Changli Institute
of Pomology (China) on modified MS medium at 2 to 4°C in the light (10 µE m^{-2} s^{-1}, 15
hour photoperiod) with yearly transfer (Reed and Chang, 1997).

Morus

Morus nigra L. shoot tips survived for only six months on multiplication medium at 4°C with a 16 hour photoperiod, but with activated charcoal in the medium (see Lynch, Chapter 4, this volume), survival could be increased to 42 per cent after nine months at 25°C (Wilkins *et al.*, 1988). Sharma and Thorpe (1990) stored 15 genotypes of *Morus alba* L. for six months at 4°C in the dark on shoot proliferation medium (80 per cent viability).

Prunus

Marino *et al.* (1985) stored shoot cultures of three *Prunus* (peach, cherry) genotypes at 8°C, 4°C or –3°C for up to 10 months on multiplication medium. A 16 hour photoperiod was important for successful 4°C and 8°C storage for 90 days, but ten-month, dark storage at –3°C was better than under lights for some genotypes. Cultures stored 14 days after subculture survived better than those stored immediately, and lower temperatures increased storage times. Druart (1985) stored 12 *Prunus* species and cultivars on basal medium at 2°C in the dark for up to four years; dimethyl sulphoxide or glycerol in the medium was toxic to the cultures. Survival of topped and partially submerged shoots was genotype dependent. Wilkins *et al.* (1988) stored five *Prunus* genotypes on multiplication medium at 4°C with a 16 hour photoperiod for nine to 18 months.

Punica

Pomegranate, *Punica granatum* L., shoot cultures died at 4°C, but survived for 18 months at 10°C with a 16 hour photoperiod (Wilkins *et al.*, 1988).

Pyrus

Pyrus communis L. subsp. *caucasica* (Fed.) Browicz shoot tips grown on basal medium at 4°C, 8°C, and 12°C with a 16 hour photoperiod exhibited depressed growth for 12 to 18 months, with highest survival at 4°C. Adding mannitol or increasing sucrose in the medium were not successful for storage at either 4°C or 28°C (Wanas *et al.*, 1986). *Pyrus pashia* D. Don had 100 per cent survival on multiplication medium at 4°C or 10°C with a 16 hour photoperiod after 12 months (Wilkins *et al.*, 1988). *P. communis* cultivars La France, Bartlett, and La France × Bartlett had good survival at 5°C with a 16 hour photoperiod and at 1°C in darkness for 20 months, but 10°C and 15°C storage with light produced poor survival. Japanese pears *P. pyrifolia* (Burm.) Nakai. CVs. 'Shinsui', 'Nijisseiki', 'Shinchu', 'Kosui' 'Hosui' and 'Hakatado' were killed when stored at 5°C, 10°C, or 15°C with a photoperiod, but had 100 per cent survival when stored at 1°C in the dark for 12 months. ABA did not improve the survival of stored Japanese pears (Moriguchi, 1995; Moriguchi *et al.*, 1990). Pears stored at NCGR–Corvallis in 1984 were kept in 20 × 100 mm tubes at 4°C in darkness, but tubes were replaced with tissue-culture bags in 1989. *Pyrus* accessions (169) were stored for eight months to 4.6 years with a mean storage time of 2.7 years in tissue-culture bags in the dark at 4°C (Reed and Chang, 1997). Storage of 46 genotypes with three treatments (4°C upright plants, 4°C three-quarters submerged, 1°C upright) showed genotype differences for the length of storage, but few differences were noted among the three storage treatments (Reed *et al.*, 1998).

10.2.2 Long-term storage in liquid nitrogen

Cryopreservation (see Benson, Chapter 6, this volume) of temperate fruit trees began in the 1970s when dormant bud freezing was successfully applied to apple, pear, peach, plum, and cherry; now additional techniques are available. Many temperate nut seeds are dehydration sensitive, liquid nitrogen sensitive, or survive for a year or less in 4°C storage. In this respect, they are similar to recalcitrant tropical seeds as demonstrated by Marzalina and Krishnapillay (Chapter 17, this volume). Excised embryonic axes are excellent material for cryopreservation of wild populations, but cultivars require methods similar to those developed for fruit trees. The three major cryopreservation techniques – slow freezing, vitrification, and encapsulation–dehydration – are useful for these plant materials. Slow-freezing techniques developed in the 1970s by several investigators are used on many different species (Kartha, 1985). Sakai (1993) developed several plant vitrification solutions with highly concentrated cryoprotectants that allow cells to dehydrate quickly and cellular liquids to form glasses at low temperatures. Vitrification solution components, the duration of exposure, the size of plant material, the cryoprotectant toxicity, and the temperature of application are all important to plant survival. Dereuddre *et al.* (1990a, 1990b) devised a new cryopreservation system involving encapsulation of shoot tips in alginate beads followed by dehydration and direct exposure to LN. In addition, combinations of these techniques are also used in certain situations.

Carya

Pence (1990) found most *Carya* embryonic axes dried to 5–10 per cent moisture content before being exposed to LN germinated or partially germinated, with some callus following thawing. Subsequent *in vitro* growth and development was best for fresh seed and declined from shoots to callus to no growth as the seed aged.

Castanea

Castanea axes dried to about 8 per cent moisture before freezing produced callus upon recovery in initial testing (Pence, 1990). Chestnut embryonic axes desiccated to 20–30 per cent moisture before LN exposure had improved survival and some shoot formation (Pence, 1992).

Corylus

Early studies by Pence (1990) found that cryopreserved embryonic axes and control axes of *Corylus* seeds produced only callus. Embryonic axes from fresh seed of *Corylus avellana* L. 'Morell' exhibited maximum recovery (85 per cent) when axes were frozen at 12 per cent moisture content, while 'Butler' required 11 per cent moisture to obtain 50 per cent recovery (González-Benito and Perez, 1994). Whole seeds of *C. avellana* 'Barcelona' did not survive LN exposure following desiccation pretreatment, but embryonic axes were excised from the thawed seed and regrown in culture (Normah *et al.*, 1994). Axes from stratified 'Barcelona' seed had improved shoot growth for both control and LN exposed treatments. Axes from stored, stratified seed dried to 8 per cent moisture were cryopreserved with 85 per cent viability and 70 per cent shoot growth, while only 30 per cent of unstratified axes produced shoots (Reed *et al.*, 1994). Embryonic axes from seeds of *Corylus colurna* L., *C. americana* Marsh., and *C. sieboldiana* var. mandshurica (Maxim.) C. Schneider were stored in LN using this technique at NCGR–Corvallis and the National Seed Storage Laboratory, Ft Collins, CO. Regrowth of the thawed axes was 75–80 per cent for all three species (Reed, unpublished).

Juglans

Dried axes of *Juglans* seeds (5 per cent moisture) germinated or partially germinated producing shoots and/or roots *in vitro* following cryopreservation (Pence, 1990). Cryoprotectant with 5 M 1,2-propanediol and 20 per cent sucrose produced 75–91 per cent survival and regrowth of *Juglans* embryonic axes (de Boucaud *et al.*, 1991). Slow freezing *in vitro* grown shoot tips of walnut was also successful (de Boucaud and Brison, 1995). Modified PVS2 cryoprotectant treatment combined with slow freezing (0.5°C/min) of shoot tips produced 34 per cent survival (Brison *et al.*, 1991). Encapsulation–dehydration and slow freezing methods were successful with isolated walnut somatic embryos (de Boucaud *et al.*, 1994).

Malus

Apple and pear winter buds exposed to subzero temperatures at slow freezing rates retained their viability after immersion in liquid nitrogen and apple buds taken from the shoots grew after being grafted onto rootstocks in the greenhouse (77 per cent regrowth) (Sakai and Nishiyama, 1978). Dormant vegetative *Malus* buds cryopreserved using a combined dehydration–encapsulation technique had 80–100 per cent viability (Stushnoff, 1987; Stushnoff and Seufferheld, 1995; Tyler and Stushnoff, 1988a, 1988b; Tyler *et al.*, 1988). In related studies winter-dormant buds had moisture contents ranging from 48 to 60 per cent. They required desiccation to 20–30 per cent moisture to survive LN exposure. Very few tolerant species, however, could be desiccated below 10 per cent. At maximum hardiness most buds were tolerant of desiccation (Stushnoff, 1987, 1991; Tyler and Stushnoff, 1988a, 1988b). Genotypes that naturally tolerate desiccation and freezing to −30°C or colder at maximum hardiness would survive this procedure best. Cryopreserved dormant buds were either thawed slowly in room temperature air or rapidly in 40°C water (Sakai, 1985; Sakai and Nishiyama, 1978; Tyler and Stushnoff, 1988a; Tyler *et al.*, 1988). Pretreating dormant apple buds with sugars and other cryoprotectants enhanced the survival of less cold hardy taxa or those that do not sufficiently acclimate (Seufferheld *et al.*, 1991). Slow freezing below −10°C and immersion in LN without a cryoprotectant was successful with good regrowth *in vitro* after thawing for dormant-shoot tips from winter apple buds (Katano *et al.*, 1983). Dormant buds of 500 apple genotypes are stored in the vapour phase of LN at the National Seed Storage Laboratory (NSSL) in Ft Collins, Colorado (Forsline *et al.*, 1993). Single-bud sections from cold-hardened, dormant apple shoots are dried to 30 per cent moisture, cooled at 1°C/h to −30°C, held for 24 hours, and then stored in the LN vapour phase.

The first tests of cryopreservation on *in vitro* grown shoot tips of apple recovered only callus (Kuo and Lineberger, 1985). Caswell *et al.* (1986) found that high sucrose concentrations in the culture medium improved the hardiness of *in vitro* grown apple shoots, but that the cold hardening response was genotype dependent. The effects of sucrose *in vitro* were applied to *Malus* cryopreservation by Niino and Sakai (1992) who used the encapsulation–dehydration method to cryopreserve *in vitro* grown apple shoot tips and obtained about 80 per cent regrowth after thawing. Decreases in moisture content of *in vitro* grown plants were also obtained through extended culture duration (Chang *et al.*, 1992). Plants cultured for 70 days without transfer before cold acclimatization (CA) had more shoot formation following slow freezing and LN exposure than those cultured for 35 days. These results are attributed to lower meristem moisture contents and slowed shoot growth. A freezing rate of 0.1 to 0.2°C/min was suitable for *in vitro* grown apple shoot tips (Chang *et al.*, 1992). Meristems of more than 70 cultivars are stored in LN at Changli Institute of Pomology, Hebei Academy of Agricultural and Forestry Sciences. Samples removed from LN after one month and one, two, and three years were recultured with no change in survival or plantlet regrowth (Chen *et al.*, 1994; Reed and Chang, 1997).

Vitrification is also a successful technique for *Malus* shoot tips. Apple shoot tips dehydrated with PVS2 (30 per cent glycerol, 15 per cent ethylene glycol and 15 per cent DMSO in MS medium containing 0.4 M sucrose) at 25°C for 80 minutes produced 80 per cent shoot formation following vitrification (Niino *et al.*, 1992c). Zhao *et al.* (1995) studied the effects of plant vitrification solutions PVS1 to PVS5 on apple meristems; plants treated in PVS3 (50 per cent sucrose, 50 per cent glycerol) for 80 minutes before exposure to liquid nitrogen had the best regrowth.

Morus

Shoot tips of prefrozen winter buds of *Morus bombycis* Koidz. cv. Kenmochi survived immersion in liquid nitrogen, but grafts and cuttings did not survive (Yakuwa and Oka, 1988; Yokoyama and Oka, 1983). Wang *et al.* (1988) regenerated plants of *M. multicaulis* Loud. Cv. Lusang through shoot tip culture from frozen winter buds using a similar method. Niino *et al.* (1992b) demonstrated that excised shoot tips from winter buds of *M. bombycis* cv. Kenmochi, prefrozen to −20°C at 5°C/day produced more shoots than buds prefrozen at 10°C/day. Partially dehydrating the buds to about 38.5 per cent moisture content at 25°C prior to prefreezing to −20°C, improved the recovery rates. Alginate-coated, winter-hardened shoot tips of several *Morus* species had maximum shoot formation (81 per cent) when dehydrated to 22–25 per cent water content before freezing (Niino *et al.*, 1992b). Thirteen mulberry cultivars tested for cryopreservation as *in vitro* grown shoot tips produced survival ranging from 40 to 81.3 per cent with all methods tested: slow freezing (0.5°C/min to −42°C), vitrification (PVS2, 90 minutes), air-drying (24 per cent water content), and encapsulation–dehydration (33 per cent water content) (Niino, 1990, 1995; Niino *et al.*, 1992a).

Prunus

Survival (70 to 80 per cent) of axillary apices excised from *Prunus persica* cv. GF 305 *in vitro*-cultured plants required pretreatment on a 5 per cent DMSO and 5 per cent proline culture medium before vitrification (Paulus *et al.*, 1993). This pretreatment medium was also effective and produced 69 per cent and 74 per cent shoot formation in two *Prunus* rootstock cultivars (Brison *et al.*, 1995). Dormant buds of ten *Prunus* species were cryopreserved with up to 100 per cent recovery by grafting (Stushnoff, 1985). *In vitro*-grown *Prunus* shoots survived cryopreservation with 75 per cent regrowth when held at −30°C for 24 hours before being transferred to LN. Culturing on 14 per cent sucrose medium and chilling at 4°C enhanced the low-temperature tolerance of cryopreserved plantlets (Stushnoff, 1985).

Pyrus

Dormant hardy *Pyrus* shoots were able to survive LN after prefreezing to −40°C or −50°C (Sakai and Nishiyama, 1978). Moriguchi *et al.* (1985) found that shoot tips from dormant buds of Japanese pear required prefreezing to −40 to −70°C before being exposed to LN. Oka *et al.* (1991) and Mi and Sanada (1992, 1994) recovered whole plants from cryopreserved buds.

In vitro grown pear-shoot meristems were first successfully cryopreserved in 1990 (Dereuddre *et al.*, 1990a, 1990b; Reed, 1990). A slow-freezing method for *in vitro* grown pear meristems which incorporated cold acclimatization and slow cooling produced 55 to 95 per cent regrowth in cryopreserved shoot tips of four *Pyrus* species including a subtropical species, *P. koehnei* (Reed, 1990). Encapsulation–dehydration was applied to pear by Dereuddre *et al.* (1990a, 1990b). A 0.75 M sucrose preculture and four hour dehydration (20 per cent residual water) produced 80 per cent recovery (Scottez *et al.*, 1992). A modified encapsulation–dehydration method developed by Niino and Sakai (1992) produced 70 per cent shoot formation for three pear cultivars. They applied the vitrification method to pears and obtained 40 to 72.5 per cent regrowth (Niino *et al.*, 1992c; Suzuki *et al.*, 1997).

A comparison of slow freezing and vitrification methods using 28 *Pyrus* genotypes found that 61 per cent had better than 50 per cent regrowth following slow freezing (0.1°C/min), while only 43 per cent of the genotypes responded this well to the vitrification technique (Luo *et al.*, 1995).

10.3   Germplasm storage

Fruit and nut trees in field genebanks are at risk from severe weather, insect and animal pests, and diseases. Quarantine laws designed to prevent the spread of diseases or insects also restrict global exchange of field germplasm. Although *in vitro* cultures are not necessarily disease free, and may be virus infected or have bacterial contaminants, some countries allow *in vitro* cultures to satisfy quarantine restrictions. This makes *in vitro* collections valuable as complementary or secondary collections. Cryopreserved storage is the ultimate base storage; available for use in case of emergency but requiring little input of care or money.

10.3.1   In vitro *stored collections*

Collections of temperate tree fruit and nut crops are held at various experiment stations and plant breeding centres throughout the world (Ashmore, 1997; Brettencourt and Konopka, 1989). *In vitro* stored collections are sometimes used as secondary collections but may also be the primary collection. Complete listings of *in vitro* stored germplasm collections are difficult to find, but germplasm workers in many countries use *in vitro* culture for other purposes including virus elimination (Table 10.1).

10.3.2   *Cryopreserved collections*

A few countries have initiated clonal-germplasm storage is liquid nitrogen. Research is underway in many more countries. For temperate tree fruit and nut crops, cryopreserved

Table 10.1 Some of the world-wide germplasm related *in vitro* culture work and culture collections of temperate tree fruits and nuts

Country	Collections[a]	Culture[b]
Belgium	*Prunus*	
Germany	*Prunus, Pyrus*	
Hungary	*Prunus*	
India		*Prunus*
France		*Prunus, Pyrus, Castanea, Juglans, Mauls*
Japan[c]	*Cydonia, Diospyros, Malus, Morus, Prunus, Pyrus*	
Portugal	*Prunus*	
United States	*Corylus, Cydonia, Pyrus*	*Malus, Prunus*
Uruguay		*Malus, Pyrus, Prunus, Cydonia*

[a] Brettencourt and Konopka (1989).
[b] Ashmore (1997).
[c] Personal communication (1998).

collections are held in China (*Malus in vitro* grown meristems), Japan (*Cydonia, Malus* pollen, *Malus, Morus* dormant buds) and the USA (*Corylus* embryonic axes, *Malus* dormant buds, *Pyrus in vitro* grown meristems, *Corylus, Pyrus* pollen).

10.3.3 Discussion: the role of storage technologies

In vitro *storage*

In vitro culture is an important tool for the international germplasm community (see Lynch, Chapter 4 this volume), but much remains to be learned about optimal *in vitro* storage conditions. Many of the factors mentioned in individual research reports require more investigation. Important data are available on the size and type of propagule stored. Barlass and Skene (1983) found that single-rooted *Vitis* shoots respond differently from proliferating cultures. For apple, single-shoot tips, nodal segments, and cultures with multiple shoots respond differently to various storage conditions (Orlikowska, 1991). Optimum age, size, and physiology must all be taken into account before cultures are stored.

Light quality and intensity are important for culture growth both before and during storage. In most cases experimentation into light effects is limited by lack of growth room availability in a facility. Data on culture storage in light versus darkness are available, but extensive information on the effects of light quality, duration, and intensity is not available for tree fruit and nut crops.

Culture conditions before storage, and culture time after subculture have important effects on storage time, but little has been done to study these conditions. The pre-storage culture period affected both storage length and the proliferation of *Prunus* rootstock cultivars following storage; 14 days was optimum (Marino *et al.*, 1985). In some apple genotypes, proliferation was better following cold storage than in non-stored plants, but this also varied with genotype (Orlikowska, 1992). Multiplication and storage media greatly affect the survival of cultures after storage. Storage time may be improved or limited by the growth regulators in storage media. For the most part this is a genotype dependent phenomenon (Orlikowska, 1992; Reed, 1993). Research on the effects of growth regulators in storage media and the genetic stability of stored cultures is still limited (Wilkins *et al.*, 1988). Genetic analysis of *in vitro* grown and stored plants is not a standard practice; genetic instability appears to be genotype dependent and it is difficult to generalize on its causes or probabilities (Moore, 1991). Field and molecular analyses are needed to determine genetic stability (Harding, 1994; Kumar, 1995). Adventitious shoot production may be a cause of genotype variation during *in vitro* storage. Improvements in multiplication and storage media should reduce the likelihood of adventitious shoot production. Additional research is badly needed to develop standard techniques for genetic stability testing.

Contaminants are a major difficulty for any *in vitro* system. Slow-growing contaminants may persist without being noticed for long periods, then suddenly become evident during or after *in vitro* storage (Gunning and Lagerstedt, 1985). Stored cultures may die from the debilitating effects of latent infections (Wanas *et al.*, 1986). Indexing cultures for latent bacterial and fungal infections should be a standard step in germplasm storage procedures (Reed and Tanprasert, 1995). Bacteriological media used to detect cultivable contaminants are more effective than simply examining the cultures visually (Reed *et al.*, 1995; Tanprasert and Reed, 1997). Special methods are still needed to detect non-cultivable contaminants, such as obligate parasites. Healthier cultures, longer storage times, and safer materials for distribution can be assured through improved detection of bacterial contaminants. Germplasm storage in heat-sealed tissue culture bags can nearly eliminate fungal, bacterial, or insect contamination during the storage period (Reed, 1991, 1992, 1993).

Cryopreservation

Cryopreservation techniques are now available for many forms of fruit and nut tree germplasm storage (cell suspensions, callus, shoot tips, somatic embryos, and embryonic axes). Some of these techniques are now used for long-term storage of germplasm (Reed and Chang, 1997). Future improvements in cryopreservation will require attention to several research topics. The choice of plant material is one important consideration since both growth stage and genotype affect survival following LN exposure. Response to cryopreservation techniques varies greatly with genotype; even related genotypes may have very different survival following LN exposure (Niino, 1995; Reed and Yu, 1995). The physiological status of mother plants directly impacts survival following cryopreservation (Chang *et al.*, 1992; Reed, 1988). More emphasis is needed on research into the physiological condition of plants prior to cryopreservation. Research into the physiology of cold acclimatization (CA) of cultures will be useful, especially since CA pretreatments are necessary for the success of many cryopreservation techniques. Comparison of cryopreservation techniques remains difficult due to wide variations in the temperature, light conditions, and duration of CA used in different laboratories.

Freezing rates and cryoprotectants have received much attention in the past but still require further study. Slow freezing, the first technique developed, remains an important protocol; a very slow freezing rate ($0.1°C/min$ to $-40°C$) produces the best survival for *in vitro* grown shoot tips from many species including *Pyrus* and *Malus* (Chang *et al.*, 1992; Reed, 1990). Most *in vitro* systems require cryoprotectants, and highly concentrated solutions (such as PVS2) formerly used only with direct LN exposure are now used for some slow freezing and encapsulation methods (Brison *et al.*, 1995; Niino, 1995).

Combined methods are successful for cryopreserving some difficult genotypes. In a combination of the slow-freezing technique and the encapsulation–dehydration technique, encapsulated grape axillary-shoot tips were slowly cooled before being exposed to LN, significantly increasing survival and shoot formation over encapsulation alone (Plessis et al., 1993). Dehydration of encapsulated dormant apple buds with a vitrification solution, followed by LN exposure was also successful (Seufferheld et al., 1991). Advances in cryopreservation of difficult genotypes may result from further exploration of combined techniques. No phenotypic changes have been observed in meristem derived plants of cryopreserved plant material (Harding and Benson, 1994; Reed and Hummer, 1995). Genetic abnormalities due to cryopreservation are expected to be rare; however, more studies are needed to confirm the genetic stability of plants held in LN.

10.4   Impact on the storage and distribution of germplasm

10.4.1   In vitro storage

The use of in vitro stored plants as primary or secondary collections of clonal crops reduces the land area required for field genebanks. Three replicates per genotype are considered ideal for germplasm storage and using an in vitro culture for one or two of these replicates greatly decreases field space and labour costs. In vitro cultures are not guaranteed to be pathogen free; however virus-indexed materials can be stored in vitro to keep them in virus-free condition. Bacterial and fungal indexing can detect cultivable contaminants and provide propagules in which requesters can have a high degree of confidence (Reed and Tanprasert, 1995; Reed et al., 1995). In vitro cultures obtained from virus-elimination programmes and indexed for cultivable bacterial and fungal contaminants often meet phytosanitary requirements for import and export. Plants distributed as in vitro cultures are useful for many requesters and in many cases cultures survive international shipment better than traditionally propagated plants (Bartlett, personal communication). A large percentage of the NCGR plant material is distributed as in vitro plantlets. Acclimatization of these plantlets requires the same care as other in vitro grown materials. Each plant shipment should include information on in vitro growth and acclimatization procedures.

10.4.2   Cryopreservation

Cryopreservation is best suited for base (long-term) storage of clonal collections, (see Benson, Chapter 6, this volume for an appraisal of methodology). The greatest cost of cryopreservation is in the initial storage of an accession, but very little input is needed for many years after storage. Collections in LN require little storage space: a 40 l dewar can hold as many as 3000 sample vials (i.e., five vials for each of 600 accessions). A clonal collection with representatives in the field, in vitro, and in LN would provide active, backup, and base storage for an accession for less cost and greater security over the long term than three field plants. Although plants can be distributed as cryopreserved samples, they are best kept as insurance in case of loss of actively growing accessions. Plants cryopreserved at one location can be shipped to a second location in a specially designed travel dewar. Pyrus, Ribes, and Rubus meristems and Corylus embryonic axes cryopreserved in Oregon were shipped to Colorado by air freight in a travel dewar for base storage (Reed *et al.*, 1997, 1998). For recovery the cryopreserved samples should be thawed by the recommended procedures, regrown *in vitro* into plantlets, and acclimatized to the greenhouse by techniques used for the specific plant type.

10.5 Conclusions

Plant conservation and germplasm exchange using *in vitro* methods have increased over the past decade, mirroring perhaps the advances in research in this field. Improved global transportation and communication have led to a wider exchange of ideas as well as to the exchange of plant materials. More institutions are now taking advantage of improved techniques to provide *in vitro* base, primary, or secondary collections to protect their germplasm collections. The advantages of *in vitro* conservation of important plant collections are the same as in the past, both in terms of phytosanitary considerations and plant security, but the willingness of curators to provide alternative storage for crops has increased.

Further improvements to these techniques are of course always needed; research is needed to improve *in vitro* culture, storage and cryopreservation, including a myriad of aspects in each of these fields. Fortunately, *in vitro* culture and cryopreservation have progressed to the point where they can be used routinely in many laboratories. *In vitro* stored plantlets are used as primary or duplicate collections in several facilities. Cryopreserved samples for long-term (base) storage of important collections are now a reality as well. They provide an important, previously missing, form of germplasm storage (base storage) for vegetatively propagated plants.

References

ASHMORE, S.E., 1997, *Status Report on the Development and Application of In Vitro Techniques for the Conservation and Use of Plant Genetic Resources*, p. 67, Rome: International Plant Genetic Resources Institute.

BARLASS, M. and SKENE, K.G.M., 1983, Long-term storage of grape *in vitro*, *FAO/IBPGR Plant Genetic Resources Newsletter*, **53**, 19–21.

BRETTENCOURT, E.J. and KONOPKA, J., 1989, *Directory of Germplasm Collections. 6. II. Temperate Fruits and Tree Nuts*, p. 296, Rome: International Board for Plant Genetic Resources.

BRISON, M., PAULUS, V. and DE BOUCAUD, M.T., 1991, Cryopreservation of walnut and plum shoot tips, *Cryobiology*, **28**, 738.

BRISON, M., DE BOUCAUD, M.T. and DOSBA, F., 1995, Cryopreservation of *in vitro* grown shoot tips of two interspecific *Prunus* rootstocks, *Plant Science*, **105**(2), 235–242.

CASWELL, K.L., TYLER, N.J. and STUSHNOFF, C., 1986, Cold hardening of *in vitro* apple and Saskatoon shoot cultures, *HortScience*, **21**, 1207–1209.

CHANG, Y., CHEN, S., ZHAO, Y. and ZHANG, D., 1992, Studies of cryopreservation of apple shoot tips, in *China Association for Science and Technology First Academic Annual Meeting of Youths Proceedings (Agricultural Sciences)*, pp. 461–464, Beijing: Chinese Science and Technology Press.

CHEN, S., CHANG, Y., ZHAO, Y. and ZHANG, D., 1994, Studies on the cryopreservation of *in vitro*-grown shoot tips, *Abstracts VIIIth International Congress of Plant Tissue and Cell Culture*, M-37, 274.

DE BOUCAUD, M.T. and BRISON, M., 1995, Cryopreservation of germplasm of walnut (Juglans species), in *Biotechnology in Agriculture and Forestry*, Vol. 32, edited by Y.P.S. BAJAJ, pp. 129–147, Berlin, Heidelberg: Springer-Verlag.

DE BOUCAUD, M., BRISON, M., LEDOUX, C., GERMAIN, E. and LUTZ, A., 1991, Cryopreservation of embryonic axes of recalcitrant seed: *Juglans regia* L. cv. Franquette, *Cryo-Letters*, **12**, 163–166.

DE BOUCAUD, M.T., BRISON, M. and NEGRIER, P., 1994, Cryopreservation of walnut somatic embryos, *Cryo-Letters*, **15**, 151–160.

DEREUDDRE, J., SCOTTEZ, C., ARNAUD, Y. and DURON, M., 1990a, Effects of cold hardening on cryopreservation of axillary pear (*Pyrus communis* L. cv Beurre Hardy) shoot tips of *in vitro* plantlets, *C. R. Acad. Sci. Paris*, **310**, 265–272.

DEREUDDRE, J., SCOTTEZ, C., ARNAUD, Y. and DURON, M., 1990b, Resistance of alginate-coated axillary shoot tips of pear tree (*Pyrus communis* L. cv Beurre Hardy) *in vitro* plantlets to dehydration and subsequent freezing in liquid nitrogen, *C. R. Acad. Sci. Paris*, **310**, 317–323.

DRUART, P., 1985, *In vitro* germplasm preservation techniques for fruit trees, in *In Vitro Techniques. Propagation and Long Term Storage*, edited by A. SCHAFER-MENUHR, pp. 167–171, Dordrecht: Martinus Nijhoff.

ECKHARD, A., 1989, Untersuchungen zur Entwicklung einer rationaellen Mehtode der in vitro-Depothaltung von Kern und Steinobst unter Kuhlbedingungen, *Archives Gartenbau*, **2**, 131–140.

FORSLINE, P.L., STUSHNOFF, C., TOWILL, L.E., WADDELL, J. and LAMBOY, W., 1993, Pilot project to cryopreserve dormant apple (*Malus* sp) buds, *HortScience*, **28**, 118.

GONZÁLEZ-BENITO, M.E. and PEREZ, C., 1994, Cryopreservation of embryonic axes of two cultivars of hazelnut (*Corylus avellana* L.), *Cryo-Letters*, **15**, 41–46.

GUNNING, J. and LAGERSTEDT, H.B., 1985, Long-term storage techniques for *in vitro* plant germplasm, *Proceedings International Plant Propagators Society*, **35**, 199–205.

HARDING, K., 1994, The methylation status of DNA derived from potato plants recovered from slow growth, *Plant Cell Tissue and Organ Culture*, **37**, 31–38.

HARDING, K. and BENSON, E.E., 1994, A study of growth, flowering, and tuberisation in plants derived from cryopreserved potato shoot-tips: implications for *in vitro* germplasm collections, *Cryo-Letters*, **15**, 59–66.

KARTHA, K.K., 1985, *Cryopreservation of Plant Cells and Tissues*, edited by K.K. KARTHA, p. 276, Boca Raton, Florida: CRC Press.

KATANO, M., ISHIHARA, A. and SAKAI, A., 1983, Survival of dormant apple shoot tips after immersion in liquid nitrogen, *HortScience*, **18**(5), 707–708.

KUMAR, M.B., 1995, Genetic stability of micropropagated and cold stored strawberries, MS Thesis, Oregon State University, Corvallis.

KUO, C.C. and LINEBERGER, R.D., 1985, Survival of *in vitro* cultured tissue of 'Jonathan' apples exposed to −196°C, *HortScience*, **20**, 764–767.

LUNDERGAN, C. and JANICK, J., 1979, Low temperature storage of *in vitro* apple shoots, *HortScience*, **14**, 514.

LUO, J., DENOMA, J. and REED, B.M., 1995, Cryopreservation screening of *Pyrus* germplasm, *Cryobiology*, **32**, 558.

MARINO, G., ROSATI, P. and SAGRATI, F., 1985, Storage of *in vitro* cultures of *Prunus* rootstocks, *Plant Cell Tissue and Organ Culture*, **5**, 73–78.

MI, W. and SANADA, T., 1992, Cryopreservation of pear winter buds and shoot tips, *China Fruits*, 20–22 and 38.

MI, W. and SANADA, T., 1994, Cryopreservation of pear shoot tips *in vitro*, *Advances in Horticulture*, 86–88.

MOORE, P., 1991, Comparison of micropropagated and runner propagated strawberry, *Fruit Varieties Journal*, **45**, 119.

MORIGUCHI, T., 1995, Cryopreservation and minimum growth storage of pear (*Pyrus* species), in *Biotechnology in Agriculture and Forestry*, Vol. 32, edited by Y.P.S. BAJAJ, pp. 114–128, Berlin, Heidelberg: Springer-Verlag.

MORIGUCHI, T., AKIHAMA, T. and KOZAKI, I., 1985, Freeze-preservation of dormant pear shoot apices, *Japanese Journal of Breeding*, **35**, 196–199.

MORIGUCHI, T., KOZAKI, I., YAMAKI, S. and SANADA, T., 1990, Low temperature storage of pear shoots *in vitro*, *Bulletin of the Fruit Tree Research Station*, **17**, 11–18.

NIINO, T., 1990, Establishment of the cryopreservation of mulberry shoot tips from in-vitro grown shoot, *J. Seric. Sci. Jpn.*, **59**, 135–142.

NIINO, T., 1995, Cryopreservation of germplasm of mulberry (*Morus* species), in *Biotechnology in Agriculture and Forestry*, Vol. 32, edited by Y.P.S. BAJAJ, pp. 102–113, Berlin, Heidelberg: Springer-Verlag.

NIINO, T. and SAKAI, A., 1992, Cryopreservation of alginate-coated *in vitro*-grown shoot tips of apple, pear and mulberry, *Plant Science*, **87**, 199–206.

NIINO, T., SAKAI, A., ENOMOTO, S., MAGOSI, J. and KATO, S., 1992a, Cryopreservation of *in vitro*-grown shoot tips of mulberry by vitrification, *Cryo-Letters*, **13**, 303–312.

NIINO, T., SAKAI, A. and YAKUWA, H., 1992b, Cryopreservation of dried shoot tips of mulberry winter buds and subsequent plant regeneration, *Cryo-Letters*, **13**, 51–58.

NIINO, T., SAKAI, A., YAKUWA, H. and NOJIRI, K., 1992c, Cryopreservation of *in vitro*-grown shoot tips of apple and pear by vitrification, *Plant Cell Tissue and Organ Culture*, **28**, 261–266.

NORMAH, M.N., REED, B.M. and YU, X., 1994, Seed storage and cryoexposure behavior in hazelnut (*Corylus avellana* L. cv. Barcelona), *Cryo-Letters*, **15**, 315–322.

OKA, S., YAKUWA, H., SATO, K. and NIINO, T., 1991, Survival and shoot formation *in vitro* of pear winter buds cryopreserved in liquid nitrogen, *HortScience*, **26**, 65–66.

ORLIKOWSKA, T., 1991, Effect of *in vitro* storage at 4°C on surviving and proliferation of apple rootstocks, *Acta Horticulturae*, **289**, 251–253.

ORLIKOWSKA, T., 1992, Effect of *in vitro* storage at 4°C on survival and proliferation of two apple rootstocks, *Plant Cell Tissue and Organ Culture*, **31**, 1–7.

PAULUS, V., BRISON, M., DOSBA, F. and DE BOUCAUD, M.T., 1993, Preliminary studies on cryopreservation of peach shoot tips by vitrification. *Comptes-Rendus-de–1'Academie-d'Agriculture-de-France*, **79**(7), 93–102.

PENCE, V.C., 1990, Cryostorage of embryo axes of several large-seeded temperate tree species, *Cryobiology*, **27**, 212–218.

PENCE, V.C., 1992, Desiccation and the survival of *Aesculus*, *Castanea* and *Quercus* embryo axes through cryopreservation, *Cryobiology*, **29**, 391–399.

PLESSIS, P., LEDDET, C., COLLAS, A., and DEREUDDRE, J., 1993, Cryopreservation of *Vitis vinifera* L. cv. Chardonnay shoot tips by encapsulation-dehydration: effects of pretreatment, cooling and postculture conditions, *Cryo-Letters*, **14**, 308–320.

REED, B.M., 1988, Cold acclimation as a method to improve survival of cryopreserved *Rubus* meristems, *Cryo-Letters*, **9**, 166–171.

REED, B.M., 1990, Survival of *in vitro*-grown apical meristems of *Pyrus* following cryopreservation, *HortScience*, **25**, 111–113.

REED, B.M., 1991, Application of gas-permeable bags for *in vitro* cold storage of strawberry germplasm, *Plant Cell Reports*, **10**, 431–434.

REED, B.M., 1992, Cold storage of strawberries *in vitro*: a comparison of three storage systems, *Fruit Varieties Journal*, **46**, 98–102.

REED, B.M., 1993, Improved survival of *in vitro*-stored *Rubus* germplasm, *Journal of the American Society of Horticultural Science*, **118**, 890–895.

REED, B.M. and CHANG, Y., 1997, Medium- and long-term storage of *in vitro* cultures of temperate fruit and nut crops, in *Conservation of Plant Genetic Resources In Vitro*, Vol. 1, edited by M.K. RAZDAN and E.C. COCKING, pp. 67–105, Enfield, NH, USA: Science Publishers, Inc.

REED, B.M. and HUMMER, K., 1995, Conservation of germplasm of strawberry (*Fragaria* species), in *Biotechnology in Agriculture and Forestry, Cryopreservation of Plant Germplasm I*, Vol. 32, edited by Y.P.S. BAJAJ, pp. 354–370, Berlin: Springer-Verlag.

REED, B.M. and TANPRASERT, P., 1995, Detection and control of bacterial contaminants of plant tissue cultures. A review of recent literature, *Plant Tissue Culture and Biotechnology*, **1**, 137–142.

REED, B.M. and YU, X., 1995, Cryopreservation of *in vitro*-grown gooseberry and currant meristems, *Cryo-Letters*, **16**, 131–136.

REED, B.M., NORMAH, M.N. and YU, X., 1994, Stratification is necessary for successful cryopreservation of axes from stored hazelnut seed, *Cryo-Letters*, **15**, 377–384.

REED, B.M., BUCKLEY, P.M. and DEWILDE, T.N., 1995, Detection and eradication of endophytic bacteria from micropropagated mint plants, *In Vitro – Plant*, **31P**, 53–57.

REED, B.M., DeNOMA, J., LUO, J., CHANG, Y. and TOWILL, L., 1997, Cryopreserved storage of a world pear collection, *In Vitro Cellular and Developmental Biology*, **33**, 51A.

REED, B.M., PAYNTER, C.L., DeNOMA, J., and CHANG, Y., 1998, Techniques for medium and long-term storage of pear (*Pyrus*) genetic resources, *FAO/IPGRI Plant Genetic Resources Newsletter*, International Plant Genetic Resources Institute, Rome.

SAKAI, A., 1985, Cryopreservation of shoot-tips of fruit trees and herbaceous plants, in *Cryopreservation of Plant Cells and Tissues*, edited by K.K. KARTHA, pp. 135–170, Boca Raton, Florida: CRC Press.

SAKAI, A., 1993, Cryogenic strategies for survival of plant cultured cells and meristems cooled to −196°C, in *Cryopreservation of Plant Genetic Resources*, Vol. 6, pp. 5–26, Tokyo: Japan International Cooperation Agency.

SAKAI, A. and NISHIYAMA, Y., 1978, Cryopreservation of winter vegetative buds of hardy fruit trees in liquid nitrogen, *HortScience*, **13**, 225–227.

SCOTTEZ, C., CHEVREAU, E., GODARD, N., ARNAUD, Y., DURON, M. and DEREUDDRE, J., 1992, Cryopreservation of cold acclimated shoot tips of pear *in vitro* cultures after encapsulation-dehydration, *Cryobiology*, **29**, 691–700.

SEUFFERHELD, M.J., FITZPATRICK, J., WALSH, T. and STUSHNOFF, C., 1991, Cryopreservation of dormant buds from cold tender taxa using a modified vitrification procedure, *Cryobiology*, **28**, 576.

SHARMA, K.K. and THORPE, T.A., 1990, *In vitro* propagation of mulberry, *Scientia Horticulturae*, **42**, 307–320.

STUSHNOFF, C., 1985, Cryopreservation of *in vitro* shoots from *Prunus pennsylvanica* and *Prunus fruticosa*, *FAO/IBPGR Plant Genetic Resources Newsletter*, **51**, 48.

STUSHNOFF, C., 1987, Cryopreservation of apple genetic resources, *Canadian Journal of Plant Science*, **67**, 1151–1154.

STUSHNOFF, C., 1991, Strategies for cryopreservation based on tissue hydration, *Cryobiology*, **28**, 745–746.

STUSHNOFF, C. and SEUFFERHELD, M., 1995, Cryopreservation of apple (*Malus* species) genetic resources, in *Biotechnology in Agriculture and Forestry*, Vol. 32, edited by Y.P.S. BAJAJ, pp. 87–101, Berlin, Heidelberg: Springer-Verlag.

SUZUKI, M., NIINO, T., AKIHAMA, T. and OKA, S., 1997, Shoot formation and plant regeneration of vegetative pear buds cryopreserved at −150 degrees C, *Journal of the Japanese Society of Horticultural Science*, **66**, 29–34.

TANPRASERT, P. and REED, B.M., 1997, Detection and identification of bacterial contaminants from strawberry runner explants, *In Vitro Cellular and Developmental Biology – Plant*, **33**, 221–226.

TYLER, N. and STUSHNOFF, C., 1988a, Dehydration of dormant apple buds at different stages of cold acclimation to induce cryopreservability in different cultivars, *Canadian Journal of Plant Science*, **68**, 1169–1176.

TYLER, N.J. and STUSHNOFF, C., 1988b, The effects of prefreezing and controlled dehydration on cryopreservation of dormant vegetative apple buds, *Canadian Journal of Plant Science*, **68**, 1163–1167.

TYLER, N., STUSHNOFF, C. and GUSTA, L.V., 1988, Freezing of water in dormant vegetative apple buds in relation to cryopreservation, *Plant Physiology*, **87**, 201–205.

WANAS, W.H., CALLOW, J.A. and WITHERS, L.A., 1986, Growth limitations for the conservation of pear genotypes, in *Plant Tissue Culture and its Agricultural Applications*, edited by L.A. WITHERS and P.G. ALDERSON, pp. 285–290, London: Butterworths.

WANG, L., ZHANG, X. and YU, Z., 1988, Cryopreservation of winter mulberry buds, *Acta Agriculturae Boreali-Sinica*, **3**, 103–106.

WILKINS, C.P., NEWBURY, H.J. and DODDS, J.H., 1988, Tissue culture conservation of fruit trees, *FAO/IBPGR Plant Genetic Resources Newsletter*, **73/74**, 9–20.

YAKUWA, H. and OKA, S., 1988, Plant regeneration through meristem culture from vegetative buds of mulberry (*Morus bombycis* Koidz) stored in liquid nitrogen, *Annals of Botany*, **62**, 79–82.

YOKOYAMA, T. and OKA, S., 1983, Survival of mulberry winter buds at super low temperatures, *J. Seric. Sci. Jpn.*, **52**, 263–264.

ZHAO, Y., CHEN, S., WU, Y., CHANG, Y. and ZHANG, D., 1995, Cryopreservation of *in vitro* shoot tips of apple by vitrification, in *China Association for Science and Technology Second Academic Annual Meeting of Youths Proceedings (Horticultural Sciences)*, pp. 406–409, Beijing: Beijing Horticultural University Press.

11

Conservation of Small Fruit Germplasm

REX M. BRENNAN AND STEPHEN MILLAM

11.1 Introduction

Small fruits are fairly recently domesticated, in comparison to the large-scale arable crops (Simmonds, 1976), and are mainly highly heterozygous woody perennial species. They are clonally propagated, and, as such, often require appropriate means of conservation rather than the use of seedbanks, although the latter have a definite role in the storage of wild accessions. The breeding of most small fruit species is increasingly dependent on the broadening of the genetic base as many of the important small fruit crops have arisen from a narrow genetic foundation. Recent molecular studies by Lanham *et al.* (1995) in *Ribes*, and Graham *et al.* (1996) in *Fragaria*, have substantiated the evidence for the narrow origins, and indicated that enhanced genetic diversity in the relevant breeding programmes may be considered essential for the introduction of certain traits of agronomic interest.

On a practical note, it must be recognized that most small fruits are regarded as minor crops, and the level of resources to fund research or maintain germplasm collections are thus correspondingly lower.

11.1.1 Small fruit germplasm collections

In the conservation of plant genetic resources, the past 30 years have seen a considerable increase in the number of collections and accessions in *ex situ* storage centres world-wide (Villalobos and Engelmann, 1995). Collections of small fruit are highly varied in their scope and nature, often determined by purpose and type of use for which the collection is designed. Many of the oldest collections are part of local or national botanic gardens, concentrating on the species comprising the fruit genus. Often, such collections are backed up by significant systematic research expertise, including extensive herbaria in many instances. In part, this can be linked to an increasing need to respond to loss of habitat caused by human activity, especially the encroachment of agriculture into native habitats. In *Ribes*, for example, various north American species are given endangered status, including *R. divaricatum* Dougl. var. *parishii* (A. Heller) Jepson. This species was found in the lower regions of southern California, and may already be extinct.

Other collections are focused more on the results of plant breeding, in terms of the cultivars produced over the past century or longer, to maintain the genetic base across breeding programmes and make them available both to breeders and other researchers, as well as other enthusiasts; one such example is the UK National Fruit Collection, at Wye in England.

Some of the most valuable collections from a genetic basis are those maintained in support of active breeding programmes. These comprise aspects of the previous examples, containing some species material and old cultivars from various geographical locations. However, they also contain advanced breeding selections which are unlikely to be duplicated elsewhere; as these genotypes often represent the key points in ongoing breeding development, they are especially valuable. Such collections are maintained at the Scottish Crop Research Institute in Dundee, Scotland and most other breeding centres for small fruits. In recent years, the number of small fruit breeding programmes has decreased, and one major disadvantage of this type of germplasm collection is that it is often lost when the programme is discontinued (Jennings *et al.*, 1990).

11.2   Small fruit breeding

The breeding of most small fruit genera is carried out in temperate areas of the globe, with the exception of strawberry (*Fragaria* spp.), where the range extends into some more tropical regions. The main centres for the breeding of *Rubus* spp. lie within northern Europe, North America and New Zealand, while *Ribes* spp. are bred almost entirely within northern Europe. Breeding programmes for blueberries and cranberries (*Vaccinium* spp.) are mainly in North America. Most programmes are supported either by government or, increasingly, by the growing and processing industries downstream of the breeding process.

For the breeding of most of these crops, the use of diverse germplasm is increasing, as breeders search for new sources of pest and disease resistance, fruit quality and tolerance of abiotic stresses. This is increasing the amount of interspecific crossing undertaken by small fruit breeders, but it must be recognized that this approach can lengthen the timespan of the breeding process to up to 30 years for the introgression of useful traits from species material (Jennings *et al.*, 1990).

There is a need to incorporate germplasm from different environments, especially in order to increase the genetic variability available to breeders in the future (Park, 1994). To do this, specific ecotypes need to be collected and conserved. Early studies by Clausen *et al.* (1940) revealed a link between environment and intraspecific variation, and in fruit species similar ecophysiological variation has been demonstrated, at both the morphological and molecular level, in *Fragaria* (Harrison *et al.*, 1997), *Vaccinium* (Woodward, 1986), *Rubus* (Graham *et al.*, 1997), and *Ribes* (Lanham *et al.*, 1995). Local ecotypes represent valuable genetic resources that are often highly sensitive to disruption, making their conservation vital for future breeding potential.

11.3   Problems related to field-based collections

The maintenance of small fruit germplasm collections at present is usually in the form of a field genebank, or clonal repository. This has the advantage of making the material readily accessible to breeders and researchers, but there are nevertheless problems to this approach. Firstly, the cost of field collections is high, with significant land, cultural and labour costs. More significantly, in terms of the conservation of the genetic resources within the collection, field genebanks are highly vulnerable to attacks by pests and diseases, some of which, such as gall mite (*Cecidophyopsis ribis*) on *Ribes* and *Phytophthora fragariae* var. *rubi* on *Rubus*, can irrevocably damage the accessions. The necessity for ensuring virus-free base collections may involve access to virus-testing facilities (see Martin and Postman, Chapter 5, this volume) and undertaking virus-indexing of small fruit can be lengthy and expensive.

A further limitation of field genebanks is the relatively low number of genotypes that can be conserved, compared with seed banks or cryo-banks (see Reed, Chapter 10, this volume). The most likely way forward at the present time is for a base collection to be maintained *in vitro*, and active collections grown in the field for exploitation by breeders and seed collections maintained to cover wild accessions. The base collection needs to be a representative sample based on ecogeographic origin and specific characteristics (Williams, 1991).

11.4 World-wide small fruit germplasm collections

The collections of genetic resources of small fruits world-wide are listed in Table 11.1. Many of the collections are linked to breeding programmes, but virtually all are extensively used by breeders. Some examples of national germplasm conservation systems are given below.

The US National Plant Germplasm System (NPGS) originated in 1943, and in 1990 legislation was passed to establish a National Genetic Resources Programme, to acquire, characterize, preserve, document and distribute germplasm to the scientific community. Base collections, including fruit, are maintained at Fort Collins, Colorado, and most of

Table 11.1 World-wide location of small fruit germplasm collections (from Bettancourt and Konopka, 1989)

Genus	Location
Major genera	
Fragaria	Australia, Austria, Belgium, China, Czech R., Denmark, France, Germany, India, Ireland, Italy, Poland, Norway, Russian Fed., S. Africa, Sweden, Switzerland, Turkey, UK, USA
Rubus	Australia, Belgium, Canada, China, Czech R., Finland, France, India, Italy, Netherlands, Norway, Poland, Russian Fed., S. Africa, Sweden, Switzerland, UK, USA
Ribes	Australia, Belgium, Canada, China, Czech R., Denmark, France, India, Netherlands, Norway, Poland, Russian Fed., Sweden, UK, USA
Vaccinium	Australia, Belgium, Poland, S. Africa, Sweden, UK, USA
Other minor genera	
Amelanchier	Belgium, Canada, Poland, UK, USA
Hippophae	Finland, Russian Fed., Sweden, UK
Sambucus	Belgium, Canada, Norway, Poland, Switzerland, UK, USA
Rosa	Belgium, Canada, China

the field collections of small fruit species, together with seed of wild accessions are kept at the National Clonal Germplasm Repository in Corvallis, Oregon.

Agriculture and Agri-Food Canada (AAFC), which has the main mandate for plant germplasm conservation, operates a seed genebank in Ottawa, which stores and documents accessions of value to Canada, and a clonal genebank in Smithfield, which concentrates on the preservation of tree and small fruits (Campbell and Fraleigh, 1995).

In Europe, the Nordic Gene Bank, based at 22 sites in Sweden, Norway, Finland, Denmark and Iceland, contains a large collection of species, commercial cultivars and many wild accessions. Other important collections are maintained at breeding centres, such as those at INRA (France) SCRI (Scotland), and Horticulture Research International (East Malling) and Wye College in England.

11.5   *In vitro* conservation and cryopreservation

The use of *in vitro* techniques for germplasm conservation (see Lynch, Chapter 4, and Benson, Chapter 6, this volume) is becoming increasingly important, particularly for conservation strategies for specific genotypes, or where alternative means (e.g. seed storage) are not appropriate (Blakesley *et al.*, 1996). The clonal nature regarding the propagation systems employed and the growing of most commercially and scientifically important fruit genera render them especially suitable to *in vitro* approaches. These include conventional tissue culture, restricted growth techniques and cryopreservation (see Reed, Chapter 10, this volume). There are increasing levels of research on the utilization of these techniques for woody plants, and furthermore, techniques for the genetic transformation of a number of such species are being rapidly developed (Oliveira *et al.*, 1996) enabling the introgression of characters from outwith the limits of sexual compatibility.

11.5.1   *Tissue culture*

Storage of plant germplasm may include, as a relatively short-term measure, its simple introduction into axenic culture. This facilitates a non-seasonal, easily maintained and transportable source of clonal plant material. However, for medium-term (more than six months) storage the use of growth inhibitors of a plant growth regulator or osmotic nature may be required for certain species, to alleviate the economic and technical problems of repeated subculturing. Such 'slow-growth' systems may, however, result in the generation of variable phenotypes. For example, the effect of using mannitol as an osmotic inhibitor often results in deleterious side-effects (Bhat and Chandel, 1993). Such symptoms include stunted shoots, reduced internodes and leaves and a propensity towards a hyperhydritic appearance. There is also evidence that such osmoticum-based storage systems affect cellular and molecular processes, notably DNA methylation patterns (Harding, 1994) and amino acid accumulation (Galiba *et al.*, 1992). The long-term effect of maintaining such material for extended periods has been relatively poorly investigated. In many cases however, the transfer of material maintained on such medium onto normal growth medium results in a phenotypically normal status being reinstated, though this may take a small number of subculturing cycles.

11.5.2 *Cold storage of tissue cultures*

A related *in vitro* approach involves using cold-storage; for example, in the case of strawberries, temperatures of 2 and 5°C, and light or dark growth conditions were employed (Jeong *et al.*, 1996). However, as is common with many cell and tissue culture investigations, significant varietal differences were recorded in this study. Three *in vitro* cold storage systems of strawberries were also compared by Reed (1992). Apple germplasm maintained *in vitro* was also successfully stored in slow growth medium at 4°C (Negri *et al.*, 1995). However, cold-storage at temperatures of 2–10°C offer only a partial solution, with related technical problems.

11.5.3 *Cryopreservation*

Fruit crops have been successfully adapted to a range of cryopreservation systems (see Benson, Chapter 6, this volume). For example, the effects of dehydration and exogenous growth regulators on dormancy quiescence and germination of cryopreserved grape somatic embryos were investigated by Gray (1989), and the cryopreservation of stem segments of kiwi fruit was also reported by Jian and Sun (1989). Initial methods were refined in a report on cryopreservation of dormant buds from apple, using a modified vitrification procedure (Seufferheld *et al.*, 1991). The viability of banana meristem cultures following cryopreservation in liquid nitrogen was cited as up to 42 per cent (Panis *et al.*, 1996), and the development of methods for successful cryopreservation of jackfruit and litchi were described by Chaudhury *et al.* (1996).

For small fruits, the cryopreservation of *Rubus* floral buds in liquid nitrogen was reported by Gu *et al.* (1990) and the same species was investigated by Reed (1990) in a study on cold hardening versus abscisic acid as a pretreatment for meristem cryopreservation. Reed and colleagues (Luo and Reed, 1997; Reed and Yu, 1995) have performed several studies on the cryopreservation of *Ribes* shoot-tip germplasm, including one approach which involved the application of proteins to enhance post-cryopreservation survival (Luo and Reed, 1997). Genotype differences in responses can however occur; thus shoot tips of *Ribes nigrum* cultivars 'Ben Tron' and 'Ben More' were found to be differentially capable of surviving and regenerating new shoots following cryopreservation, and recovery responses were also dependent upon the vitrification and encapsulation–dehydration methodology employed (Benson *et al.*, 1996) (see Figure 11.1). Further developments in this particular species are expected in the near future following ongoing research at the University of Abertay Dundee in collaboration with SCRI and Dr Barbara Reed, at the USDA National Clonal Germplasm Repository, Oregon. Cryopreservation of *in vitro* grown meristems of strawberry (*Fragaria* × *ananassa* Duch.) by encapsulation–vitrification was reported by Hirai *et al.* (1998). In their work, alginate-coated meristems from four cultivars of *in vitro* grown strawberry were successfully cryopreserved following dehydration by a vitrification solution. Successfully encapsulated vitrified meristems remained green and then developed shoots within one week after plating, without intermediary callus formation. The average rate of shoot formation of encapsulated vitrified meristems amounted to nearly 90 per cent. It was also reported that encapsulated vitrified meristems cooled to −196°C produced higher shoot formation than encapsulated dried meristems, and the authors stated that the encapsulation–vitrification method was relatively easy to handle and produced high levels of shoot formation. Cryopreservation of

Figure 11.1 Growing meristem of *Ribes nigrum* cv. Ben More following cryopreservation using an encapsulation–dehydration protocol (see Benson, Chapter 6, this volume and Benson *et al.*, 1996)

strawberry cell suspension cultures was also reported by Yongjie *et al.* (1997). Cryo-preservation is, in general, an accessible technology, and can be applied to both pilot scale and substantial collections of material.

11.6   The use of pollen storage for small fruit germplasm conservation

The storage of pollen is practised routinely by plant breeders, as a means of enabling the hybridization of non-synchronous genotypes, as well as providing material for germplasm exchanges (Barnabás and Kovács, 1997). However, there is increasing interest in the development of suitable protocols for the preservation of genetic resources of various species, including fruit. For most woody perennial fruit crops, the predominant means of germplasm storage has been as clonal material, but the use of pollen, particularly in terms of the preservation of species and ecotypic material, has great potential.

Pollen of most berry fruits of the Saxifragaceae and Rosaceae is relatively long-lived (Harrington, 1970), remaining viable for six months to one year in normal circumstances. Loss of viability is strongly affected by environmental factors, especially temperature and humidity, and viability is best retained in low humidity conditions (Holman and Brubaker, 1926). The most suitable means of short-term storage is therefore in unsealed glass containers placed in dessicators containing dried silica gel.

Longer term storage requires temperatures of less than 0°C, to prevent the loss of viability. This falls into two categories, namely lyophilization and cryopreservation. Lyophilization has proved effective for storing pollen of various species (King, 1965), and can be divided into: freeze-drying, where pollen is rapidly frozen to ca. −60 to −80°C and water is gradually removed thereafter; vacuum drying, where pollen is simultaneously subjected to vacuum; and cooling and evaporative drying.

Cryopreservation of pollen in liquid nitrogen has great potential for the maintenance of genetic resources (Bajaj, 1987), with the successful long-term storage of pollen of several fruit genera reported, including *Vitis vinifera* for five years (Ganeshan and Alexander, 1988). It is to be hoped and expected that protocols for other species will be developed in the near future.

11.7 Conservation of transgenic plant small fruit plant germplasm

Since the first report of the creation of a transgenic plant in 1983 the number of plant species successfully transformed now totals over 120, with small fruits represented by blueberry, cranberry, strawberry, raspberry and blackcurrant to date. As such, the number of transgenic germplasm collections is set to rapidly rise. In the light of this, there has been a perceived need to address the area of long-term maintenance of such material and to identify potential problems. Identified areas of concern include the need for containment of transgenic pollen to avoid release and possible introgression into the wild. The guidelines involving the handling and release of transgenic material vary considerably from state to state.

Assessments of the risks associated with the long-term maintenance are related to the guidelines concerning the work and handling of transgenic material set out by National authorities (in the UK, the Advisory Committee on Genetic Manipulation). Key features of the granting of a licence to undertake work with transgenics are the areas of documentation and validation. Regarding documentation, as much detail as possible concerning constructs and sites of insertion needs to be recorded, with details of selectable markers used for the transformation. In as much as it is known that certain tissue culture procedures may induce genetic variation, the effect of gene transfer methods could be reasonably concluded to be increasingly prone to variation. This area of research requires detailed investigation on a case by case basis.

11.8 Genetic stability aspects

The potential use of molecular genetic techniques for conservation of crop plants and wild relatives has been widely recognized (Hodgkin and Debouck, 1991) as biochemical markers, though easily assayed can be affected by environment and may show poor levels of polymorphism. Molecular genetic techniques (see also Harris, Chapter 2, this volume) have made a significant impact on plant genetic resources, conservation and use (Hodgkin, 1995). Though initial studies focused on the analysis of specific genes, increasing sophistication and methods of discrimination have allowed studies of phylogeny and species evolution to be undertaken. Molecular techniques for the analysis, characterization and conservation of plant genetic resources with particular application to extent and distribution of genetic diversity were reviewed by Karp and Edwards (1995). Analysis of variation in plastid DNA also permits analysis of the maternal and paternal contributions in evolutionary strides.

Variation arising from plant tissue culture is a topic of great debate. Material derived from shoot tips, maintained on media lacking growth regulators and not subject to osmotic stress will generally be genetically stable. However, the prolonged maintenance of shoots in culture may increase the frequency of spontaneous somatic mutations and much of this variation may be of a cryptic nature (Cassells *et al.*, 1987). The topic of genetic instability of regenerated and transgenic plants was reviewed by Karp (1993) but with the development of more discriminating molecular assay systems in recent years further elucidation is expected (see Harding, Chapter 7, this volume).

11.9   DNA banking

The storage of individual DNA samples in appropriate form, with 'identifiers' for later retrieval, has enormous potential for a number of studies. This technology is generically known as DNA banking, and has foreseeable applications in the conservation of small fruit germplasm. Probably the leading exponent in the area of botanical DNA banking is the Missouri Botanic Garden which holds material in its DNA bank to support studies of plant relationships. More details are available via the WWW at http://www.mobot.org. The subject was also reviewed by Adams (1997).

11.10   Conclusions

The maintenance and expansion of germplasm collections of small fruit, and in particular the integration of the various collections already in existence, is vital to the future development of improved cultivars for commercial fruit production. Further additions of wild germplasm into collections and more widespread use of *in vitro* techniques for base collection maintenance should be adopted. The increased diversification of horticulture will result in an expanded demand for novel plant material, and the increasingly important technologies of functional genomics and gene transfer offer wide scope for the enlarged use of small fruit conservation programmes.

Acknowledgements

The authors thank the Scottish Office Agriculture, Environment and Fisheries Department for continuing financial support.

References

ADAMS, P.R., 1997, Conservation of DNA:DNA banking, in CALLOW, J.A., FORD-LLOYD, B.V. and NEWBURY, H.J. (Eds), *Biotechnology in Agriculture Series, 19, Biotechnology and Plant Genetic Resources: Conservation and Use*, pp. 163–174, Wallingford: CAB International.

BAJAJ, Y.P.S., 1987, Cryopreservation of pollen and pollen embryos, and the establishment of pollen banks, *International Review of Cytology*, **107**, 397–420.

BARNABÁS, B. and KOVÁCS, G., 1997, Storage of pollen, in SHIVANNA, K.R. and SAWNHEY, V.K. (Eds), *Pollen Biotechnology for Crop Production and Improvement*, pp. 293–314, Cambridge: Cambridge University Press.

BENSON, E.E., REED, B.M., BRENNAN, R.M., CLACHER, K.A. and ROSS, D.A., 1996, Use of thermal analysis in the evaluation of cryopreservation protocols for *Ribes nigrum* L. germplasm, *Cryo-Letters*, **17**, 347–362.

BETTANCOURT, E.J. and KONOPKA, J., 1989, *Temperate Fruits and Tree Nuts*, IPGRI Directory of Germplasm Collections, Rome: IPGRI.

BHAT, S.R. and CHANDEL, K.P.S., 1993, *In vitro* conservation of *Musa* germplasm: effects of mannitol and temperature on growth and storage, *Journal of Horticultural Science*, **69**, 841–846.

BLAKESLEY, D., PASK, N., HENSHAW, G.G. and FAY, M.F., 1996, Biotechnology and the conservation of forest genetic resources: *in vitro* strategies and cryopreservation, *Plant Growth Regulation*, **20**, 11–16.

CAMPBELL, K.W. and FRALEIGH, B., 1995, The Canadian Plant Germplasm System, *Canadian Journal of Plant Science*, **75**, 5–7.

CASSELLS, A.C., AUSTIN, S. and GOETZ, E.M., 1987, Variation in tubers in single-cell derived clones of potato in Ireland, in BAJAJ, Y.P.S. (Ed.), *Biotechnology in Agriculture and Forestry, Volume 3, Potato*, pp. 375–391, Berlin: Springer-Verlag.

CHAUDHURY, R., CHANDEL, K.P.S. and MALIK, S.K., 1996, Development of cryopreservation techniques for conservation of jackfruit, lichi and cocoa, *Cryobiology*, **33**, 653–654.

CLAUSEN, J., KECK, D.D. and HIESEY, W.M., 1940, Experimental studies on the nature of species. I. Effect of varied environment on western North American Plants, *Carnegie Institution of Washington*, Publ. No. 520.

GALIBA, G., SIMON-SARKADI, L., KOCSY, G., SALGO, A. and SUTKA, J., 1992, Possible chromosomal location of genes determining the osmoregulation of wheat, *Theoretical and Applied Genetics*, **85**, 415–418.

GANESHAN, S. and ALEXANDER, M.P., 1988, Fertilizing ability of cryopreserved grape (*Vitis vinifera* L.) pollen, *Genome*, **30**, Suppl. 1, 464.

GRAHAM, J., MCNICOL, R.J. and MCNICOL, J.W., 1996, A comparison of the methods for the estimation of genetic diversity in strawberry cultivars, *Theoretical and Applied Genetics*, **93**, 402–406.

GRAHAM, J., SQUIRE, G.R., MARSHALL, B. and HARRISON, R.E., 1997, Spatially dependent genetic diversity within and between colonies of wild *Rubus idaeus* detected using RAPD markers, *Molecular Ecology*, **6**, 1001–1008.

GRAY, D.J., 1989, Effects of dehydration and exogenous growth regulators on dormancy quiescence and germination of grape somatic embryos, *In Vitro Cell Developmental Biology*, **25**, 1173–1178.

GU, J., WARMUND, M. and GEORGE, M., 1990, Cryopreservation of *Rubus* floral buds in liquid nitrogen, *HortScience*, **25**, 1083.

HARDING, K., 1994, The methylation status of DNA derived from potato plants recovered from slow growth, *Plant Cell, Tissue and Organ Culture*, **37**, 31–38.

HARRINGTON, J.F., 1970, Seed and pollen storage for conservation of plant gene resources, in FRANKEL, O.H. and BENNET, E. (Eds), *Genetic Resources in Plants: Their Exploration and Conservation*, pp. 501–521, Oxford and Edinburgh: Blackwell.

HARRISON, R.E., LUBY, J.J., FURNIER, G.R. and HANCOCK, J.F., 1997, Morphological and molecular variation among populations of octoploid *Fragaria virginiana* and *F. chiloensis* (Rosaceae) from North America, *American Journal of Botany*, **84**, 612–620.

HIRAI, D., SHIRAI, K., SHIRAI, S. and SAKAI, A., 1998, Cryopreservation of *in vitro*-grown meristems of strawberry (*Fragaria x ananassa* Duch.) by encapsulation-vitrification, *Euphytica*, **101**, 108–115.

HODGKIN, T. and DEBOUCK, G.D., 1991, Some possible applications of molecular genetics in the conservation of wild species for crop improvement, in ADAMS, R.P. and ADAMS, J.E. (Eds), *Conservation of Plant Genes: DNA Banking and In Vitro Biotechnology*, pp. 153–182, San Diego: Academic Press .

HOLMAN, R.M. and BRUBAKER, F., 1926, On the longevity of pollen, *University of California Publications of Botany*, **13**, 179–204.

JENNINGS, D.L., DAUBENY, H.A. and MOORE, J.N., 1990, Blackberries and raspberries (*Rubus*), in MOORE, J.M and BALLINGTON, J.R. (Eds), *Genetic Resources of Temperate Fruit and Nut Crops*, pp. 329–390, Wageningen: International Society of Horticultural Science.

JEONG, H.-B., HA, S.-H. and KANG, K.-Y., 1996, *In vitro* preservation method of culturing shoot tip at low temperature in strawberry, *Journal of Agricultural Science and Biotechnology*, **38**, 284–289.

JIAN, L.-C. and SUN, L.-H., 1989, Cryopreservation of the stem segments of kiwi fruit, *Acta Botanica Sinica*, **31**, 66–68.

KARP, A.,1993, Are your plants normal? Genetic instability in regenerated and transgenic plants, *Agro Food Industry Hi-Tech*, **5**, 7–12.

KARP, A. and EDWARDS, K.J., 1995, Molecular techniques in the analysis of the extent and distribution of genetic diversity, in AYAD, W.G., HODGKIN, T., JARADAT, A. and RAO, V.R. (Eds), *Molecular Genetic Techniques for Plant Genetic Resources, Report of IPGRI Workshop*, pp. 11–22, Rome: IPGRI.

KING, J.R., 1965, The storage of pollen particularly by the freeze-drying method, *Bulletin of the Torrey Botanical Club*, **92**, 270–288.

LANHAM, P.G., BRENNAN, R.M., HACKETT, C. and MCNICOL, R.J., 1995, RAPD fingerprinting of blackcurrant (*Ribes nigrum* L.) cultivars, *Theoretical and Applied Genetics*, **90**, 166–172.

LUO, J. and REED, B.M., 1997, Abscisic acid-responsive protein, bovine serum albumin and proline pretreatments improve recovery of *in vitro* currant shoot-tips meristems and callus by vitrification, *Cryobiology*, **34**, 240–250.

NEGRI, V., HAMMAD, A.H.A. and STANDARDI, A., 1995, A space-saving storage technique for *in vitro* cultures of apple germplasm, *Journal of Genetics and Breeding*, **49**, 127–131.

OLIVEIRA, M.M., MIGUEL, C.M. and RAQUEL, M.H., 1996, Transformation studies in woody fruit species, *Plant Tissue Culture and Biotechnology*, **2**, 76–93.

PANIS, B., TOTTE, N., VAN NIMMEN, K., WITHERS, L.A. and SWENNEN, R., 1996, Cryopreservation of banana (*Musa* spp.) meristem culture after preculture on sucrose, *Plant Science*, **121**, 95–106.

PARK, Y.G., 1994, Strategy for biodiversity and genetic conservation of forest resources in Korea, *Journal of the Korean Forestry Society*, **83**, 191–204.

REED, B.M., 1990, Cold hardening versus ABA as a pretreatment for meristem cryopreservation, *Hortscience*, **25**, 1086.

REED, B.M., 1992, Cold storage of strawberries *in-vitro*: a comparison of three storage systems, *Fruit Varieties Journal*, **46**, 98–102.

REED, B.M. and YU, X., 1995, Cryopreservation of *in vitro*-grown gooseberry and currant meristems, *Cryo-Letters*, **16**, 131–136.

SEUFFERHELD, M.J., STUSHNOFF, C., FITZPATRICK, J. and FORSLINE, P., 1991, Cryopreservation of dormant buds from cold-tender taxa using a modified vitrification procedure, *HortScience*, **27**, 1262.

SIMMONDS, N.W., 1976, *Evolution of Crop Plants*, London: Longman.

VILLALOBOS, V.M. and ENGELMANN, F., 1995, *Ex situ* conservation of plant germplasm using biotechnology, *World Journal of Microbiology and Biotechnology*, **11**, 375–382.

WILLIAMS, J.T., 1991, Plant genetic resources: some new directions, *Advances in Agronomy*, **45**, 61–91.

WOODWARD, F.I., 1986, Ecophysiological studies on the shrub *Vaccinium myrtillis* L. taken from a wide altitudinal range, *Oecologia*, **70**, 580–586.

YONGJIE, W., ENGELMANN, F., FRATTARELLI, A., DAMIANO, C. and WITHERS, L.A., 1997, Cryopreservation of strawberry cell suspension cultures, *Cryo-Letters*, **18**, 317–324.

12

Biotechnological Advances in the Conservation of Root and Tuber Crops

ALI M. GOLMIRZAIE, ANA PANTA AND JUDITH TOLEDO

12.1   Introduction

The Andean region shelters a wide variety of root and tuber species that, over hundreds of years, have developed a broad diversity. The varied ecosystems of this region and cultural inheritance of conservation from the Inca civilization, have enhanced this biodiversity. In recent years, however, these crops have been grown in environments that are increasingly undergoing changes in farming and land use, and changes caused by rural development. Such changes placed Andean root and tuber species at risk, and exposed them to genetic erosion. For this reason, in 1971 the International Potato Centre (CIP) assumed a mandate to safeguard these genetic resources. In the early years, CIP scientists and researchers worked on potato, then on sweet potato. In recent years the centre has assumed the responsibility of protecting nine other Andean root and tuber species. CIP is working on the establishment of a complete genetic pool for each species, through germplasm collections. Cultivated and wild species are maintained in the field, *in vitro*, and as botanical seed, including those of potato (*Solanum tuberosum*), sweet potato (*Ipomoea batatas)*, ulluco (*Ullucus tuberosus*), oca (*Oxalis tuberosa*), mashua (*Tropaeolum tuberosum*), yacon (*Polymnia sonchifolia*), arracacha (*Arracacia xanthorhiza*), maca (*Lepidium meyenii*), achira (*Canna edulis*), mauka (*Mirabilis expansa*) and ajipa (*Pachyrhyzus ahipa*). In this way biodiversity is protected against dangerous agents and at the same time an extensive collection of germplasm is available as a source of desirable traits for breeding.

At CIP, *in vitro* collections are grouped by species and utilization. Each crop is divided into three collections: world, pathogen tested, and research (Table 12.1).

The world collection contains wild and native cultivars that are under continual evaluation to identify desirable genes. The pathogen tested collection contains selected accessions that have been cleaned up and evaluated under rigorous procedures for virus detection and genetic stability (see Martin and Postman, Chapter 5, this volume). This material is available for germplasm distribution. The research collection contains accessions of advanced material and breeding lines.

CIP maintains, cleans, and distributes germplasm from its *in vitro* collections. The centre employs a wide range of tissue culture methods to preserve this germplasm (Golmirzaie

Table 12.1 Number of *in vitro* accessions in genebank at CIP

Crop	Species	Collection	Accession number
Potato	*Solanum tuberosum*	World	4218
		Pathogen tested	938
		Research	2247
		Total	7403
Sweet potato	*Ipomoea batatas*	World	4996
		Pathogen tested	414
		Research	7
		Total	5417
Oca	*Oxalis tuberosa*	World	476
Ulluco	*Ullucus tuberosus*	World	439
Mashua	*Tropaelum tuberosum*	World	48
Yacon	*Polymnia sonchifolia*	World	38
Arracacha	*Arracacia xanthorhiza*	World	9
Achira	*Canna edulis*	World	36

and Panta, 1997b). Tissue culture techniques have advanced considerably in recent years, and methods such as cryopreservation provide an alternative to continuous plant culture. Other techniques, such as medium- and long-term storage are routinely applied for all CIP mandate crops.

For many years, CIP technologies in tissue culture (see Lynch, Chapter 4, this volume) have been applied to *in vitro* maintenance of germplasm. Since 1975 (Roca, 1975) *in vitro* potato collections and, later sweet potato (Sigueñas, 1987) have been maintained by clonal propagation of nodes, securing genetic stability, using many propagation methods over time. However, some of these techniques are being improved (slow growth in sweet potato), and the establishment of *in vitro* techniques for new crops (Andean tuber and root crops) is being undertaken by using *in vitro* tuberization for germplasm maintenance.

The following sections will describe tissue culture technologies for germplasm maintenance of potato, sweet potato, and other Andean root and tuber crops. The principles of these techniques have been presented by Lynch (Chapter 4, this volume).

12.2   Establishment of aseptic cultures

The establishment of aseptic cultures is feasible in vegetatively propagated crops, by the use of actively growing buds or nodes. Explants are chosen from young plants growing in the greenhouse, or from tuber sprouts (in tuberous plants). Plants received for *in vitro* introduction must be healthy and strong and devoid of symptoms induced by fungi, bacteria, viruses or viroids. Plant tissues are disinfected using various sterilizing agents, with high efficiency for large collections (Espinoza *et al.*, 1992; Lizarraga *et al.*,

1990; Espinoza *et al.*, 1992). However viroids, viruses or bacteria produce systemic infections that require special treatment.

12.2.1 *Infection produced by viroids*

Viroids are the most infective group of pathogens and are easily disseminated during *in vitro* manipulation. There are *in vitro* procedures to eliminate viroids from plants (Lizarraga, 1980). However, handling infected plants is considered a risk, and it is necessary to incinerate them (Singh *et al.*, 1989), thereby avoiding their introduction into the laboratory.

For safety (and before they are included into CIP germplasm collections) *in vitro* plants obtained from other laboratories are evaluated by NASH (nucleic acid spot hybridization test; Salazar and Querci, 1992) for viroid detection.

12.2.2 *Infection produced by viruses*

Vegetatively propagated plants are frequently infected with viral diseases. The introduction of large numbers of accessions from field collections makes it difficult to eradicate viruses in germplasm collections. This process is time consuming, expensive, and difficult to handle in large collections. The aseptic procedures of micropropagation prevent virus contamination between accessions, maintaining non-cleaned accessions without any risk. Clean-up procedures have been developed with a high level of technique efficiency for the entire collection of potato and sweet potato. Three meristems per accession are used to obtain at least one clean plant. This technique is performed as part of CIP's long-term germplasm conservation activities.

The international distribution of potato and sweet potato germplasm requires use of clonal material that is free from disease. A combination of thermotherapy of plants and meristem culture is applied to obtain pathogen-free material which can be distributed worldwide (see also Martin and Postman, Chapter 5, this volume).

The *in vitro* plantlets are propagated in magenta vessels and exposed to high temperatures (32–34°C for potato and 35–37°C for sweet potato); after one month, meristem tips are excised and cultured to obtain plants free of pathogens. This procedure permits a 99 per cent cleaning efficiency rate in potato plants (Lizarraga *et al.*, 1991) and a 90 per cent cleaning efficiency rate in sweet potato (Lizarraga *et al.*, 1990). The plants obtained are evaluated once again under rigorous procedures for virus detection using ELISA tests for detection of known viruses and indexing by indicator plants for detection of unknown viruses.

12.2.3 *Infection produced by systemic bacteria*

Lack of disease symptoms in a plant does not necessarily mean that there is an absence of pathogen infection. In some species, like sweet potato, bacteria (*Bacillus circulans, Pseudomona cepacia* and *Erwinia herbicola*) produce systemic infections without ever showing symptoms in field plants. Procedures developed at CIP (Egusquiza, 1996; Toledo *et al.*, 1998a) use specific antibiotics (200 mg/l of streptomycin and penicillin) that are introduced into culture media by filtration. Infected nodes grow in the antibiotic medium. After one month, apical buds are extracted from the plants that are healthy.

12.3 *In vitro* maintenance

12.3.1 *Short-term storage*

The choice of a particular medium for *in vitro* establishment, depends on the plant species. In general, the medium should contain minerals (Murashige and Skoog, 1962), a carbon source, vitamins, and low concentration of growth regulators (Table 12.2) (see also Lynch, Chapter 4, this volume). In this medium, plants can grow in *in vitro* conditions for 2–3 months 'short-term storage'. Environmental conditions for *in vitro* plantlet growth are simulated to be similar to field environmental parameters. These are 18–22°C for potato and Andean root and tuber crops (ARTCs) and 23–25°C for sweet potato. Additionally, 3000 lux and 16 light hours are applied for all species (Espinoza *et al.*, 1992; Lizarraga *et al.*, 1990; Toledo *et al.*, 1994).

12.3.2 *Long-term storage*

In vitro collections maintained under short-term storage require a large amount of manpower. The maintenance of *in vitro* plants for long periods or 'long-term storage', is possible by reducing the growth rate through changing the environmental conditions, and modifying some media components. Thus, slow growth can be achieved by temperature reduction, light intensity reduction, using growth regulators, limiting mineral supply, adding osmotic stress agents, or by combining any of these procedures. Details of these growth limiting procedures are as follows:

- *Temperature reduction*: To minimize the growth rate of species, room temperatures can be reduced to near zero; for tropical crops a moderate reduction is applicable (George and Sherrington, 1993).

- *Light intensity*: In vitro plants use sugar as a carbon source through heterotrophic absorption in the culture vessels. These plants still maintain their photosynthetic ability however, using CO_2 as a carbon source. By using low light intensity, carbon support autotrophically obtained is reduced, which results in delayed growth (Hughes, 1981).

- *Growth regulators*: Applying abscisic acid (ABA) to media reduces the overall growth rate of oca plantlets; however, some symptoms of glassiness (vitrification) in leaves occur during periods of long maintenance, which must therefore be avoided. Although this method is used for several crops, mutations can appear which can threaten germplasm genetic stability (Wescott, 1981).

- *Osmotic stress agents*: The inclusion of sugar in media increases the osmotic potential, thus reducing the uptake of minerals by cells. As a consequence, plant growth is delayed. Mannitol and sorbitol are sugar alcohols and are used extensively in germplasm conservation. Osmotic agents may only enter the cell slowly and produce osmotic effects without being metabolized (Dodds and Roberts, 1985).

Long-term storage is the most important goal in germplasm conservation and the efficiency of a technique depends on the periods between subcultures. Thus, the cultivar must be maintained with these methods for more than two years.

In vitro conservation of potato, sweet potato, and some Andean root and tuber crops, is maintained by using a combination of osmotic stress agents, low temperatures, and low light intensities. Most potato accessions can be maintained for up to four years, without the need to subculture (Espinoza *et al.*, 1992; Golmirzaie and Toledo, 1998; Toledo

Table 12.2 Media and incubation conditions of genebank crops maintained *in vitro* at CIP

Crop	Culture media and incubation conditions
Potato	*Introduction* MS[a] + 2.5% sucrose + stock 3[d] + 0.5 mg/l gibberellic acid + 0.35% phytagel, 18–22°C, 3000 lux and 16 hours light *Propagation* MS[a] + 2.5% sucrose + stock 1[b] + 0.25 mg/l gibberellic acid + 0.35% phytagel, 18–22°C, 3000 lux and 16 hours light *Conservation* MS[a] + 2.5% sucrose + stock 3[d] + 4% sorbitol + 2% sucrose + 0.35% phytagel, 6–8°C, 1000 lux and 16 hours light
Sweet potato	*Introduction and propagation* MS[a] + 3% sucrose + stock 2[c] + 0.35% phytagel, 23–26°C, 3000 lux and 16 hours light *Conservation* MS[a] + 2% sorbitol + stock 3[d] + 2 mg/l spermidine + 2% dextrose + 0.35% phytagel, 18–21°C, 2000 lux and 16 hours light
Oca	*Introduction* MS[a] + 3% sucrose + stock 3[d] + 2 mg/l calcium-d-pantothenate + 1 mg/l gibberellic acid + 1% charcoal activated + 0.7% agar, 18–22°C, 3000 lux and 16 hours light *Propagation* MS[a] + 2% sucrose + stock 3[d] + 2 mg/l calcium-d-pantothenate + 0.25 mg/l gibberellic acid + 10 mg/l putrescine + 0.8% agar, 18–22°C, 3000 lux and 16 hours light *Conservation* MS[a] + 4% sorbitol + 2% sucrose + stock 3[d] + 2 mg/l calcium-d-pantothenate + 0.8% agar, 18–22°C, 3000 lux and 16 hours light
Ulluco	*Introduction* MS[a] + 3% sucrose + stock 3[d] + 2 mg/l calcium-d-pantothenate + 0.25 mg/l gibberellic acid + 0.7% agar, 18–22°C, 3000 lux and 16 hours light *Propagation* MS[a] + 2% sucrose + stock 3[d] + 2 mg/l calcium-d-pantothenate + 0.8% agar, 18–22°C, 3000 lux and 16 hours light *Conservation* MS[a] + 2% sucrose + 3% mannitol + stock 3[d] + 2 mg/l calcium-d-pantothenate + 0.8% agar, 18–22°C, 3000 lux and 16 hours light
Mashua	*Introduction* MS[a] + 3% sucrose + stock 3[d] + 2 mg/l calcium-d-pantothenate + 1 mg/l gibberellic acid + 1% charcoal activated + 0.7% agar, 18–22°C, 3000 lux and 16 hours light *Propagation* MS[a] + 2% sucrose + stock 3[d] + 2 mg/l calcium-d-pantothenate + 0.5 mg/l naphthalene acetic acid + 0.8% agar, 18–22°C, 3000 lux and 16 hours light *Conservation* MS[a] + 3% sorbitol + 2% sucrose + stock 3[d] + 2 mg/l calcium-d-pantothenate + 0.8% agar, 18–22°C, 3000 lux and 16 hours light
Yacon	*Introduction* MS[a] + 3% sucrose + stock 3[d] + 2 mg/l calcium-d-pantothenate + 10 mg/l gibberellic acid + 5% charcoal activated + 0.8% agar, 18–22°C, 3000 lux and 16 hours light *Propagation* MS[a] + 3% sucrose + stock 3[d] + 2 mg/l calcium-d-pantothenate + 0.05 mg/l naphthalene acetic acid + 2 mg/l gibberellic acid + 0.8% agar, 18–22°C, 3000 lux and 16 hours light *Conservation* MS[a] + 4% mannitol + 3% sucrose + stock 3[d] + 2 mg/l calcium-d-pantothenate + 0.8% agar, 18–22°C, 3000 lux and 16 hours light
Achira	*Propagation* MS[a] + 2.5% sucrose + stock 3[d] + 0.1 mg/l naphthalene acetic acid + 0.8 mg/l benzylaminopurine + 0.7% agar, 18–22°C, 3000 lux and 16 hours light
Arracacha	*Propagation* MS[a] + 2.5% sucrose + stock 3[d] + 0.1 mg/l naphthalene acetic acid + 1.5 mg/l benzylaminopurine + 0.7% agar, 18–22°C, 3000 lux and 16 hours light

[a] MS: basal medium, Murashige and Skoog (Murashige and Skoog, 1962).
[b] Stock 1: 2 mg/l glycine, 0.5 mg/l nicotinic acid, 0.5 mg/l pyridoxine, 0.4 mg/l thiamine HCl, 0.1 mg/l gibberellic acid.
[c] Stock 2: 0.1 g/l arginine, 0.02 g/l putrescine, 0.01 g/l gibberellic acid, 0.2 g/l ascorbic acid, 0.002 g/l calcium-d-panthotenate.
[d] Stock 3: 2 mg/l glycine, 0.5 mg/l nicotinic acid, 0.5 mg/l pyridoxine, 0.4 mg/l thiamine HCl.

Figure 12.1   Culture tubes containing *in vitro* plantlets of five Andean root and
tuber crops: mashua (*Tropaeolum tuberosum*), ulluco (*Ullucus tuberosus*), oca (*Oxalis
tuberosa*), yacon (*Polymnia sonchifolia*), and arracacha (*Arracacia xanthorhiza*)

and Golmirzaie, 1998). Sweet potato accessions can be maintained for 12 to 18 months
(Cubillas, 1997; Lizarraga *et al.*, 1990). Other crops (see Table 12.2), like ulluco and
mashua, are maintained for 24 months; oca and yacon are maintained for six 6 months
(Borda *et al.*, 1998; Toledo *et al.*, 1994).

12.3.3   In vitro *tuberization for long-term conservation*

Microtubers of potato have been developed at CIP for a wide range of genotypes for the
purpose of international germplasm distribution, seed production, and, as an alternative
for germplasm conservation. Microtubers are produced in 2–3 months and can be stored
at 10°C for 21 months (Kwiatkowski *et al.*, 1988) after harvest. Additionally, tuber
dormancy can be controlled by environmental changes (Estrada *et al.*, 1986; Tovar *et al.*,
1985) or sprout growth can be retarded by storage of sprouted tubers embedded in conservation medium, thus permitting 24 months of conservation.

12.4   Other operational considerations for germplasm maintenance

For germplasm conservation, facilities play an important role. An adequate room and
well-trained personnel that can efficiently handle plant material prevent contamination.
The operational areas should be organized as a restricted zone. This restriction permits a
better control of sanitary conditions, and prevents surface contamination. Additionally,
frequent disinfecting of surfaces and shelves will prevent mite contamination.

Figure 12.2 *In vitro* germplasm room of sweet potato at CIP

Electronic systems allow an efficient control of environmental conditions in storage rooms, which are essential for germplasm maintenance.

Maintenance of large collections involves the use of numbers or names to identify accessions. Databases are essential for monitoring the *in vitro* genebanks; CIP uses several computer programs such as Fox pro for Windows, MS Access queries, or MS SQL-server. They are also helpful for labelling.

12.5 Advances in germplasm utilization

Germplasm from pathogen tested collections is distributed and utilized by scientists worldwide. The utilization of clean material for production has been widely proved, and most potato programmes are currently using this type of genetic resource. *In vitro* laboratory culture is necessary for the seed programme, in order to maintain clean material and propagate large numbers of plantlets.

Since CIP has the responsibility for distributing material, CIP scientists develop training activities related to *in vitro* propagation and virus eradication directed at technicians and professionals from national programmes, NGO, and universities, etc. from developing countries.

Assisting tissue culture laboratories involved in seed programmes has helped to improve seed production. Some media modifications for reducing costs have been suggested. These are: the replacement of agar by maize starch (9 per cent), reducing MS basal salts to half strength, and using common sugar instead of sucrose (Toledo *et al.*, 1998b). These modifications have reduced the costs for *in vitro* plantlets for seed production by more than 70 per cent.

12.6 Cryopreservation

Cryo-conservation must be considered an alternative method for long-term conservation of plant genetic resources. By this method, plant material is frozen at ultra-low temperature ($-70°C$ to $-196°C$) and stored for indefinite periods without genetic erosion. For this reason, cryopreservation is adopted as one of the best alternatives for the long-term conservation of several crops, especially those that are vegetatively propagated. (See also Benson, Chapter 6, this volume, for the principles of cryopreservation methodology.)

Potato cryopreservation of shoot tips (meristems with some leaf primordia) is now a workable technique at CIP (Golmirzaie and Panta, 1997a). Application of this method for sweet potato, cassava, and yam cryopreservation is also being studied at CIP, CIAT, and IITA, respectively (Escobar *et al.*, 1997; Ng and Ng, 1997). In the case of other Andean root and tuber crops, basic research has been carried out only on ulluco (*Ullucus tuberosum*) (Estrada, unpublished observations). At CIP, further experiments will take place on cryopreservation of shoot tips of ulluco by using the vitrification method.

With all cryopreservation methods, success has been obtained using shoot tips or seeds. Shoot tip cryopreservation has an advantage over other tissues, in that they can be regenerated into plants that are faithfully identical to mother plants (Sakai, 1993). Due to the success of the application of this method, a workable protocol for meristem and/or shoot tips culture is needed. In the case of potato, sweet potato, cassava, yam, and ulluco, different protocols for meristem culture have already been developed. This approach constitutes a base for applying cryopreservation to these crops. Additional funds are needed to reinforce this work and produce a common effort within international and national genebanks.

Cryopreservation research on higher plants was performed over 40 years ago. This technology followed the conventional procedures applied to other living materials. Early protocols developed for plants were based on chemical cryoprotection and dehydrative freezing. The new improved methods are based on the vitrification phenomenon. Vitrification is defined as the transition of water directly from the liquid phase into an amorphous phase or glass, while avoiding the formation of crystalline ices (Fahy *et al.*, 1984). (See Benson, Chapter 6, this volume, for details of the principles of cryopreservation.)

12.6.1 *Potato cryopreservation*

Bajaj began potato cryopreservation work in 1977 (see Bajaj, 1987 for a review) using excised meristems, pollen, shoot tips, cell cultures, and protoplast-derived cell colonies. Entire plants capable of undergoing normal tuberization have been obtained from cultures cryopreserved for four years. These findings justify the application of this technique in the long-term conservation of potato (Bajaj, 1987).

There are currently three protocols used for potato cryopreservation. One is the vitrification method, developed by Peter Steponkus and collaborators at Cornell University, USA (Schneibel-Preikstas *et al.*, 1992; Steponkus *et al.*, 1992). In this method, shoot tips are dehydrated by exposure to concentrated solutions of sugars and cryoprotectants, and then rapidly frozen, to prevent intracellular ice formation. The other methods which have been more recently developed are the dehydration and encapsulation process which has been tested for several potato species (Benson *et al.*, 1996) and the droplet method (Schafer-Menuhr, 1996) which has been applied for conserving potato germplasm in Germany.

At CIP, after three years of testing the vitrification method (Steponkus *et al.*, 1992), around 200 potato genotypes have been frozen and stored in liquid nitrogen. These include: diploid, triploid, and tetraploids genotypes (*Solanum tuberosum* subsp. *andigena*, *S. chaucha*, *S. phureja*, *S. stenotomum*, *S. goniocalyx*, and natural hybrids of *S. goniocalyx* × *S. stenotomum* and *S. stenotomum* × *S. goniocalyx*). By making the necessary modifications to the original protocol, 75 per cent of the tested genotypes have been successfully recovered. On thawing, after three months storage, the survival of 80 genotypes was statistically the same as it was after one day of storage. From this result, the feasibility of applying cryopreservation techniques for the long-term storage of a wide range of potato genotypes has been shown. A 130 capacity liquid nitrogen tank based at CIP is currently used for cryopreserving more than 150 potato accessions. For each genotype 250 shoot tips are stored. Assays to test their genetic stability are underway. For increasing the survival rate, CIP is working to improve the vitrification method and will also test other methods, such as dehydration, encapsulation, and the droplet technique. The goal is to cryopreserve the 4000 accessions that have a very low request rate by users.

The vitrification method that has been applied at CIP is illustrated in Figure 12.3. (Golmirzaie and Panta, 1997a). This method, based on Steponkus's protocol, involves removing apical shoot tips (1.5 mm long) from plantlets grown *in vitro* for 30–45 days. The shoot tips consist of 4–5 leaf primordia and the apical dome. They are first precultured, in a modified Murashige–Skoog medium (supplemented with 0.04 mg/l kinetin, 0.5 mg/l indole acetic acid, and 0.2 mg/l gibberellic acid) containing 0.09 M sucrose for 24 hours under the incubation conditions used for micropropagation, and then in the same medium containing 0.6 M sucrose for five hours at room temperature. The shoot tips are dehydrated by placing them in a vitrification solution containing ethylene glycol:sorbitol:bovine serum albumin (50:15:6 wt per cent) for 50 minutes at room temperature. The shoot tips are then transferred to 4.0 cm propylene straws with 60 µl of vitrification solution; straws are rapidly quenched in liquid nitrogen. Ten shoot tips are loaded per straw. Following storage in liquid nitrogen, the shoot tips are thawed, expelled from the straws into a hypertonic (1.5 osm) sorbitol solution at room temperature, rinsed twice and incubated for 30–45 minutes. The shoot tips are then plated onto semisolid potato meristem medium containing Murashige–Skoog salts supplemented with 0.04 mg/l kinetin, 0.1 mg/l gibberellic acid, and 25 g/l sucrose, and maintained under normal incubation conditions for micropropagation. After 4–6 weeks, survival is evaluated by counting the plantlets growing from shoot tips.

The lowest survival rate (29 per cent) was found in *S. tuberosum* subsp. *andigena* and *S. goniocalyx* genotypes, and the highest (47 per cent) was for the natural hybrid of *S. goniocalyx* × *S. stenotomum*. More research is needed to confirm whether this variation is due to differences between species. Working with 80 genotypes, in the first survival evaluation (with material thawed one day after freezing), the average survival was 46 per cent, and after three months of storage it was 40 per cent. This difference was not statistically significant. Theoretically, the survival rate should not change, even if the plant material was stored for many years. To confirm this hypothesis, a third evaluation after 1–2 years of freezing will be carried out.

Attempts to optimize potato cryopreservation procedures

Optimizing potato cryopreservation procedures is a continuous aspect of research work at CIP and during 1997 the following assays were carried out:

POTATO CRYOPRESERVATION

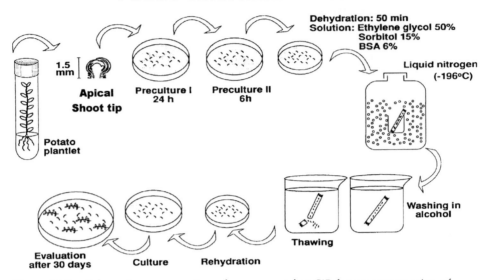

Figure 12.3   Schematic representation of process used at CIP for cryopreservation of potato shoot tips by vitrification (method based on Steponkus *et al.*, 1992)

- *Plant vigour and use of apical shoot tips*: After freezing about 100 accessions, it was evident that plant vigour is a bottleneck in the cryopreservation process and that the survival rate obtained by using axillary shoot tips varies in some genotypes. With apical shoot tips from vigorous plants, seven genotypes tested showed higher survival and the average survival rate was thus increased from 31 to 67 per cent.

- *Post-thaw culture medium improvement*: In some genotypes evaluated 45 days after thawing, the survival rate was lower than it was in preliminary evaluations. This suggested that shoot tip death could occur during the first weeks of post-thaw culture. One assay to improve the recovering medium was also carried out. Seven potato genotypes were used in this experiment. Medium supplementation with 100 mg/l vitamins (0.5 mg/l nicotic acid, 0.5 mg/l pyridoxine-HCl, 0.1 mg/l thiamine-HCl, and 2 mg/l glycine) increased the average survival rate by 10 per cent.

- *Structural observations on potato shoot tips after thawing from liquid nitrogen*: This assay was undertaken to define types of damage caused by cryopreservation, using the vitrification method. Knowledge of the basis and cause of cell damage may help increase the survival rate by targeting modifications in the freezing and thawing process. For this experiment, four genotypes of *Solanum stenotonum* were used. After one day of storage in liquid nitrogen, shoot tips were thawed; these samples were processed for transmission electron microscopy following the protocol used at CIP (Ureña, 1986). The main cell danage symptoms observed were abnormal cytoplasm, different grades of cell plasmolysis and a large number of small vesicles. No such damage was ever observed in the non-frozen control material. More studies are needed to determine if the damage suffered is non-repairable in post-thaw culture, and if the function of these cells is necessary for shoot tip regrowth. A minor reduction in the rate of the rehydration process (after thawing) could reduce this damage. Assays to test this hypothesis are in process.

12.6.2 *Sweet potato cryopreservation*

Sweetpotato cryopreservation is being studied by L. Towill (USDA), H. Takagi (JIRCAS, Japan) and CIP. Using six genotypes, different protocols for cryopreserving shoot tips and somatic embryos were tested at CIP. Shoot tips vitrification was defined as the most promising method, but more research is needed before starting to apply this method on a large scale.

Shoot tip cryopreservation of the cultivars, María Angola, Jonathan, Chugoku, Morada Inta, Tib 10, and Jewel, was tested using the vitrification method developed for Towill and Jarret (1992). After two months of thawing, shoot tip regrowth of only four genotypes was observed: María Angola had 24 per cent survival, Jonathan 17 per cent, Morada Inta 8 per cent and Jewel 26 per cent. Since survival rate of frozen sweet potato shoot tips has been low, somatic embryos were used as an alternative explant for sweet potato cryopreservation. Five cultivars (María Angola, Jonathan, Chugoku, Morada Inta, and Tanzania) were tested using the dehydration of encapsulated and non-encapsulated embryo methods, developed by Blakesley *et al.* (1995, 1996). Unfortunately, no survival was observed (Díaz, 1998).

12.7 Genetic stability

Morphological criteria are being applied to identify one accession and to determine if genetic changes in plants after *in vitro* procedures is taking place. Nevertheless, these have to be supported by molecular techniques in order to increase the number of markers used to define that both accessions (cryopreserved and control) are similar. Genetic stability has to be monitored after meristem culture, long-term storage, and cryopreservation, through assessing regenerated plants and comparing them with field-derived material. (See also Harding, Chapter 7, this volume).

Meristem culture can produce variants when cells become callus. When these callus regenerate shoots, the plant produced could be modified. Long-term storage exposes the plantlet to stress over many years. However, from CIP's experience no genetic changes have been detected up to now.

12.8 Conclusions

It is now widely recognized that the *in vitro* maintenance of germplasm prevents losses from natural disasters and pathogen damage. CIP currently maintains the largest collections of potato (*Solanum*) and sweet potato (*Ipomoea*) in active *in vitro* genebanks, and is working to develop *in vitro* collections of *Ullucus tuberosus* (ulluco), *Oxalis tuberosa* (oca), *Tropaeolum tuberosum* (mashua), *Polymnia sonchifolia* (yacón), *Arracacia xanthorrhiza* (arracacha), *and Canna edulis* (achira).

Many techniques of *in vitro* tissue culture have been incorporated in the research of different crops and they have a direct impact on conservation and seed production, such as pathogen elimination, micropropagation, conservation, and international exchange. The incorporation of these techniques in the maintenance of germplasm has made possible the conservation of resources such as potato, sweet potato, and other ARTCs under safe conditions. This method has also made these species available to all breeding programmes and National Agricultural Research Centres.

Cryopreservation techniques can assist the long-term conservation of plant genetic resources. The application of these techniques by CIP in its mandate crops will help the management of these genetic resources, in combination with other tissue culture techniques.

References

BAJAJ, Y.P.S., 1987, Cryopreservation of potato germplasm, in *Biotechnology in Agriculture and Forestry, Vol. 3: Potato*, edited by Y.P.S. BAJAJ, pp. 472–486, Berlin: Springer-Verlag.

BENSON, E.E., WILKINSON, M., TODD, A., OKURA, U. and LYON, J., 1996, Developmental competence and ploidy stability in plants regenerated from cryopreserved potato shoot-tips, *Cryo-Letters*, **17**, 119–128.

BLAKESLEY, D., AL-MAZROOEI, S. and HENSHAW, G.G., 1995, Cryopreservation of embryogenic tissue of sweetpotato (*Ipomoea batatas*): use of sucrose and dehydration for cryoprotection, *Plant Cell Reports*, **15**, 259–263.

BLAKESLEY, D., AL-MAZROOEI, S., BHATTI, M.H. and HENSHAW, G.G., 1996, Cryopreservation of non-encapsulated embryogenic tissue of sweetpotato (*Ipomoea batatas*), *Plant Cell Reports*, **15**, 873–876.

BORDA, C., TOLEDO, J. and GOLMIRZAIE, A., 1998, Evaluación del efecto del sorbitol, manitol y ABA en el crecimiento de *Oxalis tuberosa*, in *Proceedings of the International Symposium of Biotechnology – Andean Region*, La Molina, Lima, Peru, p. 27.

CUBILLAS, L.A., 1997, Uso del estrés osmótico como método de conservación *in vitro* del germoplasma de *Ipomoea batatas* (L.) Lam. [Use of osmotic stress as method of in vitro conservation of *Ipomoea batatas* (L.) Lam germplasm], *Thesis*, San Marcos National University, Lima, Peru.

DÍAZ, S., 1998, Criopreservación de embriones somáticos y apices de yeas de camote (*Ipomoea batatas* L. (Lam). [Cryopreservation of somatic embryos and shoot tips of sweetpotato (*Ipomoea batatas* L. (Lam)], *Thesis*, Ricardo Palma University, Lima, Peru.

DODDS, J.H. and ROBERTS, L.W., 1985, *Experiments in Plant Tissue Culture*, 2nd edn, Cambridge, NY: Cambridge University Press.

EGUSQUIZA, V., 1996, Erradicación bacteriana del germoplasma de camote (*Ipomoea batatas* (L.) Lam) in vitro. [Bacterial eradication from in vitro sweetpotato germplasm (*Ipomoea batatas* (L.)Lam)], Thesis, Ricardo Palma University, Lima, Peru.

ESCOBAR, R.H., MAFLA, G. and ROCA, W.M., 1997, A methodology for recovering cassava plants from shoot tips maintained in liquid nitrogen, *Plant Cell Reports*, **16**, 474–478.

ESPINOZA, N., LIZÁRRAGA, R., SIGUEÑAS, C., BUITRÓN, F., BRYAN, J. and DODDS, J.H., 1992, Tissue culture: micropropagation, conservation, and export of potato germplasm, *CIP Research Guide 1*, International Potato Center, Lima, Peru.

ESTRADA, R., TOVAR, P. and DODDS, J.H., 1986, Induction of in vitro tubers in a broad range of potato genotypes, *Plant Cell, Tissue and Organ Culture*, **7**, 3–10.

FAHY, G.M., MACFARLANDE, D.R., ANGELL, A.A. and MERYMAN, H.T., 1984, Vitrification as an approach to cryopreservation, *Cryobiology*, **21**, 407–426.

GEORGE, E.F. and SHERRINGTON, P.D., 1993, Minimal growth techniques: storing and distributing clonal material, in *Plant Propagation by Tissue Culture (Part 1) The Technology*, 2nd Edition, Exegetics Ltd. Edington, Wetsbury, Wilts, England, pp. 164–165.

GOLMIRZAIE, A.M. and PANTA, A., 1997a, Advances in potato cryopreservation by vitrification, *CIP Program Report*, International Potato Center, Lima, Peru. pp. 71–76.

GOLMIRZAIE, A.M. and PANTA, A., 1997b, Tissue culture methods and approaches for conservation of root and tuber crops, in *Conservation Plant Genetic Resources In Vitro, Vol. 1: General Aspects*, edited by M.K. RAZDAN, and E.C. Cocking, pp. 123–152, USA: Science Publishers, Inc.

GOLMIRZAIE, A. and TOLEDO, J., 1998, Non-cryogenic, long-term germplasm storage, in *Plant Cell Culture Protocols*, edited by R. HALL, Netherlands: Humana Press.

HUGHES, K.W., 1981, In vitro ecology: exogenous factors affecting growth and morphogenesis in plant culture system, *Environmental and Experimental Botany*, **21**, 281–288.

KWIATKOWSKI, S., MARTIN, M., BROWN, C. and SLUIS, C., 1988, Serial microtuber formation as a long-term conservation method for in vitro potato germplasm, *American Potato Journal*, **65**, 369–375.

LIZARRAGA, R.E., 1980, Elimination of potato spindle tuber viroid by low temperature and meristem culture, *Phytopathology*, **70**, 754–755.

LIZARRAGA, R., PANTA, A., ESPINOZA, N. and DODDS, J.H., 1990, Tissue culture of *Ipomoea batatas*: micropropagation and conservation, *CIP Research Guide 32*, International Potato Center, Lima, Peru, p. 21.

LIZARRAGA, R., PANTA, A., JAYASINGHE, U. and DODDS, J.H., 1991, Tissue culture for elimination of pathogens, *CIP Research Guide 3*, International Potato Center, Lima, Peru, p. 21.

MURASHIGE, T. and SKOOG, F., 1962, A revised medium for rapid growth and bioassays with tobacco tissue culture, *Physiologia Plantarum*, **15**, 473–497.

NG, S.Y. and NG, N.Q., 1997, Germplasm conservation in food yams (*Discorea* spp.): constraints, application and future prospects, in *Conservation Plant Genetic Resources in Vitro. Vol. 1: General Aspects*, edited by M.K. RAZDAN and E.C. COCKING, pp. 257–286, USA: Science Publishers, Inc.

ROCA, W.M., 1975, Tissue culture research at CIP, *American Potato Journal*, **52**, 281.

SAKAI, A., 1993, Cryogenic strategies for survival of plant cultured cells and meristems cooled to −196°, in *Cryopreservation of Plant Genetic Resources Projects*, Ref. 6 JICA, Tokyo, pp. 5–21.

SALAZAR, L.F. and QUERCI, M., 1992, Detection of viroids and viruses by nucleic acids probes, in *Techniques for the Rapid Detection of Plant Pathogens*, edited by J.M. DUNCAN and L. TORRANCE, pp. 129–144, Oxford: Blackwell Scientific.

SCHAFER-MENUHR, A., 1996, Protocol for the cryopreservation of potato varieties, in *Refinement of Cryopreservation Techniques for Potato*, Final project report, edited by Deutsche Sammlung von Mikroorganismen und Zell-kulturen GmbH (DMSZ), Braunschweig, Germany.

SCHNEIBEL-PREIKSTAS, B., EARLE, E.D. and STEPONKUS, P., 1992, Cryopreservation of potato shoot tips by vitrification, Abstracts of the 29th Annual Meeting of the Society for Cryobiology, June 14–19, 1992, Cornell University, Ithaca, NY, *Cryobiology*, **29**, 747.

SIGUEÑAS, C., 1987, Propagacion y conservacion *in vitro* de dos cultivares de camote (*Ipomoea batatas* Lam). [Propagation and *in vitro* conservation of two sweet potato cultivars (*Ipomoea batatas* Lam)], *Thesis*, La Molina, National Agrarian University, Lima, Peru.

SINGH, R.P., BOUCHER, A. and SOMERVILLE, T.H., 1989, Evaluation of chemicals for disinfection of laboratory equipment exposed to potato spindle tuber viroid, *American Potato Journal*, **66**, 239–245.

STEPONKUS, P.L., LANGIS, R. and FUJIKAWA, S., 1992, Cryopreservation of plant tissues by vitrification, *Advances in Low-Temperature Biology*, **1**, 1–16.

TOLEDO, J. and GOLMIRZAIE, A., 1998, Conservación in vitro de *Solanum* sp. bajo condiciones de estres osmótico y ambiental. [In vitro conservation of *Solanum* sp. under environmental and osmotic stress], *Abstracts of III Latin-American Meeting on Plant Biotechnology*, La Habana, Cuba, p. 192.

TOLEDO, J., ARBIZU, C. and HERMANN, M., 1994, Coleccion internacional *in vitro* de ulluco, oca, mashua y yacon. [*In vitro* international collection of ulluco, oca, mashua and yacon], *Abstracts of VIII International Congress of Andean Agriculture Systems*, Valdivia, Chile, p. 1.

TOLEDO, J., EGUZQUISA, V. and GOLMIRZAIE, A., 1998a, Limpieza de bacterias sistémicas en plántulas in vitro de camote *I. batata*. [Elimination of systemic bacterial infection in sweetpotato plantlets], *Abstracts of III Latin-American Meeting on Plant Biotechnology*, La Habana, Cuba, p. 146.

TOLEDO, J., ESPINOZA, N. and GOLMIRZAIE, A., 1998b, Cultivo de tejidos: Manejo de plántulas in vitro en la producción de semilla de papa. [Tissue culture: use of *in vitro* plants in seed potato production]. *CIP Training Guide*, International Potato Center, Lima, Peru.

TOVAR, P., ESTRADA, R., SCHILDE-RENTSCHLER, L. and DODDS, J.H., 1985, Induction of in vitro potato tubers, *CIP Circular*, International Potato Center, Lima, Peru, **13**(4): 1–4.

TOWILL, L. and JARRET, R., 1992, Cryopreservation of sweetpotato (*Ipomoea batatas* (L.) Lam.) shoot tips by vitrification, *Plant Cell Reports*, **11**, 175–178.

UREÑA, F. 1986, Manual de técnicas básicas para microscopía electrónica de trasmisión y barrido. [Basic techniques of electronic mycroscope of transmission and scanning], *Training Guide*, Electronic microscopic unit (UME), Costa Rica University, Costa Rica.

WESCOTT, R.J., 1981, Tissue culture storage of potato germplasm 2. Use of growth retardants, *Potato Research*, **24**, 343–352.

13

Biotechnology in Germplasm Management of Cassava and Yams

S.Y.C. NG, S.H. MANTELL AND N.Q. NG

13.1 Introduction

Yam (*Dioscorea* spp.) and cassava (*Manihot esculenta* Crantz) are major food crops for millions of people in Africa, Latin America, the Caribbean, Asia and the Pacific. Yam tubers are a source of carbohydrate. They are also high in amino acids and rich in minerals. The crop is especially important in the yam zone of West Africa, stretching from Cote d'Ivoire to Cameroon. The storage roots of cassava form the basic carbohydrate source in Africa and Latin America. Cassava leaves, rich in vitamins, minerals and proteins, are consumed as leafy vegetables in many parts of Africa. Cassava storage roots are also used as animal feed, and raw materials for alcohol production, food and starch industries.

Yam is propagated by planting a portion of the tuber or bulbils and cassava is propagated by woody stem cuttings. Yams are dioecious; male and female flowers are produced in the leaf axils of separate plants. Many yam species do flower; however, some germplasm of cultivated species does not flower regularly (Asiedu *et al.*, 1997). Thus it is a challenge to crop improvement. Cassava is monoecious and predominantly outcrossing which leads to a high degree of heterozygosity in plants.

13.1.1 Origin, distribution and cultivation

Dioscorea is the largest genus in the family Dioscoreaceae which consists of about 600 species. The majority of the species are distributed in the tropics except for *D. japonica* Thumb. and *D. opposita* Thumb. which can tolerate frosty conditions (Coursey, 1972). Within the genus there are about ten species cultivated for food and six species for pharmaceutical use. The major cultivated food yams are *D. rotundata* Poir (white Guinea yam, white yam), *D. alata* L. (water yam), *D. cayenensis* Lam. (yellow yam), *D. dumetorum* (Knuth) Pax (bitter yam), *D. esculenta* (Lour.) Burk. (Chinese yam), and *D. bulbifera* L. (aerial yam). These species account for over 90 per cent of all the food yams grown in the tropics.

Centres of origin of cultivated yam species are distributed in three independent areas in the tropics (Alexander and Coursey, 1969; Coursey, 1967; Hahn, 1995): south-east Asia for *D. alata*, *D. esculenta* and *D. bulbifera*, Africa for *D. rotundata*, *D. cayenensis*, *D. bulbifera* and *D. dumetorum*, and the Americas for *D. trifida* L. During the domestication process, there was an east-to-west movement of yam species. The Asiatic ones (*D. alata* and *D. esculenta*) were transferred to Africa and the Americas and are now grown widely in these continents. Similarly, the African species (*D. rotundata*) were taken to the Americas and have become an important crop in that region, particularly in the Caribbean.

The cultivation of yams is widespread across the continents of Africa, Asia, and the Americas. *D. rotundata* is the dominant yam species grown in the yam zone of West Africa representing the major yam-producing region in Africa. *D. alata* is another major cultivated yam species in the Americas, Asia, and the Pacific with the bulk being produced in the Caribbean and the Pacific. FAO statistics show that world yam production in 1995 was about 33 million metric tonnes and the total area under cultivation was about 3 million ha (FAO, 1996). Africa accounted for 96 per cent of the world yam production and similarly of the area under cultivation.

Cassava belongs to the genus Manihot, family Euphorbiaceae. This genus contains nearly 100 species of herbs, shrubs, and trees among which the production of latex and cyanogenic glucosides is common. The geographic origin of cassava is in lowland tropical America. It shares the Brazilian–Paraguayan centre of origin with crops such as peanuts, cacao, and rubber. North-western South America is proposed as the heart of domestication. The Mesoamerican region extending from the north-western coast of Mexico and covering parts of Guatemala, El Salvador, and Nicaragua is also a potential area of early domestication (Rogers, 1965). Cassava was carried by the Arawak tribes of Central Brazil to the Caribbean Islands and Central America in the eleventh century, by the Portuguese to the west coast of Africa, via the Gulf of Benin and the Congo River at the end of the sixteenth century, and to the east coast via the islands of Reunion, Madagascar, and Zanzibar at the end of the eighteenth century (Jennings, 1976). The crop arrived in India about 1800 and the Spaniards took it to the Pacific. Cassava production is limited to the area between 30°N and 30°S. The total area under cultivation in 1995 was about 16 million ha with 61 per cent in Africa (FAO, 1996). Africa accounts for almost 51 per cent of world cassava production, followed by Asia.

13.1.2   *Genetic diversity*

The centres of diversity and cultivation of each yam species differ except for *D. alata*, which is quite widespread throughout the yam-growing areas in the world. The major areas of genetic diversity are in the continents where the yam species originated. Genotypes of yam exhibit wide variation in their morphology and anatomy, responses to biotic and abiotic stresses, biochemical qualities and yields (Hamon and Toure, 1991; Okoli, 1991), as well as in their cytology (Hahn, 1995), due to the different ploidy levels within single species. The current diversity in traditional landraces has been as a result of the availability of wild yams with cropping potential, different selection pressures, successive domestication, culture-derived modifications and somatic mutation. There is also great genetic diversity of wild yam species in West and Central Africa (Hamon *et al.*, 1995).

Gulick *et al.*, (1983) suggested the existence of three regions as the primary sources of the genetic diversity of *M. esculenta*. The most important region covers part of northeast, central-east and southeast Brazil and Paraguay. The second region covers southern Venezuela, eastern Colombia, and northwest Brazil and the third region is centred in Nicaragua, extending to northern Honduras, and southern Panama. Africa is one of the important secondary centres of genetic diversity for cassava. The wild *Manihot* genepool has a number of desirable characteristics which could contribute to the improvement of cassava (Bonierbale *et al.*, 1997a). Genes from closely related wild species are also accessible through hybridization (Asiedu *et al.*, 1994). The genetic variability in wild *Manihot* species and *M. esculenta* in particular is significant for morphological characters, resistance to biotic and abiotic stresses, and food quality aspects such as cyanogenic glucosides contents in the tuberous roots and leaves.

13.2   Integrated strategy for germplasm conservation

Germplasm of both cassava and yams is under the threat of erosion. The wild habitats and local landraces, which are the traditional sources of genetic variability, are being lost through land clearance, drought, bushfires, and the abandonment of farms by aged and retiring farmers (Acheampong, 1996a). Thus, there is a need to collect traditional cultivars and wild related species for preservation so as to prevent further loss of genetic diversity. The importance of wild species in the breeding of these two crops and the diversity that exists in the centres of origin demand immediate attention for collection and conservation (Hamon *et al.*, 1995; Hershey, 1994). In the late 1960s and early 1970s, international agricultural research centres (IARCs) and national agricultural research systems (NARs) around the world started to assemble, conserve, and use germplasm collections of cassava and yams. Several international and regional meetings on germplasm conservation and utilization have been held to create awareness, and promote and harmonize the conservation and utilization of the plant germplasm.

13.2.1   In situ *conservation*

In situ conservation functions best when genetic diversity is concentrated in relatively small areas which are not immediately subject to high pressure from human activity which threatens their existence. Creating reserves and parks is most easily accomplished in limited areas. Since the *Manihot* species are generally sporadic in their distribution which coincides with some of the areas where agriculture is rapidly expanding, creating massive natural reserves may not be possible (Hershey, 1994).

Recently, IITA initiated research on the *in situ* conservation of yams (Ng *et al.*, 1997a) at the forest reserves (more than 300 ha) in the campus where wild yam species grow. Surveys carried out in this reserve showed that six species of yam are found growing throughout the forest. These yam species have been preserved *in situ*, among trees and shrubs over the last 30 years. The population dynamics of different yam species are being monitored. This will provide useful information on the implications of *in situ* conservation of yam.

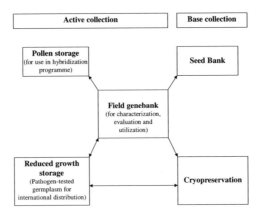

Figure 13.1   Relationships and function of the different *ex-situ* conservation methods

13.2.2   *Ex situ conservation*

To reduce the loss of valuable germplasm collections of yams and cassava, a combination of different conservation methods is necessary. Different options available for use are the field genebank, *in vitro* culture, and seed storage (Ng and Ng, 1994). Pollen storage may have potential for the conservation of yam genetic resources. The relationships and function of different conservation methods are illustrated in Figure 13.1.

Field genebanks

A field genebank has been the traditional method for preserving yams and cassava germplasm. Enormous losses from the collections have been reported by several workers (Acheampong, 1996a; Taylor, 1996a) and a yearly loss of as high as 10 per cent has been reported in yams (Okoli, 1991).

An improved field genebank management protocol for yams adopting a modified minisett technique was successfully used to conserve a wide range of species and genotypes of yams (Ng and Ng, 1994). The losses are prevented by treating the minisetts with a mixture of fungicides and insecticides before planting and protecting the plants by the application of fungicides when necessary. Since the planting sett is a minisett, the harvested tubers are relatively small. These are easier to harvest, thereby reducing the damage caused during harvesting and handling.

For the field genebank maintenance of cassava, stem cuttings are planted in the field at the beginning of the rainy season. The plants are allowed to grow through the whole year. They can then either be re-planted the following year or ratooned at the beginning of the rainy season and re-planted the following year. The problem related to field genebank maintenance is the exposure of the collections to biotic and abiotic stresses. Those that are susceptible to certain biotic stresses could be totally wiped out. It is always recommended that the collections be planted at more than one location.

In vitro *cultures*

In vitro conservation can relieve some of the storage problems of cassava and yams. However, for routine use, methodologies have to be worked out to cover adequately a wide range of genotypes, including the wild related species (Withers, 1994). At the same time, problems associated with genetic instability need attention (see Harding, Chapter 7, this volume).

Different approaches in the use of *in vitro* cultures complementary to the field gene-bank were described (Ng and Ng, 1991). The choice of the different options depends on the availability of the facility and routine protocols (see Lynch, Chapter 4, this volume). The use of a normal culture medium and culture conditions for *in vitro* storage provide a short-term measure. The use of reduced growth culture medium and reduced incubation temperature and light intensity offer short- to medium-term storage of germplasm. These two approaches are now currently available for both plant species although more research is needed on the use of these protocols in their wild related species. Cryopreservation (see Benson, Chapter 6, this volume) offers long-term conservation; however, research is underway to develop protocols for both cassava and yams.

Seed storage

Seed storage is the most common and convenient method for germplasm conservation. It provides a very high degree of accessibility and security. However, this is not feasible for all crops. Fortunately, yam seeds follow the orthodox seed storage physiology pattern (Daniel *et al.*, 1995). Cassava seeds may be stored under similar conditions. Thus seed storage can be used to store and maintain gene combinations of these two species.

Pollen and DNA storage

Pollen storage provides an opportunity to store genetic materials for those yams that produce male flowers. Studies showed that pollen of yams can be stored at −80°C for more than two years (Ng *et al.*, 1997b). This finding has great potential for those yams that are male flowering.

The DNA bank is another possible option for the conservation of genes. However, strategies and procedures for the utilization of conserved DNA need to be devised. The DNA bank will be more appropriate for the conservation of collections of genes with particular values.

13.3   Tissue culture for germplasm conservation and exchange

Since cassava and yams are vegetatively propagated, tissue culture methods are particu-larly important to these two crops in the areas of the conservation and exchange of clonal germplasm. These methods include meristem/shoot-tip culture, embryo culture, micro-tuberization, somatic embryogenesis, organogenesis, reduced growth storage, and cryopre-servation (see Lynch, Chapter 4, this volume).

13.3.1   *Meristem/shoot-tip culture*

This is a technique which is well developed for both yams and cassava. Plantlets have been regenerated successfully from meristems of different yam species, including *D. alata, D. rotundata, D. cayenensis, D. esculenta, D. dumetorum* and *D. bulbifera* (Mantell *et al.*, 1980; Ng, 1986), and cassava (Kartha and Gamborg, 1975; Ng and Hahn, 1985;

Roca, 1985). This technique is commonly used for eliminating disease and establishing cultures for micropropagation and for germplasm conservation. Successful application of meristem culture in disease elimination depends on a knowledge of diseases, especially viral diseases, infecting the crop/plant and the availability of reliable testing methods. Pathogen-tested plants are particularly useful in establishing a pathogen-tested *in vitro* genebank as well as for the safe international germplasm exchange of clonal materials (Ng and Ng, 1997a).

13.3.2 *Axillary bud and nodal cultures*

Axillary buds and nodal culture are the commonly used culture systems for rapid clonal propagation and for the germplasm conservation of yams and cassava. The detailed procedures for rapid clonal propagation using nodal cultures were described by Mantell *et al.* (1979) and Ng (1992a) for yams, and by Roca (1984) and Ng and Hahn (1985) for cassava.

Plant regeneration from axillary buds or nodes has been achieved in several yam species, *D. alata, D. rotundata, D. abyssinica, D. cayenensis, D. floribunda, D. trifida, D. esculenta, D. dumetorum,* and *D. mangenotiana* (Chaturvedi, 1975; Lauzer *et al.*, 1992; Mantell *et al.*, 1979; Ng, 1992a). Nodal cuttings used were 5 to 15 mm lengths of vine with one to two axillary buds and a 2 to 8 mm long petiole. Plantlets obtained were further subdivided into nodal cuttings for subculturing. It was evident that the growth and development of nodal cuttings were faster when cultured in liquid medium, temporary immersion in liquid medium, and aeration, than in an unaerated semi-solid medium (Ng and Mantell, 1996). Mineral salts, carbon sources, and cytokinins in the culture medium as well as daylength also have dramatic influences on the growth of yam nodal cuttings (see Lynch, Chapter 4, this volume).

Using axillary buds or nodal cultures, plant regeneration was obtained from cassava (both local and improved cultivars), from many countries around the world (Chishimba and Lingumbwanga, 1997; Guo and Liu 1995; Zok *et al.*, 1993). The effects of auxins and cytokinins on plantlet growth were evaluated (Guo and Liu, 1995) and the optimal concentration to support shoot and root growth was determined. Based on these findings, 800 000 cassava plantlets were produced in two years in rural villages using simple structures for an incubation facility without electricity.

13.3.3 *Embryo culture*

Embryo culture technique is an important tool for rescuing immature embryos and increasing the recovery of zygotic embryos from germination of seeds (also see Lynch, Chapter 4 and Marzalina and Krishnapillay, Chapter 17, this volume). Embryo culture techniques and culture media for both mature and immature embryo culture of cassava and some other *Manihot* species (Ng, 1992b; Roca, 1984) and several yam species, *D. opposita, D. rotundata,* and *D. abyssinica* (Ng *et al.*, 1994b; Okezie *et al.*, 1984) have been reported.

According to Okezie *et al.* (1984), the embryos of yam seeds continue to grow, changing from a heart- to a fan-shape, long after seeds have been harvested. Immature embryo/seed culture in *D. alata* is particularly important to rescue hybrids from the crosses aimed at transferring anthracnose resistance genes to susceptible genotypes which have other desirable characteristics. Culture media for supporting the further growth and development of immature seeds at eight weeks after anthesis were identified (Ng, unpublished results). However, efforts need now to be focused on increasing the percentage of plant regeneration and to develop media suitable for culturing even younger seeds.

13.3.4 *Microtuberization*

The ability of yam explants to produce tubers *in vitro* has been observed in many yam species, *D. alata* (Alhassan and Mantell, 1994; Jean and Cappadocia, 1991), *D. bulbifera* (Ammirato, 1984; Forsyth and Van Staden, 1982), *D. cayenensis* (Ng and Mantell, 1996), *D. opposita* (Mantell and Hugo, 1986), *D. rotundata* (Ng, 1988), and *D. abyssinica* and *D. mangenotiana* (Lauzer *et al.*, 1992). However, this phenomenon was not observed in cassava. Observations on cassava only showed a thickening of the roots. Since cassava is not propagated through tuberous roots, research in this area has not been explored.

The formation of microtubers was influenced by several factors, both culture medium composition and incubation conditions. Carbon sources, auxins, growth inhibitors, cytokinins, and nitrogen source in the culture medium can determine the frequency of microtuber formation, the sizes, and numbers. These parameters are also affected by photoperiod and light intensity. The responses to these various factors differed among the species studied (Ng and Mantell, 1996). Differential responses among genotypes within the same species were also recorded. Microtubers of yams are normally formed at the node where roots also initiate. Sucrose, fructose, mannose, and glucose are suitable for use as a carbon source, both for shoot growth and the induction and growth of microtubers. Carbon sources at 5 per cent (w/v) gave a higher percentage microtuber formation than at 3 per cent (Ng, 1988; Ng and Mantell, 1997). Microtubers normally formed at about two to three months after culturing in the induction medium. They are ready for harvest two to four months after they are induced. The tubers can be harvested aseptically by detaching them from the roots and shoots (Ng and Ng, 1997a). Microtubers often have a longer dormancy period than the tubers harvested from the field and germination is also irregular. Alhassan (1991) reported stimulatory responses of *D. alata* microtubers to 6 per cent 2-chloroethanol treatment and strong interaction of cytokinin type and concentration used for microtuber induction on germination responses. Once more consistent tuber germination is achieved, microtubers offer great potential for germplasm exchange as well as for germplasm conservation.

13.3.5 *Adventitive regeneration*

Advances in developing adventitive regeneration systems provide efficient large-scale micropropagation of plant materials and an opportunity for gene transfer through genetic engineering. Although callus cultures can be readily established from different explants of yam, plant regeneration via either embryogenesis or organogenesis is achieved only in certain genotypes of a few food yam species. The most successful results obtained so far have been from the use of zygotic embryos of *D. rotundata* (Osifo, 1988), as well as *D. composita* and *D. cayenensis* (Viana and Mantell, 1989). Direct multiple bud formation was achieved in *D. opposita* using immature leaves (Kohmura *et al.*, 1995) and callus formation and plant from young leaf petioles of *D. alata* (Fautret *et al.*, 1985). Somatic embryogenesis and plant regeneration were also achieved from callus cultures derived from the roots of *in vitro* plants (Twyford and Mantell, 1996). Somatic embryos have also been obtained from the callus and suspension cultures of *D. rotundata* (Ng and Mantell, 1996) which were derived from young leaves *in vitro*. Most recently, plant regeneration via somatic embryogenesis and/or organogenesis was reported from several medicinal yams and *D. alata* (Lee, 1997).

The development of the adventitive regeneration system in cassava is well advanced. Leaves and shoots have occasionally been regenerated from callus grown from stem sections (Tilquin, 1979) and calli derived from leaf mesophyll protoplasts (Shahin and Shepard, 1980). Direct somatic embryogenesis and plant regeneration were first reported on the use of cotyledonary explants from mature cassava seeds (Stamp and Henshaw, 1982). In the late 1980s, direct somatic embryogenesis and plant regeneration were achieved from leaf explants obtained from *in vitro* plantlets (Stamp and Henshaw, 1987a) and secondary embryogenesis was also achieved (Stamp and Henshaw, 1987b). Since then, successes in direct embryogenesis and plant regeneration from leaf lobes of cassava of various origins, America, Asia and Africa, have been reported (Li *et al.*, 1995; Mroginski and Scocchi, 1993; Ng and Adeniyi, 1994; Raemakers *et al.*, 1995; Sudarmonowati and Bachtiar, 1995). A few years later, a plant regeneration system was reported from embryogenic suspension culture (Taylor *et al.*, 1995, 1997). Using both regeneration systems, genetic transformation systems are now being developed (Puonti-Kaerlas *et al.*, 1997; Schopke *et al.*, 1997).

13.3.6   *Reduced growth storage*

Reduced growth storage of germplasm is considered as a short- to medium-term storage method for germplasm conservation (see Lynch, Chapter 4, and Reed, Chapter 10, this volume). This can be achieved most commonly by the reduction of incubation temperature, manipulation of culture media, or the combination of both (Ng and Ng, 1991). The effects of sucrose in combination with mannitol in culture medium drastically reduced the plant growth of *D. alata in vitro* (Ng and Ng, 1991) and at the same time gave a reasonably high percentage recovery after 13 months of storage. This indicated that simply adding mannitol in the culture media can effectively reduce plant growth, thus extending the storage period. Lower incubation temperature (18–22°C) slowed down the shoot growth of *D. rotundata* compared to the normal incubation temperature (25–30°C) and the percentage recovery after 12 months storage at reduced temperature was higher than in cultures stored at normal incubation temperature (Ng and Ng, 1997a). This showed that a lower incubation temperature could also prolong the culture storage period.

Research on reduced growth storage has provided a means for maintaining cassava germplasm *in vitro*. Studies indicated that storage temperature, light intensity, media manipulation (addition of osmoticum, growth hormone, and nutrient level) as well as culture vessels influenced the growth of cassava plantlets and viability (CIAT, 1980). Low incubation temperature (20°C) reduced the growth of cassava shoots by one-fifth compared to those cultures stored at 25–30°C (Roca, 1984). An increasing concentration of benzyl amino purine (BAP) as well as sucrose in the culture media slowed down shoot elongation. Growth of cultures was also reduced by the reduction of nitrogen content in the medium. The addition of mannitol in combination with sucrose (Roca, 1984) and mannitol or absisic acid (Acedo, 1995) also reduced shoot growth without significantly reducing viability. Media for culturing wild *Manihot* species were investigated (Mafla *et al.*, 1993; Ng and Ng, 1997b). However, the wild *Manihot* species studied represent a rather small portion of the whole range of wild *Manihot* species.

13.3.7 Cryopreservation

Cryopreservation offers long-term conservation of clonal germplasm. This is equivalent to the base collection in seed conservation. Recent developments in cryopreservation (see Benson, Chapter 6, this volume), especially those reducing cryodamage, offer an improvement of survival after freezing and the list of successfully cryopreserved species is also increasing. A recent review of the different cryopreservation protocols was given by Ashmore (1997). The classical cryopreservation protocols have normally been by slow cooling with the use of cryoprotectants and these require sophisticated and expensive programmable freezing equipment. The new techniques recently developed offer an opportunity to use fast freezing, and direct immersion in liquid nitrogen.

Cryopreservation in yam is in its infancy. Successful cryopreservation of cell suspension cultures of *D. deltoidea* (a medicinal yam species) in liquid nitrogen was reported (Butenko *et al.*, 1984). More recently a report showed that successful cryopreservation of two *Dioscorea* species in liquid nitrogen was achievable by encapsulation–dehydration of shoot-tips (Mandal and Chandel, 1995). Research in developing cryopreservation protocols in food yams is underway at ORSTOM, Montpellier (Engelmann, personal communication), in Japan (Takagi, personal communication) and at IITA.

Early work carried out by Bajaj (1983) showed that cassava meristems frozen directly in liquid nitrogen with cryoprotectants resumed growth and formed plants after thawing. A survival percentage of 26 was obtained after three years storage in liquid nitrogen. Other successful approaches involving controlled cooling regimes have been described for shoot meristem (Escobar *et al.*, 1993), seeds and zygotic embryos (Marin *et al.*, 1990) and somatic embryos (Sudarmonowati and Henshaw, 1990). Pregrowth of shoot-tips in proliferation medium, followed by treatment with cryoprotectants and dehydration before direct immersion of shoot-tips in liquid nitrogen resulted in a plant recovery rate as high as with programmable freezing (Escobar *et al.*, 1995) and this protocol was successfully tested in several cassava genotypes. It was suggested that pretreatment and dehydration of explants were critical factors determining the success of cassava cryopreservation. Preliminary results obtained at IITA showed that shoot meristems, dissected from plantlets grown in medium containing dimethyl sulphoxide, directly immersed in liquid nitrogen formed buds after being removed from liquid nitrogen and placed on the normal meristem culture medium (IITA, 1998a).

13.4 Biochemical and molecular fingerprinting for auditing of germplasm collection

Characterization and evaluation of a germplasm collection is normally carried out in the field, using agrobotanical characters and their reactions to biotic and abiotic stresses. Biochemical and molecular fingerprinting are added tools for germplasm characterization (see Harris, Chapter 2, this volume). Molecular markers for specific traits could be used to select germplasm that has specific characteristics such as resistance to biotic and abiotic stresses as well as some quality attributes of the plant. Biochemical and molecular markers are listed in the recent revised descriptor list for *Dioscorea* spp. (IPGRI/IITA, 1997).

13.4.1 Range of methods available

Genetic variation can be identified and quantified by measuring the variation in DNA composition and also by analysing the primary products of the genes – protein.

Biochemical markers

Hamon and Toure (1990a, 1990b) reported the development of isozyme electrophoresis using starch gel in yam and used such a system for the characterization of yam germplasm. Procedures for protein extraction from different yam tissues, leaves, and tubers, were studied. Isozyme electrophoresis studies using acrylamide gel on extracts of yam leaf, tuber or *in vitro* plant materials of three yam species were conducted (IITA, 1993). Of eight enzymes studied, five isozymes have given consistent results with good separation and resolution of bands. These isozymes were glutamate oxaloacetate transaminase (GOT), alchohol dehydrogenase (ADH), esterase (EST), 6-phosphogluconate dehydrogenase (6-PGD), and peroxidase (PER).

A method for the extraction of crude protein extracts from the root tissues of cassava plants for electrophoresis was developed (Ocampo *et al.*, 1993). The crude extract was used directly for isozyme electrophoresis using polyacrylamide gel electrophoresis. Preliminary studies indicated that αß-esterase is ideal for use in the characterization of cassava germplasm. This isozyme provided high polymorphism and a high number of bands were detectable in cassava clones.

Molecular markers

Procedures for the extraction of chloroplast DNA and nuclear ribosomal DNA from yam tissues were developed by Terauchi *et al.* (1989, 1992). A fast, simple, and efficient mini-scale method for the preparation of DNA from tissues of yam has also been described by Asemota (1995). Using this method, good quality genomic DNA was extracted from both fresh and lyophilized organs and tissues of field grown yam plants as well as from that of *in vitro* plantlets. The amount of plant samples required to extract a reasonable quantity of DNA ranged from 25 to 100 mg of fresh materials. Among the different tissues used, leaf samples gave a higher yield of DNA than other organs tested. Different methods, i.e. RAPD–PCR, amplified fragment length polymorphism (AFLP) and restriction fragment length polymorphism (RFLP), were also developed for yam (Mignouna *et al.*, 1997; Ng and Mantell, 1996).

A method for the extraction of total and mitochondrial DNA from young leaves of both cassava and related species (Carvalho *et al.*, 1993) and chloroplast DNA from cassava has been described by Joseph and Yeoh (1995). This extraction method resulted in the production of a high quality DNA as well as a yield of up to 730 ng/g fresh weight of leaf tissue. RAPD–PCR, RFLP, AFLP technology and a microsatellite technique have recently been developed in cassava by Fregene *et al.* (1995) and Agyare-Tabbi *et al.* (1997).

13.4.2 *Isozyme fingerprinting*

Applications of isozyme analysis to *Dioscorea* species and *M. esculenta* are summarized in Table 13.1. A variety of isozymes were used to classify or distinguish different *Dioscorea* species, and different cultivars within the same species. Results of a study conducted on 393 accessions of yam collected in Cote d'Ivoire using five isozymes, malate dehydrogenase (MDH), isocitrate dehydrogenase (ICD), PGD, shikimate dehydrogenase (SDH), and phosphoglucoseisomerase (PGI), revealed four major clusters (Hamon and Toure, 1990a). One consisted of 15 varietal groups having fast migrating bands. The remaining three

Table 13.1 Summary information on the application of isozymes in *Dioscorea* and *Manihot* germplasm collections

Species	Isozymes*	Application	Reference
D. cayenensis–rotundata complex	MDH, ICD, PGD, SDH and PGI	Used to classify 393 accessions of yam germplasm collected mainly from Cote d'Ivoire	Hamon and Toure (1990a)
D. cayenensis–rotundata complex	MDH, ICD, PGD, SDH and PGI	Used to classify 453 accessions of yams collected in West Africa	Hamon and Toure (1990b)
Dioscorea spp.	Peroxidase and acid phosphatase	Identification of four yam species, *D. alata*, *D. cayenensis/ D. rotundata*, *D. esculenta* and *D. bulbifera*	Twyford *et al.* (1990)
Dioscorea spp.	GOT, ADH, EST, PGD and peroxidase	Distinguish different yam species, *D. rotundata*, *D. cayenensis* and *D. alata*	IITA (1993)
M. esculenta	EST and GOT	Genetic studies of these two isozyme loci	Sarria *et al.* (1993)
M. esculenta	EST	Fingerprinting of 4304 cassava accessions of germplasm collection, selection of core collection, identification of possible duplicates and genetic diversity	Ocampo *et al.* (1993)
		Fingerprinting of elite cassava cultivars	Laminski *et al.* (1997)

* MDH, malate dehydrogenase; SDH, shikimate dehydrogenase; ICD, isocitrate dehydrogenase; PGI, phosphoglucoseisomerase; PGD, 6-phosphogluconate dehydrogenase; GOT, glutamate oxaloacetate transaminase; ADH, alchohol dehydrogenase; EST, esterase.

clusters included five varietal groups which have slow migrating bands. Another study conducted on 453 yam accessions collected in West Africa using the range of the same enzyme systems showed that the germplasm collection can be classified to 62 genotypes (Hamon and Toure, 1990b). Twelve cultivar groups could be identified using these five isozyme systems. These results correspond closely to the classifications based on the morphological characters.

Isoelectric focusing characterization of different *Dioscorea* species by Twyford *et al.* (1990), based on two isozyme systems (peroxidases and acid phosphatases), showed that these can be used to distinguish the four species (*D. alata*, *D. esculenta*, *D. bulbifera*, and *D. cayenensis/D. rotundata*). Peroxidases were also useful in identifying cultivars of *D. alata* with similar morphological characters.

Polyacrylamide gel electrophoresis was used by Sarria *et al.* (1993) to study the genetics of two isozyme loci, esterase and GOT in cassava. The esterase locus has five multiple alleles, including one null allele, and behaves as a monomer, with a diploid inheritance pattern with two alleles presenting in each individual. The GOT locus has three alleles, also with diploid inheritance. The diploid inheritance of these two loci supported the allotetraploid nature of cassava. A total of 4304 accessions of the cassava collection were analysed for esterase isozyme patterns by Ocampo *et al.* (1993). A total of 2146 different banding patterns were found among this collection. Of these banding patterns, 1407 were represented by one clone and the rest had 2 to 39 clones with the same banding pattern. These data indicate the presence of duplicates in the germplasm collection which has serious implications for conservation. Esterase characterization can therefore be used as one of the criteria to eliminate such duplicates.

13.4.3   *Use of RAPD–PCR, microsatellites and other molecular markers*

Molecular markers (see Harris, Chapter 2, this volume) have often been used to study genetic diversity, phylogeny and fingerprinting as well as to construct physical genetic maps in cassava and yam. A summary of some of the applications of molecular markers in germplasm management is presented in Table 13.2. The range of marker systems includes RAPD, AFLP, microsatellites, and RFLP. Studies were mostly carried out on cultivated species of these two crops. However, in a few cases wild related species were included, especially in phylogenetic studies.

The origin and phylogeny of Guinea yams, *D. rotundata* and *D. cayenensis*, have been investigated by means of RFLP analysis of chloroplast DNA and nuclear ribosomal DNA using seven restriction endonucleases and various heterologous probes (Terauchi *et al.*, 1992). Fourteen cultivars of Guinea yams were used together with 12 accessions of seven wild yam species in this particular study. Results of both chloroplast and ribosomal DNA surveys indicated that *D. rotundata* was probably domesticated from either *D. abyssinica*, *D. liebrechtsiana* or *D. praehensilis* or their hybrid and that *D. cayenensis* was obtained from the hybridization between a male plant of either *D. burkilliana*, *D. minutiflora* or *D. smilacifolia* and a female plant of either *D. rotundata*, *D. abyssinica*, *D. liebrechtsiana* or *D. praehensilis*. This was also supported by RAPD–PCR using two random sequence primers, OPB1 and OPB2 (Mignouna *et al.*, 1993). Based on these results, it was concluded that *D. rotundata* is a distinct species and that *D. cayenensis* is a variety of *D. rotundata*. Thus, this data has helped clarify the so-called *D. rotundata–cayenensis* complex. Mignouna *et al.* (1995) further suggested that *D. abyssinica* and *D. praehensilis* are probably progenitors of *D. rotundata* and *D. cayenensis*.

RAPD markers generated by ten oligoprimers were used to assess the intraspecific variability of aerial yam from different geographic regions (Ramser *et al.*, 1996). Results showed that the African collections formed a distinct group, whereas Asian and Polynesian materials were more heterogeneous, suggesting the classification of some accessions should perhaps be re-examined. This study indicated that RAPD markers provide an additional cultivar identification method in *Dioscorea* spp. thus assisting in germplasm management and in the identification of bulbils of unknown origin.

RAPD, AFLP and RFLP have been used to study the genetic diversity of cultivated cassava (Mignouna and Dixon, 1997; Tonukari *et al.*, 1997) and the genus *Manihot* (Beeching *et al.*, 1995) as well as for various phylogenetic studies (Carvalho *et al.*, 1993; Schaal *et al.*, 1997; Second *et al.*, 1997). Results of a study on the genetic diversity of limited accessions of cassava and a few *Manihot* spp. using both RAPDs and RFLPs revealed that both marker systems are comparable in sensitivity (Beeching *et al.*, 1995). Genetic variation among 28 cassava varieties collected from the Republic of Benin was investigated using RAPD. Results showed that these 28 accessions of cassava varieties can be assigned to six groups. However, the grouping did not necessarily relate to the locations where they were collected (Tonukari *et al.*, 1997). A large number of plants of

Table 13.2 Summary information on the application of molecular markers in *Dioscorea* and *Manihot* germplasm collections

Species	Methods/systems	Application	Reference
D. bulbifera	RFLP	Linkage (physical) map	Terauchi *et al.* (1989)
D. rotundata and *D. cayenensis*	RFLP AFLP	Origin and phylogeny study Genetic diversity study	Terauchi *et al.* (1992) Mignouna *et al.* (1997)
D. bulbifera	RAPD	Genetic variability and relationship within the species	Ramser *et al.* (1996)
Dioscorea spp.	RAPD	Genetic variability study	Narayanaswamy *et al.* (1994)
Manihot species	RFLP and RAPD	Phylogeny study and cultivar identification	Carvalho *et al.* (1993)
	RFLP, RAPD, microsatellite	Phylogeny study and germplasm characterization	Carvalho *et al.* (1995)
	AFLP	Study on the genetic structure, domestication and taxonomy	Second *et al.* (1997)
M. esculenta × *M. aesculifolia*	RFLP and RAPD	Construction of genetic map	Angel *et al.* (1993)
M. esculenta × *M. esculenta*	RFLP and RAPD	Production of linkage map	Fregene *et al.* (1995) Fregene *et al.* (1997)
Manihot genus	RAPD	Study the origin of cassava, phylogenetic relationships and diversity of the genus	Schaal *et al.* (1995) Schaal *et al.* (1997)
	RFLP	Phylogenetic relationships of species within the genus	Haysom *et al.* (1995)
	RAPD and RFLP AFLP	Genetic diversity Genetic diversity	Beeching *et al.* (1995) Bonierbale *et al.* (1997b)
Manihot interspecific hybrids	RAPD	Isolation of true hybrids	Buso *et al.* (1995)
M. esculenta	RAPD	Characterization of cultivated cassava clones	Bonierbale *et al.* (1995) Laminski *et al.* (1997)
Manihot intraspecific crosses	RAPD	Analysis of three F1 populations obtained from intraspecific crosses of cassava	Gomez *et al.* (1995)
M. esculenta	RAPD	Genetic diversity of cultivated cassava Genetic variation study	Tonukari *et al.* (1997) Mignouna and Dixon (1997)

wild *Manihot* species (276) and cassava (86) were analysed using AFLP (Second *et al.*, 1997). Results indicated that cassava is closely related to several other species, including the hybrids between *M. glazovii* and cassava. It also suggested that the domestication of cassava not only involves *M. esculenta* spp. *flabellifolia* and *peruviana*, but some other species also have made a genetic contribution at some stage in their evolution. The genetic diversity of cassava is high. However, a nearly equivalent diversity is found in a single Amazonian field based on AFLP analysis. This suggested that a dynamic conservation strategy for the wild species of *Manihot* is a practical approach.

Phylogenetic studies using RAPD on nuclear ribosomal DNA indicated that cassava originated from South America and that *M. esculenta* spp. *flabellifolia* and *M. esculenta* spp. *peruviana* were the nearest wild species (Schaal *et al.*, 1997). Several gene sequences are identical in these two species and in cassava, indicating their close affinities and, probably, a recent common ancestry. Similar results were obtained using RFLP and RAPD except that *M. pilosa* was found to have a high similarity with *M. esculenta* (Carvalho *et al.*, 1993).

The presence of duplicates in a germplasm collection increases the cost of conservation. Suspected duplicates identified by morphological and isozyme patterns were subjected to DNA fingerprinting (Ocampo *et al.*, 1995). Southern blot hybridization was used to screen 85 clones that formed 35 groups of possible duplicates. Fingerprint analysis indicated that 29 groups (72 clones) gave an identical hybridization pattern corresponding to 83 per cent of the groups. This confirmed that these groups were genetically duplicate. These results can be used systematically along with morphological characters and isozyme patterns to eliminate duplicates.

13.5 Applications to germplasm conservation, exchange, and improvement

Various biotechnological techniques developed for yam and cassava are applied to the conservation, exchange, and improvement of these two crop species. These techniques include meristem and nodal cultures, embryo culture, microtuberization, reduced growth storage, and DNA-based molecular tools.

13.5.1 Germplasm exchange

The success of a breeding and selection programme depends on the availability of diverse germplasm (see Hummer, Chapter 3, this volume). Cassava and yam are vegetatively propagated; germplasm exchange in vegetative form was a main constraint until the development of meristem culture and virus indexing methods. Although cassava and yam seeds can be obtained from a large proportion of the germplasm, seeds are highly heterozygous. The development of meristem culture and the virus indexing techniques in cassava and yam contributed to the production of virus-tested germplasm. This allows the safe movement of clonal germplasm across national boundaries (see Martin and Postman, Chapter 5, this volume).

Meristem culture and virus indexing have been applied to eliminate viruses from cassava (Ng and Hahn, 1985, Roca 1985) and from yam (Mantell *et al.*, 1979; Ng, 1986). Virus-tested germplasm is then distributed to collaborators (Asiedu *et al.*, 1997; Bonierbale *et al.*, 1997a; Ng and Ng, 1997a, 1997b; Taylor, 1996b). Using meristem culture techniques,

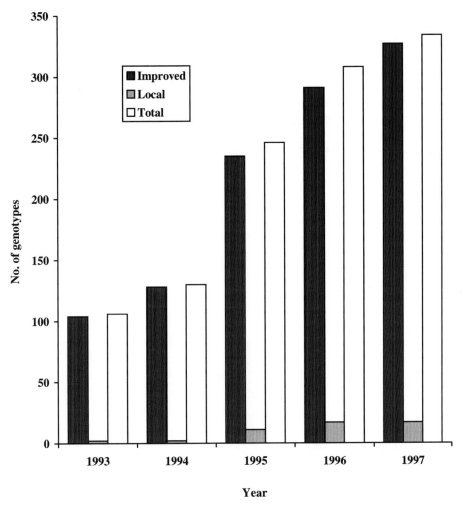

Figure 13.2 Number of virus-tested cassava germplasm accessions available for distribution (1993–1997)

IITA now has 344 and 45 virus-tested cassava and yams genotypes, respectively, available for distribution (Figures 13.2 and 13.3). The number of virus-tested cassava and yam genotypes available for distribution has steadily increased over the years.

A number of local selected cassava cultivars are also available for distribution. The number of virus-tested yam genotypes available in 1997 has almost tripled those available in 1993 and 1994. For the first time, two virus-tested *D. alata* genotypes are available for distribution. Over the years, virus-tested cassava and yam were distributed by IITA to collaborating NARs in more than 40 countries in Africa and worldwide, respectively. Summaries of the distributions of virus-tested cassava germplasm and yam germplasm are presented in Tables 13.3 and 13.4, respectively. From 1993 to 1997, a total of 41 818 *in vitro* plantlets of cassava had been distributed by IITA to its collaborators. More than 10 000 plantlets were distributed in 1996 and in 1997. In addition, IITA in collaboration with World Vision International, produced and delivered 18 090 cassava virus-tested plantlets to Angola under the 'Seed of Freedom Project' during 1996 and 1997.

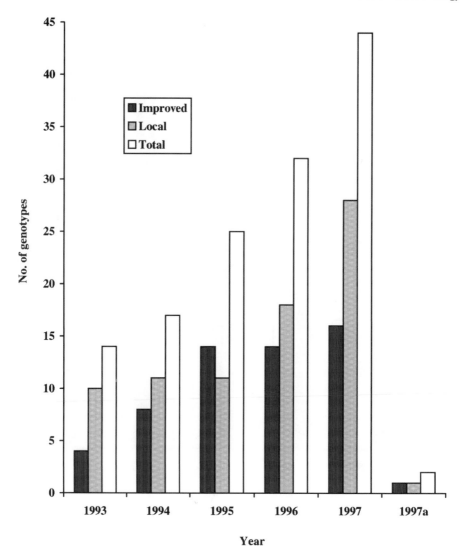

Figure 13.3 Number of virus-tested yam genotypes available for distribution (1993–1997 for *D. rotundata* and 1997a for *D. alata*)

Table 13.3 Distribution of virus-tested cassava plantlets by IITA (Nigeria) to collaborators during 1993 to 1997

Year	No. of countries	No. of plantlets
1993	18	3007
1994	18	2385
1995	24	8291
1996	20	11 271
1997	11	16 864
	Total	41 818

Table 13.4 Distribution of virus-tested yam propagules by IITA (Nigeria) to collaborators during 1993 to 1997

Year	In vitro plantlets		Minitubers	
	No. of countries	No. of plantlets	No. of countries	No. of minitubers
1993	14	400	7	1400
1994	12	1772	3	1001
1995	14	4559	5	3417
1996	9	3689	7	5500
1997	7	6385	8	6886
Total		16 805		18 204

In the case of yam, both virus-tested *in vitro* plantlets and minitubers derived by planting virus-tested plantlets in the screenhouse, were used for distribution by IITA. From 1993 to 1997, a total of 16 805 virus-tested *in vitro* yam plantlets and 18 204 minitubers were distributed to its collaborators. These materials are now being used by the collaborators either as genetic stocks for use in the improvement of the plant or for releasing to farmers after evaluation.

Meristem or nodal cultures were also used at IITA to transfer germplasm collections of cassava from collaborating NARs in Ghana and the Republic of Benin, to the *in vitro* genebank at IITA. A simple collapsible portable transfer hood made of plywood and installed with UV light was used as the transfer hood for the inoculation of cultures where a tissue culture facility was not available (Ng and Ng, 1997b). Culture media and sterile water were prepared at IITA and transported to the location where the field genebank was located. Cultures were shipped to IITA within 10–14 days for incubation and cultures upon receipt were inspected by the Nigerian Plant Quarantine Services (NPQS). The embryo culture technique has been applied to germinate isolated embryos of *Manihot* species, *M. flabellifolia, M. alutacea, M. peruviana, M. gabrielensis* and *M. quinquepartita* (Ng and Ng, 1997b).

Procedures developed and adopted by IITA for the exchange of clonal germplasm of cassava and yam are presented in Figure 13.4.

At IITA, selected genotypes of cassava (woody cuttings) and yams (tubers) are planted in pots and maintained in a screenhouse. Approximately four weeks after sprouting, buds are collected and used for meristem culture. For cassava, only apical buds are used whereas both apical and axillary buds are used for yams. Cassava buds are treated with 7 per cent sodium hypochlorite solution with Tween 20 for 20 minutes and rinsed with three changes of sterile distilled water, whereas buds of yams are disinfected by a double disinfection procedure. Meristems with two leaf primordia are excised and inoculated on a meristem culture medium. The medium used for the meristem culture of cassava is basically Murashige and Skoog (MS) basal medium (Murashige and Skoog, 1962) supplemented with 3 per cent sucrose, 80 mg l^{-1} adenine sulphate, 0.5 mg l^{-1} benzyl aminopurine (BAP), 0.2 mg l^{-1} naphthalene acetic acid (NAA), 0.04 mg l^{-1} gibberellic acid (GA$_3$), and 0.6 per cent agar.

The yam meristem culture medium is the same as that used for cassava meristem culture except that GA$_3$ is doubled and L-cysteine at 40 mg l^{-1} is added. Cultures are incubated in a culture room set at 28–30°C under 12 hour photoperiods. Plantlets regenerated from

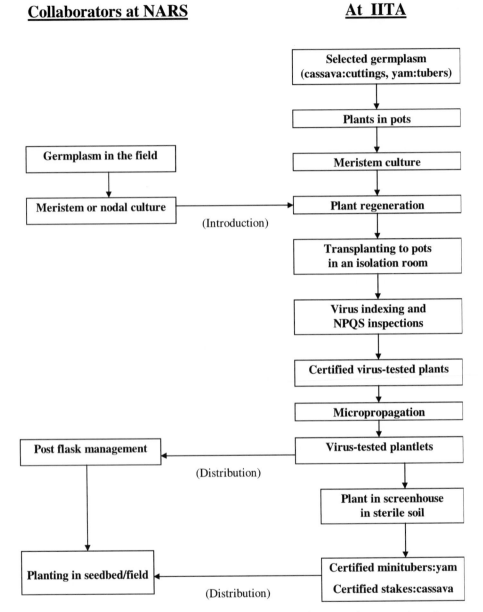

Figure 13.4 The procedures adopted at IITA for the application of tissue culture in germplasm exchange

meristems are subcultured (cloned) using nodal culture. The medium for nodal culture of cassava consists of MS basal medium supplemented with 3 per cent sucrose, 0.01 mg l^{-1} NAA, 0.05 mg l^{-1} BAP, and 0.7 per cent agar. For yams, the culture medium consists of MS basal medium with 0.5 mg l^{-1} kinetin, 20 mg l^{-1} L-cysteine, and 0.7 per cent agar. Cultures are incubated under similar conditions to those described above. After one or two months, about two to three of the plantlets obtained from each clone (clones of the same meristem) are transplanted to an insect-free isolation room. Plants are allowed to grow under these conditions for at least four months and symptomless plants are indexed for viruses. For cassava, tests for African cassava mosaic virus are carried out. This is done by sap inoculation to *N. benthamiana*, enzyme linked immunosorbent assay (ELISA) using monoclonal and polyclonal antibodies and PCR (see Martin and Postman, Chapter 5, this volume). As for yams, tests are made for several viruses. These are yam mosaic virus, yam potyvirus 1, and *D. alata* badnavirus. Virus indexing methods include sap inoculation to *N. benthamiana*, ELISA, electron microscopy, reverse transcriptase PCR (RT-PCR) for yam mosaic virus, and immunocapture PCR (IC-PCR) for *D. alata* badnavirus. NPQS inspection is also carried out during the virus indexing process. Negatively indexed plants are certified and are micropropagated using nodal cultures. Virus-tested plants are then distributed to NARs or used to produce minitubers for distribution. Upon arrival, the plantlets are transplanted to a humidity chamber for hardening and then planted in nursery beds. With respect to minitubers, once sprouted they can be planted in a nursery or field for evaluation. Germplasm collections introduced to IITA from collaborators in different NARs are first micropropagated and then transplanted for virus indexing and inspection by NPQS before being released and planted out in the field for evaluation.

13.5.2 *Germplasm conservation*

The application of *in vitro* methods to the germplasm conservation of yam and cassava at present is mainly by reduced growth storage (see Lynch, Chapter 4, this volume). Progress is being made in the area of cryopreservation in cassava. However, the methodology has yet to be tested properly on a wide range of genotypes. By contrast, in the case of yam cryopreservation is still at the developing stage. According to a recent survey in 107 countries around the world, 64 countries reported using an *in vitro* facility for germplasm conservation (Ashmore, 1997). This indicated that 59.8 per cent of the total countries surveyed had an *in vitro* facility. Only 3.7 per cent of the countries surveyed are using cryopreservation for germplasm conservation. In Africa, only 36.7 per cent of countries surveyed are practising *in vitro* conservation of germplasm and none are using cryopreservation.

Table 13.5 provides a summary of the current status of the application of *in vitro* conservation to the maintenance of cassava and yam germplasm. The number of accessions maintained *in vitro* ranges from 5 to over 5000. Most of the cultures are maintained under reduced temperature coupled with reduced light intensity. The storage period ranges from 3 to 21 months.

IITA has a collection of *Dioscorea* species from the major yam-growing countries in Africa and a collection of *Manihot* species. The yam collection consists of six cultivated species and wild species which make up to 2682 accessions (Ng and Ng, 1997a). The *Manihot* collection includes local cultivars from Africa, germplasm introduced from Latin America, the hybrids between Nigerian and Latin American germplasm and wild *Manihot* species, to a total of 1863 accessions (Ng *et al.*, 1994a). The collections of *Dioscorea* and *Manihot* germplasm are also maintained as living collections in the field. About 56 per cent of the *Dioscorea* collection and 26 per cent of the *Manihot* collection are duplicated in the *in vitro* genebank as a complementary method for the conservation of the germplasm.

In vitro cultures are incubated under reduced temperature (18–22°C). Cultures of yams can be stored for 8 to 36 months depending on the accessions and species. Studies drawn from the past five years' results on the subculture intervals of different accessions of

Table 13.5 Summary of yam and cassava maintained *in vitro*

Country/ institution	Crop	No. of accessions	Culture condition	Storage duration (months)	Reference
Philippines	Yam	13	Reduced growth media, 8 h light	12	Zamora and Paet (1996)
	Cassava	13	Normal medium, 8 h light	2	Zamora and Paet (1996)
Brazil	Yam	12	–	3–14	De Goes *et al.* (1996)
	Cassava	1307	–	3–14	De Goes *et al.* (1996)
South Pacific	Yam	5	20°C	9	Taylor (1996b)
	Cassava	19	20°C	9	Taylor (1996b)
Cuba	Yam	93	–	–	Morales (1996)
	Cassava	440	Normal medium, 20–22°C, 500 lux	–	Morales (1996)
CIAT	Cassava	5714	23–24°C, 1000 lux, low sucrose	10–18	Guevara and Mafla (1996)
Ghana	Yam	29	28°C	–	Acheampong (1996b)
	Cassava	9	28°C	–	Acheampong (1996b)
Caribbean	Yam	32	18–22°C	–	Bateson (1996)
	Cassava	28	18–22°C	–	Bateson (1996)
IITA	Cassava	727	18–22°C	8–12	Ng and Ng (1997b)
	Yam	1504	18–22°C	8–21	Ng and Ng (1997a)
UK, Wye College	Yam	140	20°C	–	Mantell (unpublished)

D. rotundata (500), *D. alata* (100), and *D. bulbifera* (60) showed that there are genotypic differences in storability within the same species as well as among the three species (Ng and Ng, 1997a). However, a higher proportion of the germplasm collections of these three species fell into the subculture interval of 11 to 15 months, which is the intermediate storage duration for yam germplasm (Figure 13.5). Cultures of cassava can be kept for 8 to 12 months only and some of the wild *Manihot* species are difficult to micropropagate and maintain.

The procedures of *in vitro* conservation of cassava and yam adopted by IITA for the transfer of germplasm from the field to an *in vitro* genebank and its maintenance are shown in Figure 13.6.

The living collection of germplasm maintained in the field or screenhouse is the source of plant materials. Buds are collected from plants at their active growing stages and then surface disinfected by a double disinfection method. Meristems with 1–2 leaf primordia are dissected, excised and placed in culture media as described for the meristem culture of cassava and yam in the previous section. Cultures are incubated under normal culture conditions (25–30°C under 12 hour photoperiods) to achieve plant regeneration. Regenerated plants are subcultured (cloned) using nodal cuttings, placed in multiplication medium and incubated in a culture room held at 25–30°C. Regenerated plantlets are then moved to

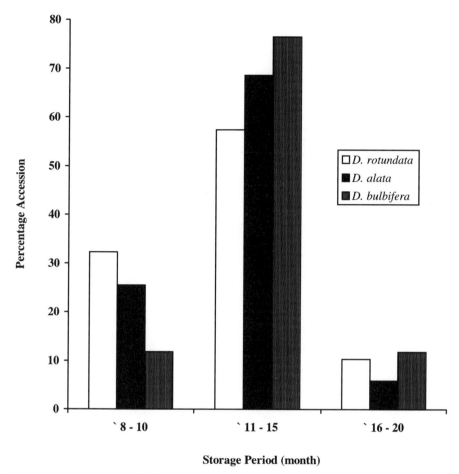

Figure 13.5 Storage period of some yam germplasm maintained *in vitro*

a lower incubation temperature (18–22°C) for storage. Cultures are checked monthly and those showing deterioration are subcultured using nodal culture. Theoretically, this cycle can be repeated indefinitely and plantlets can also be transplanted to the isolation room for virus indexing. The indexing methods used are basically the same as those described in Section 13.5.1. Through this process, the health status of stored germplasm can be documented and negatively tested germplasm can also be used for distribution. The germplasm introduced as *in vitro* cultures also follows the same process for storage and virus indexing. Eventually a pathogen-tested germplasm bank can be established, though this will be a long-term goal.

13.5.3 *Conservation biotechnology and germplasm improvement*

The contribution of conservation biotechnology to the improvement of cassava and yam germplasm is in several areas. The impact of applications to germplasm exchange is contributing to the more rapid adoption and release of improved cultivars and to the genetic conservation of these two crops. Phylogeny studies using isozyme and molecular

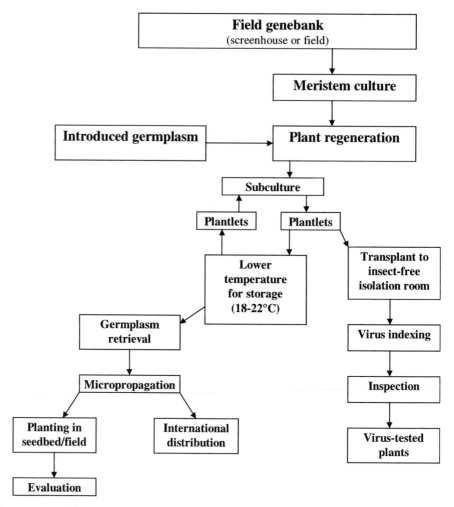

Figure 13.6 Schematic representation of reduced growth storage for germplasm conservation at IITA

markers provide information on species relationships and contribute to the development of appropriate breeding strategies to tap useful genes in wild species. Marker-assisted selection for particular traits will begin to contribute immensely to yam and cassava breeding and selection programmes to reliably screen progenies at early stages of plant growth.

The IITA-selected cassava germplasm distributed as *in vitro* virus-tested plantlets is now established, evaluated and in some cases selected for further testing in multi-locational trials or released and cultivated by farmers in countries such as Angola, Botswana, Gambia, Malawi, Mozambique, Senegal, Tanzania, Uganda, Zambia, and Zimbabwe (Teri *et al.*, 1997). One of the recent examples is the contribution of IITA cassava germplasm to the release of five cassava varieties to farmers in Uganda. These varieties are resistant to the devastating cassava mosaic disease (CMD) in the country (Gatsby Charitable Foundation, 1997) and cover 100 000 ha (IITA, 1998b). The released cassava cultivars resistant to CMD are multiplied and distributed to the farmers to combat this disease.

In a similar manner, *D. rotundata* germplasm has been sent in the form of virus-tested plantlets and minitubers are established and are being evaluated in countries such as Benin, Burkina-Faso, the Gambia, Ghana, Guinea, Malawi, Togo, Uganda, and Zambia. Five genotypes of *D. rotundata* are currently maintained and distributed by the South Pacific Commission's Tissue Culture Laboratory which had been sent by IITA in the form of virus-tested plantlets (Devi, 1997). The selected genotypes are cultivated by farmers in the region (Taylor, personal communication).

Embryo culture technique and micropropagation have been applied to establish four mapping populations of cassava, aiming at developing markers for screening gene resistance to CMD (Akano *et al.*, 1997). Through this approach, the time it takes to multiply sufficient planting materials for the evaluation of these mapping populations is considerably shortened (from at least two growing cycles to one only). The mapping populations can also be sent to other countries for similar evaluation and at the same time also meet the phytosanitary regulations. A similar approach using meristem/nodal culture for the micropropagation of yam mapping populations is carried out at IITA (IITA, 1998a, 1998b) to multiply *D. alata* mapping populations, for mapping anthracnose resistance genes, and *D. rotundata* mapping populations, for mapping yam mosaic virus and nematode resistance genes (IITA, 1997). Along with other morphological traits, the results of these studies will contribute to the construction of a genetic linkage map in the Guinea yam. Similarly, a genetic linkage map of cassava has been constructed using 132 RFLPs, 30 RAPDs, three microsatellites, and three isozyme markers segregating from the heterogeneous female parent of an intraspecific cross (Fregene *et al.*, 1997) and this map will be saturated with additional markers.

13.6 Conclusions and future prospects

Different biotechnology techniques have been or are being developed and applied to both yam (Mantell and Boccon-Gibod, 1997) and cassava in the areas of germplasm exchange, conservation, molecular markers, and the development of genetic linkage maps. Among these techniques, meristem culture, virus indexing, micropropagation, and reduced growth storage are routinely used. The application of biotechnology in germplasm exchange has contributed significantly to the safe exchange of clonal germplasm and to allow testing and evaluation, leading to the selection and release of disease-resistant and high-yielding varieties. The reduced growth storage has contributed to the safeguarding of germplasm collections. Isozyme and molecular markers can be used to confirm duplicates which are identified using morphological characters, thus increasing confidence in the process of eliminating duplicate germplasm and reducing the conservation cost. Different molecular markers are used to study genetic diversity, identify cultivars and hybrids, study phylogeny and develop genetic linkage maps.

It is expected that biotechnology will be increasingly important in the conservation and use of the genetic resources of cassava and yams. Its application in germplasm exchange and conservation will continue to play an important role. Cryopreservation will be the main focus of research in order to develop reliable protocols for routine use in long-term conservation of cassava and yam. This will minimize the problems associated with genetic stability using reduced growth storage and will reduce the costs involved in maintaining the germplasm. Emphasis will be given to the development of molecular markers in yam and cassava for traits which are difficult to screen. Such traits are cyanogenesis and biodeterioration in cassava, and disease and pest resistance in cassava and yams. The development of genetic linkage maps will contribute significantly to the genetic studies of cassava and yams.

References

ACEDO, V.Z., 1995, Meristem culture and *in vitro* maintenance of Philippines cassava, in *Proceedings of the 2nd International Scientific Meeting of the Cassava Biotechnology Network*, Vol. 1., pp. 202–209, Colombia: CBN/CRIFC/AARD/CIAT.

ACHEAMPONG, E., 1996a, 'Issues affecting the field genebank management in Ghana', Presentation at the Consultation Meeting on the Management of Field and In Vitro Genebank, Colombia, January.

ACHEAMPONG, E., 1996b, '*In vitro* genebank management of clonally propagated crops under minimal conditions', Presentation at the Consultation Meeting on the Management of Field and In Vitro Genebank, Colombia, January.

AGYARE-TABBI, A., PEREIRA, L.F. and ERICKSON, L.R., 1997, Isolation and characterization of repetitive and microsatellite DNA sequences in cassava, in *Proceedings of the 3rd International Scientific Meeting of Cassava Biotechnology Network*, edited by A.M. THRO and M.O. AKORODA, *African Journal of Root and Tuber Crops*, 2(1/2), 135–137.

AKANO, A.O., NG, S.Y.C., DIXON, A.G.O. and THOTTAPPILLY, G.T., 1997, Application of embryo culture in establishing African cassava mosaic disease resistant gene mapping population, in *Book of Abstract of the 11th Symposium of the International Society for Tropical Root Crops*, Trinidad and Tobago, p. 61.

ALEXANDER, J. and COURSEY, D.G., 1969, The origin of yam cultivation, in *Domestication and Exploitation of Plants and Animals*, edited by P.J. UCKO and G.W. DIMBELSBY, pp. 405–426, London: Gerald Duckworth.

ALHASSAN, A.Y., 1991, Cultural factors affecting rapid clonal multiplication and microtuberization in shoot culture systems of Dioscorea alata L. food yam, MPhil Thesis, Wye College, University of London.

ALHASSAN, A.Y. and MANTELL, S.H., 1994, Manipulation of cultural factors to increase microtuber size and frequency in shoot cultures of food yam, *Dioscorea alata* L. c.v. Oriental Lisbon, in *Proceedings of 9th Symposium of International Society of Tropical Root Crops*, edited by F. OFORI and S.K. HAHN, pp. 342–348, Nigeria: ISTRC/Ghana/IITA.

AMMIRATO, P.V., 1984, Yams, in *Handbook of Plant Cell Culture*, Vol. 3: Crop Species, edited by P.V. AMMIRATO, D.A. EVANS, W.R. SHARP and Y. YAMADA, pp. 327–354, New York: Macmillan.

ANGEL, F., GIRALDO, F., GOMEZ, R., IGLESIAS, C., TOHOME, J. and ROCA, W.M., 1993, Use of RFLPs and RAPDs in cassava genomic studies, in *Proceedings of the 1st International Scientific Meeting of Cassava Biotechnology Network*, edited by W.M. ROCA and A.M. THRO, pp. 62–68, Colombia: CBN/CRIFC/AARD/CIAT.

ASEMOTA, H.N., 1995, A fast, simple, and efficient miniscale method for the preparation of DNA from tissues of yam (*Dioscorea* spp.), *Plant Molecular Biology Reporter*, 13(3), 214–218.

ASHMORE, S.E., 1997, *Status Report on the Development and Application of In Vitro Techniques for the Conservation and Use of Plant Genetic Resources*, Italy: IPGRI.

ASIEDU, R., HAHN, S.K., BAI, K.V. and DIXON, A.G.O., 1994, Interspecific hybridization in the genus *Manihot*: progress and prospects, in *Proceedings of the 9th Symposium of the International Society of Tropical Root Crops*, edited by F. OFORI and S.K. HAHN, pp. 110–113, Nigeria: ISTRC/Government of Ghana/IITA.

ASIEDU, R., WANYERA, N.M., NG, S.Y.C. and NG, N.Q., 1997, Chapter 5. Yams, in *Biodiversity in Trust*, edited by D. FUCCILLO, L. SEARS and P. STAPLETON, pp. 57–66, UK: Cambridge University Press.

BAJAJ, Y.P.S., 1983, Cassava plant from meristem cultures freeze-preserved for three years. *Field Crop Research*, 7, 161–167.

BATESON, J.M., 1996, '*In vitro* genebank management at CARDI tissue laboratory', Presentation at the Consultation Meeting on the Management of Field and In Vitro Genebank, Colombia, January.

BEECHING, J.R., MARMEY, P., HUGHES, M.A. and CHARRIER, A., 1995, Evaluation of molecular approaches for determining genetic diversity in cassava germplasm, in *Proceedings of the 2nd International Scientific Meeting of the Cassava Biotechnology Network*, Vol. 1., pp. 80–89, Colombia: CBN/CRIFC/AARD/CIAT.

BONIERBALE, M.W., MAYA, M.M., CLAROS, J.L. and IGLESIAS, C., 1995, Application of molecular markers to describing the genetic structure of cassava gene pools, in *Proceedings of the 2nd Scientific Meeting of the Cassava Biotechnology Network*, Vol. 1, pp. 106–112, Colombia: CBN/CRIFC/AARD/CIAT.

BONIERBALE, M., GUEVARA, C., DIXON, A.G.O., NG, N.Q., ASIEDU, R. and NG, S.Y.C., 1997a, Chapter 1, Cassava, in *Biodiversity in Trust*, edited by D. FUCCILLO, L. SEARS and P. STAPLETON, pp. 1–20, UK: Cambridge University Press.

BONIERBALE, M., ROA, A.C., MAYA, M.M., DUQUE, M.C. and THOME, J., 1997b, Assessment of genetic diversity in *Manihot* species with AFLPs (Abstract), in *Proceedings of the 3rd International Scientific Meeting of Cassava Biotechnology Network*, edited by A.M. THRO and M.O. AKORODA, *African Journal of Root and Tuber Crops*, **2**(1/2), 139.

BUSO, G.S.C., CARVALHO, L.J.C.B. and NASSAR, N.M.A., 1995, Analysis of genetic relationships among *Manihot* interspecific hybrids and their parental species, using RAPD assay, in *Proceedings of the 2nd Scientific Meeting of the Cassava Biotechnology Network*, Vol. 1, pp. 101–105, Colombia: CBN/CRIFC/AARD/CIAT.

BUTENKO, R.G., POPOV, A.S., VOLKOVA, I.A., CHEMYAK, N.D. and NOSOV, A.M., 1984, Recovery of cell cultures and their biosynthetic capacity after storage of *Dioscorea deltoidea* and *Panax genseng* cells in liquid nitrogen, *Plant Science Letter*, **33**, 285–292.

CARVALHO, L.J.C.B., CASCARDO, J.M.C., LIMEIRA, P.S., RIBEIRO, M.C.M., ALLEM, A.C. and FIALHO, J.F., 1993, Study of DNA polymorphism in *Manihot esculenta* Crantz and related species, in *Proceedings of the 1st International Scientific Meeting of Cassava Biotechnology Network*, edited by W.M. ROCA and A.M. THRO, pp. 56–61, Colombia: CBN.

CARVALHO, L.J.C.B., BUSO, G.S.C., BRANDONI, A., ALLEM, A.C., FUKUDA, W.M.G. and SAMPALO, M.J.A.M., 1995, Study on interspecific evolutionary relationships and intraspecific characterization of cassava germplasm at EENARGEN/EMBRAPA, in *Proceedings of the 2nd International Scientific Meeting of the Cassava Biotechnology Network*, Vol. 1, pp. 163–173, Colombia: CBN/CRIFC/AARD/CIAT.

CHATURVEDI, H.C., 1975, Propagation of *Dioscorea floribunda* from *in vitro* culture of single node stem segments, *Current Science*, **44**, 839–841.

CHISHIMBA, W.K. and LINGUMBWANGA, E.S., 1997, *In vitro* propagation of six Zambian clones of cassava (*Manihot esculenta* Crantz), in *Proceedings of the 3rd International Scientific Meeting of Cassava Biotechnology Network*, edited by A.M. THRO and M.O. AKORODA, *African Journal of Root and Tuber Crops*, **2**(1/2), 212–213.

CIAT, 1980, *CIAT Annual Report 1980*, pp. 33–37, Colombia: CIAT.

COURSEY, D.H., 1967, *Yam*, London: Longman.

COURSEY, D.G., 1972, The origin and domestication of yams in Africa, in *The Origin of the African Plant: Domestication*, edited by J.R. HARLEN, pp. 383–408, The Hague: Mouton Publishers.

DANIEL, I.O., NG, N.Q. and TAYO, T.O., 1995, 'Potential of seed storage for long-term conservation of yam germplasm', Presentation at the 6th Triennial Symposium of International Society of Tropical Root Crops – Africa Branch, Malawi, October.

DE GOES, M., MENDES, R.A., LABUTO, L.B.D. and CARDOSO, L.D., 1996, 'Present status of germplasm conservation in vitro at CENARGEN/EMBRAPA, Brazil', Presentation at a Consultation Meeting on the Management of Field and In Vitro Genebanks, Colombia, January.

DEVI, S., 1997, *Role of the South Pacific Commission's Tissue Culture Laboratory in support of Pacific Agriculture*, Suva, Fiji: South Pacific Commission.

ESCOBAR, R.H., ROCA, W.M. and MAFLA, G., 1993, Cryopreservation of cassava shoot tips, in *Proceedings of the 1st International Scientific Meeting of the Cassava Biotechnology Network*, edited by W.M. ROCA and A.M. THRO, pp. 116–119, Colombia: CBN.

ESCOBAR, R.H., MAFLA, G. and ROCA, W.M., 1995, Cryopreservation for long-term conservation of cassava genetic resources, in *Proceedings of the 2nd Scientific Meeting of the Cassava Biotechnology Network*, Vol. 1, pp. 190–193, Colombia: CBN/CRIFC/AARD/CIAT.

FAO, 1996, *FAOSTAT Database Gateway*, Italy: FAO.

FAUTRET, A., DUBLIN, P. and CHAGVARDIEFF, P., 1985, 'Callus formation and neoformation with two edible *Dioscorea* species: *D. alata* and *D. trifida*' Presentation at *7th Symposium International Society of Tropical Root Crops*, Guadeloupe, July.

FORSYTH, C. and VAN STADEN, J., 1982, An improved method of *in vitro* propagation of *Dioscorea bulbifera*, Plant Cell, Tissue and Organ Culture, **1**, 275–281.

FREGENE, M., ANGEL, F., GOMEZ, R., RODRIGUEZ, F., MAYA, M., BONIERBALE, M., TOHME, J., IGLESIAS, C. and ROCA, W., 1995, A linkage map of cassava (*Manihot esculenta* Crantz) based on RFLP and RAPD markers, in *Proceedings of the 2nd Scientific Meeting of the Cassava Biotechnology Network*, Vol. 1, pp. 49–61, Colombia: CBN/CRIFC/AARD/CIAT.

FREGENE, M., ANGEL, F., GOMEZ, R., RODRIGUEZ, F., CHAVARRIAGA, P., BONIERBALE, M., ROCA, W. and TOHME, J., 1997, A molecular genetic map of cassava (*Manihot esculenta* Crantz) (Abstract), in *Proceedings of the 3rd International Scientific Meeting of Cassava Biotechnology Network*, edited by A.M. THRO and M.O. AKORODA, *African Journal of Root and Tuber Crops*, **2**(1/2), 150–151.

GATSBY CHARITABLE FOUNDATION, 1997, *Mastering Mosaic: The Fight for Cassava Production in Uganda*, UK: Gatsby Charitable Foundation.

GOMEZ, R., ANGEL, F., BONIERBALE, M.W., RODRIGUEZ, F., TOHME, J. and ROCA, W.M., 1995, Selecting heterozygous parents and single-dose markers for genetic mapping in cassava, in *Proceedings of the 2nd International Scientific Meeting of The Cassava Biotechnology Network*, Vol. 1, pp. 113–124, Colombia: CBN/CRIFC/AARD/CIAT.

GUEVARA, C.L. and MAFLA G., 1996, '*Manihot in vitro* collections held at CIAT', Presentation at a Consultation Meeting of the Management on Field and In Vitro Genebanks, Colombia, January.

GULICK, P., HERSHEY, C.H. and ESQUINAS-ALCAZAR, J., 1983, *Genetic Resources of Cassava and Wild Relatives*, Italy: IBPGR Secretariat.

GUO, J.Y. and Y.O. LIU, 1995, Rapid propagation of cassava by tissue culture and its application in rural districts in China, in *Proceedings of the Second Scientific Meeting of the Cassava Biotechnology Network*, Vol. 1, pp. 183–189, Colombia: CBN/CRIFC/AARD/CIAT.

HAHN, S.K., 1995, Yams, *Dioscorea* spp. (Dioscoreaceae), in *Evolution of Crop Plants*, edited by J. SMARRT and N.W. SIMMONDS, pp. 112–120, London: Longman.

HAMON, P. and TOURE, B., 1990a, Characterization of traditional yam varieties belonging to the *Dioscorea cayenensis–rotundata* complex by their isozymic patterns, *Euphytica*, **46**, 101–107.

HAMON, P. and TOURE, B., 1990b, The classification of the cultivated yams (*Dioscorea cayenensis–rotundata* complex) of West Africa, *Euphytica*, **47**, 179–187.

HAMON, P. and TOURE, B., 1991, New trends for yam improvement in the *Dioscorea cayenensis–rotundata* complex, in *Crop Genetic Resources of Africa*, Vol. II, edited by N.Q. NG, P. PERRINO, F. ATTERE and H. ZEDAN, pp. 119–125, Nigeria: IITA/IBPGR/UNEP/CNR.

HAMON, P., DUMONT, R., ZOUNDJIHEKPON, J., TIO-TOURE, B. and HAMON, S., 1995, *Wild Yams in West Africa: Morphological Characteristics*, France: ORSTOM.

HAYSOM, H.R., CHAN, T.L.C., LIDDLE, S. and HUGHES, M.A., 1995, Phylogenetic relationships of *Manihot* species revealed by restriction fragment length polymorphism, in *Proceedings of the 2nd Scientific Meeting of the Cassava Biotechnology Network*, Vol. 1, pp. 125–134. CBN/CRIFC/AARD/CIAT.

HERSHEY, C.H., 1994, *Manihot* genetic diversity, in *Report of 1st Meeting of the International Network for Cassava Genetic Resources, International Crop Network*, Series No. 10, pp. 111–134, Italy: IPGRI.

IITA, 1993, *Root and Tuber Improvement Program Archival Report (1989–1993), Part III, Yams (Dioscorea spp.)*, pp. 13–14, Nigeria, IITA.

IITA, 1997, *Project 15, Recombinant DNA, Molecular Diagnostics and Cellular Biotechnology for Crop Improvement, Annual Report 1996*, p. 28, Nigeria: IITA.

IITA, 1998a, *Project 15, Recombinant DNA, Molecular Diagnostics and Cellular Biotechnology for Crop Improvement, Annual Report 1997*, Nigeria: IITA.

IITA, 1998b, *Project 14, Cassava Productivity in the Lowland and Mid-altitude Agroecologies of Sub-Saharan Africa, Annual Report 1997*, Nigeria: IITA.

IPGRI/IITA, 1997, *Descriptor for Yam (Dioscorea spp.)*, Italy: International Institute of Tropical Agriculture/International Plant Genetic Resources Institute.

JEAN, M. and CAPPADOCIA, M., 1991, In vitro tuberization in *Dioscorea alata* L. 'Brazo fuerte' and 'Florido', and *D. abyssinica* Hoch., *Plant Cell, Tissue and Organ Culture*, **26**, 147–152.

JENNINGS, D.L., 1976, Cassava, *Manihot esculenta* (Euphorbiaceae), in *Evolution of Crop Plants*, edited by N. SIMMONDS, pp. 81–84, London: Longman.

JOSEPH, S.S. and YEOH, H.H., 1995, Structural variation in cassava chloroplast DNA, in *Proceedings of the 2nd Scientific Meeting of the Cassava Biotechnology Network*, Vol. 1, pp. 135–140, CBN/CRIFC/AARD/CIAT.

KARTHA, K.K. and GAMBORG, O.L., 1975, Elimination of cassava mosaic disease by meristem culture, *Phytopathology*, **65**, 826–828.

KOHMURA, H., ARAKI, H. and IMOTO, M., 1995, Micropropagation of 'Yamatoimo' Chinese yam (*Dioscorea opposita*) from immature leaves, *Plant Cell, Tissue and Organ Culture*, **40**, 271–276.

LAMINSKI, S., ROBINSON, E.R. and GRAY, V.M., 1997, Application of molecular markers to describe South African elite cassava cultivars, in *Proceedings of the 3rd International Scientific Meeting of Cassava Biotechnology Network*, edited by A.M. THRO and M.O. AKORODA, *African Journal of Root and Tuber Crops*, 2(1/2), 132–134.

LAUZER, D., LAUBLIN, G., VINCENT, G. and CAPPADOCIA, M., 1992, *In vitro* propagation and cytology of wild yams, *Dioscorea abyssinica* Hoch. and *D. mangenotiana* Miege, *Plant Cell Tissue and Organ Culture*, **28**, 215–223.

LEE, R.H., 1997, Cell–cell interactions during the induction of asexual embryogenic determination in yam (*Dioscorea* spp.) leaf tissues, PhD thesis, Wye College, University of London.

LI, H.Q., HUANG, Y.W., LIANG, C.Y. and GUO, J.Y., 1995, Improvement of plant regeneration from secondary somatic embryos of cassava, in *Proceedings of the 2nd Scientific Meeting of the Cassava Biotechnology Network*, Vol. 1, pp. 289–302, CBN/CRIFC/AARD/CIAT.

MAFLA, G., ROCA, W., REYES, R., ROA, J.C., MUNOZ, L., BACA, A.E. and IWANAGA, M., 1993, In vitro management of cassava germplasm at CIAT, in *Proceedings of the 1st International Scientific Meeting of Cassava Biotechnology Network*, edited by W.M. ROCA and A.M. THRO, pp. 168–174, Colombia: CBN.

MANDAL, B.B. and CHANDEL, K.P.S., 1995, Cryopreservation of encapsulated shoot-tips of yams (*Dioscorea* spp.) for long-term conservation, *IPGRI Newsletter for Asia, the Pacific and Oceania*, **19**, 14.

MANTELL, S.H. and BOCCON-GIBOD, J., 1997, 'Progress towards a realisation of biotechnological tools for genetic improvement of *Dioscorea* spp.', Presentation at Seminaire International sue l'Igname: plante seculaire et culture d'avenir-acquis et perspective de la recherche, France, June.

MANTELL, S.H. and HUGO, S.A., 1986, International germplasm transfer using micropropagules, in *Proceedings of Training Workshop and Symposium on Micropropagation and Meristem Culture/Vegetative Propagation*, CSC Technical Publication Series No. 205, pp. 88–98.

MANTELL, S.H., HAQUE, S.Q. and WHITEHALL, A.P., 1979, A rapid propagation system for yams, *Yam Virus Project Bulletin*, **1**, Barbados: CARDI.

MANTELL, S.H., HAQUE, S.Q. and WHITEHALL, A.P., 1980, Apical meristem tip culture for eradication of flexous rod viruses in yam (*Dioscorea alata*), *Tropical Pest Management*, **26**, 170–179.

MARIN, M.L., MAFLA, G., ROCA, W.M. and WITHERS, L.A., 1990, Cryopreservation of cassava zygotic embryos and whole seeds in liquid nitrogen, *Cryo-Letters*, **1**, 251–264.

MIGNOUNA, H.D. and DIXON, A.G.O., 1997, Genetic relationships among African cassava clones with varying levels of resistance to African cassava mosaic disease using RAPD markers, in *Proceedings of the 3rd International Scientific Meeting of Cassava Biotechnology Network*, edited by A.M. THRO and M.O. AKORODA, *African Journal of Root and Tuber Crops*, **2**(1/2), 28–32.

MIGNOUNA, H.D., ASIEDU, R., THOTTAPPILLY, G. and HAHN, S.K., 1993, 'Study on the genetic diversity and phylogenetic relationships in the genus *Dioscorea* using PCR-RAPD', Presentation at Symposium on Africa and the Challenge of Plant Biotechnology: Case of Yam, Cite d'Ivoire, April.

MIGNOUNA, H.D., ASIEDU, R. and THOTTAPPILLY, G., 1995, 'Molecular taxonomy of cultivated and wild yams in West Africa', Presentation at the 6th Triennial Symposium of ISTRCAB, Malawi, October.

MIGNOUNA, H.D., ASIEDU, R., NG, N.Q., KNOX, M. and ELLIS, N.T.H., 1997, Analysis of genetic diversity in Guinea yams (*Dioscorea* spp.) using AFLP finger printing, in *Book of Abstracts of the 11th Symposium of the International Society for Tropical Root Crops*, p. 70, Trinidad and Tobago.

MORALES, S.R., 1996, 'Conservation "In vitro" del germplasma de raices y tubercules tropicales, platano y bananos en la Republica de Cuba'. Presentation at the Consultation Meeting on the Management of Field and in vitro Genebank, Colombia, January.

MROGINSKI, L.E. and SCOCCHI, A., 1993, Somatic embryogenesis of Argentine cassava varieties, in *Proceedings of the 1st International Scientific Meeting of Cassava Biotechnology Network*, edited by W.M. ROCA and A.M. THRO, pp. 175–179, Colombia: CBN.

MURASHIGE, T. and SKOOG, F., 1962, A revised medium for rapid growth and bioassay with tobacco tissue culture, *Physiologia Plantarum*, **15**, 473–497.

NARAYANASWAMY, P., THANGAVELU, M., BAILEY, T., TRUKSA, M. and MANTELL, S.H., 1994, 'Genetic variability studies in yams (*Dioscorea* spp.) using RAPD markers', Presentation at the VIIIth International Congress of Plant Tissue and Cell Culture, Abstract S6-63, Firenze, Italy.

NG, S.Y., 1986, Virus elimination in sweet potato, yam and cocoyam, in *Proceedings of a Regional Workshop on Root and Tuber Crops Propagation*, pp. 97–102, Colombia: CIP/IITA/UNDP/CIAT.

NG, S.Y.C., 1988, *In vitro* tuberization in white yam (*Dioscorea rotundata* Poir), *Plant Cell Tissue and Organ Culture*, **14**, 121–128.

NG, S.Y.C., 1992a, Micropropagation of white yam (*Dioscorea rotundata* Poir), in *Biotechnology in Agriculture and Forestry, Vol. 19: High-tech and Micropropagation III*, edited by Y.P.S. BAJAJ, pp. 135–159, Berlin: Springer-Verlag.

NG, S.Y.C., 1992b, Embryo culture and somatic embryogenesis in cassava, in *Proceedings of the 4th Triennial Symposium of the International Society for Tropical Root Crops – Africa Branch*, edited by M.O. AKORODA and O.B. ARENE, pp. 129–131, Nigeria: ISTRC-AB/IITA/CTA.

NG, S.Y.C. and ADENIYI, O.J., 1994, Somatic embryogenesis in African-adapted cassava and evaluation of regenerates, in *Book of Abstracts of 2nd International Scientific Meeting of Cassava Biotechnology Network*, p. 56, Indonesia: CBN.

NG, S.Y. and HAHN, S.K., 1985, Application of tissue culture to tuber crops at IITA, in *Proceedings of the Inter-centre Seminar on International Agricultural Research Centers (IARCS) and Biotechnology*, pp. 29–40, Philippines: IRRI.

NG, S.Y.C. and MANTELL, S.H., 1996, *Technologies for Germplasm Conservation and Distribution of Pathogen-free Dioscorea Yams to National Root Crop Research Programs*, ODA Project R 4886 (H) Final Report, Nigeria: IITA/Wye College/SARI.

NG, S.Y.C. and MANTELL, S.H., 1997, Influence of carbon source on *in vitro* tuberization and growth of white yam (*Dioscorea rotundata* Poir.), in *Book of Abstracts of the 11th Symposium of the International Society for Tropical Root Crops*, p. 26, Trinidad and Tobago.

NG, S.Y.C. and NG, N.Q., 1991, Reduced growth storage of germplasm, in *In Vitro Methods for Conservation of Plant Genetic Resources*, edited by J.H. DODDS, pp. 11–39, London: Chapman and Hall.

NG, N.Q. and NG, S.Y., 1994, Approaches for yam germplasm conservation, in *Proceedings of the 5th Triennial Symposium of the International Society for Tropical Root Crops – Africa Branch*, edited by M.O. AKORODA, pp. 135–140, Nigeria: ISTRC-AB/IITA/CTA.

NG, S.Y.C. and NG, N.Q., 1997a, Germplasm conservation in food yams (*Dioscorea* spp.): constraints, application and future prospects, in *Conservation of Plant Genetic Resources In Vitro, Vol 1, General Aspects*, edited by M.K. RAZDAN and E.C. COCKING, pp. 257–286, USA: Science Publishers, Inc.

NG, S.Y.C. and NG, N.Q., 1997b, Cassava *in vitro* germplasm management at the International Institute of Tropical Agriculture, in *Proceedings of the 3rd International Scientific Meeting of Cassava Biotechnology Network*, edited by A.M. THRO and M.O. AKORODA, *African Journal of Root Crops*, **2**(1/2), 232–233.

NG, N.Q., ASIEDU, R. and NG, S.Y.C., 1994a, Cassava genetic resources programme at the International Institute of Tropical Agriculture, Ibadan, in *Report of the 1st Meeting of the International Network for Cassava Genetic Resources*, International Crop Network Series No. 10, pp. 71–76, Italy: IPGRI.

NG, S.Y.C, NGU, M.A. and LADEINDE, T.A.O., 1994b, Embryo culture of yams: germination and callus induction, in *Proceedings of the 5th Triennial Symposium of the International Society for Tropical Root Crops – Africa Branch*, edited by M.O. AKORODA, pp. 141–144, Nigeria: ISTRC-AB/IITA/CTA.

NG, N.Q., FAWOLE, I. and UDOH, E.A., 1997a, *In situ* conservation of yams, in *Project 16 Conservation and Genetic Enhancement of Plant Biodiversity Annual Report 1996*, p. 5, Nigeria: IITA.

NG, N.Q., TOGUN, A.O. and DANIEL, I.O., 1997b, Storage of seeds, pollen and tuber of yam, in *Project 16 Conservation and Genetic Enhancement of Plant Biodiversity Annual Report 1996*, p. 5, Nigeria: IITA.

OCAMPO, C., HERSHEY, C., IGLESIAS, C. and IWANAGA, M., 1993, Esterase isozyme fingerprinting of CIAT cassava germplasm collection, in *Proceedings of the 1st International Scientific Meeting of Cassava Biotechnology Network*, edited by W.M. ROCA and A.M. THRO, pp. 81–89, Colombia: CBN.

OCAMPO, C., ANGEL, F., JIMENEZ, A., JARAMILLO, G., HERSHEY, C., GRANADOS, E. and IGLESIAS, C., 1995, DNA fingerprinting to confirm possible genetic duplicates in cassava germplasm, in *Proceedings of the 2nd Scientific Meeting of the Cassava Biotechnology Network*, Vol. 1, pp. 145–151, CBN/CRIFC/AARD/CIAT.

OKEZIE, C.F.A., NWOKE, F.I.O. and OKONKWO, S.N.C., 1984, *In vitro* culture of *Dioscorea rotundata* embryo, in *Proceedings of 2nd Triennial Symposium of International Society for Tropical Root Crops – Africa Branch*, pp. 121–124, Nigeria: ISTRC-AB/IITA/CTA.

OKOLI, O.O., 1991, Yam germplasm diversity, uses and prospects for crop improvement in Africa, in *Crop Genetic Resources of Africa*, Vol. II, edited by N.Q. NG, P. PERRINO, F. ATTERE and H. ZEDAN, pp. 109–117, Nigeria: IITA/IBPGR/UNEP/CNR.

OSIFO, E.O., 1988, Somatic embryogenesis in *Dioscorea*, *Journal of Plant Physiology*, **133**, 378–380.

PUONTI-KAERLAS, J., LI, H.Q., SAUTTER, C. and POTYKUS, I., 1997, Production of transgenic cassava (*Manihot esculenta* Crantz) via organogenesis and *Agrobacterium*-mediated transformation, in *Proceedings of the 3rd International Scientific Meeting of Cassava Biotechnology Network*, edited by A.M. THRO and M.O. AKORODA, *African Journal of Root and Tuber Crops*, **2**(1/2), 181–186.

RAEMAKERS, C.J.J.M., SOFIARI, E., KANJU, E., JACOBSEN, E. and VISSER, R.G.F., 1995, NAA-induced somatic embryogenesis in cassava, in *Proceedings of the 2nd Scientific Meeting of the Cassava Biotechnology Network*, Vol. 1, pp. 355–363, CBN/CRIFC/AARD/CIAT.

RAMSER, J., LOPEZ-PERALTA, C., WETZEL, R. WEISING, K. and KAHL, G., 1996, Genomic variation and relationships in aerial yam (*Dioscorea bulbifera* L.) detected by random amplified polymorphic DNA, *Genome*, **39**, 17–25.

ROCA, W.M., 1984, Chapter 10, Cassava, in *Handbook of Plant Cell Culture, Vol. 2, Crop Species*, edited by W.R. SHARP, D.A. EVANS, P.V. AMMIRATO and Y. YAMADA, pp. 269–301, USA: Macmillan.

ROCA, W.M., 1985, In vitro clonal propagation to eliminate crop diseases, in *Proceedings of the Inter-Center Seminar on International Agricultural Research Centers (IARCs) and Biotechnology*, pp. 1–10, Philippines: IRRI.

ROGERS, D.J., 1965, Some botanical and ethnological considerations of *Manihot esculenta*, *Economic Botany*, **19**(4), 369–377.

SARRIA, R., OCAMPO, C., RAMIREZ, H., HERSHEY, C. and ROCA, W.M., 1993, Genetics of esterase and glutamate oxalo-acetate transaminase isozymes in cassava, in *Proceedings of the 1st International Scientific Meeting of Cassava Biotechnology Network*, edited by W.M. ROCA and A.M. THRO, pp. 75–80, Colombia: CBN.

SCHAAL, B., OLSON, P., PRINZIE, T., CARVALHO, L.J.C.B., TONUKARI, N.J. and HAYWORTH, D., 1995, Phylogenetic analysis of the genus *Manihot* based on molecular markers, in *Proceedings of the 2nd Scientific Meeting of the Cassava Biotechnology Network*, Vol. 1, pp. 62–70, CBN/CRIFC/AARD/CIAT.

SCHAAL, B., CARVALHO, L.J.C.B., PRINZIE, T., OLSEN, K., OLSON, P., CABRAL, G. and HERNANDEZ, M., 1997, Phylogenetic relationships among *Manihot* species, in *Proceedings of the 3rd International Scientific Meeting of Cassava Biotechnology Network*, edited by A.M. THRO and M.O. AKORODA, *African Journal of Root and Tuber Crops*, **2**(1/2), 147–149.

SCHOPKE, C., CARCAMO, R., BEACHY, R.N. and FAUQUET, C., 1997, Plant regeneration from transgenic and non-transgenic embryogenic suspension cultures of cassava (*Manihot esculenta* Crantz), in *Proceedings of the 3rd International Scientific Meeting of Cassava Biotechnology Network*, edited by A.M. THRO and M.O. AKORODA, *African Journal of Root and Tuber Crops*, **2**(1/2), 194–195.

SECOND, G., ALLEM, A.C., EMPERAIRE, L., INGRAM, C., COLOMBO, C., MENDES, R.A. and CARVALHO, L.J.C.B., 1997, AFLP based *Manihot* and cassava numerical taxonomy and genetic structure analysis in progress: implications for dynamic conservation and genetic mapping, in *Proceedings of the 3rd International Scientific Meeting of Cassava Biotechnology Network*, edited by A.M. THRO and M.O. AKORODA, *African Journal of Root and Tuber Crops*, **2**(1/2), 140–146.

SHAHIN, E.A. and SHEPARD, J.F., 1980, Cassava mesophyll protoplasts: isolation, proliferation, and shoot formation, *Plant Science Letter*, **17**, 459–465.

STAMP, J.A. and HENSHAW, G.G., 1982, Somatic embryogenesis in cassava, *Z. Pflanzenphysiol.*, **105**, 183–187.

STAMP, J.A. and HENSHAW, G.G., 1987a, Somatic embryogenesis from clonal leaf tissues of cassava, *Annals of Botany*, **59**, 445–450.

STAMP, J.A. and HENSHAW, G.G., 1987b, Secondary somatic embryogenesis and plant regeneration in cassava, *Plant Cell Tissue and Organ Culture*, **10**, 227–233.

SUDARMONOWATI, E. and BACHTIAR A.S., 1995, Induction of somatic embryogenesis in Indonesian cassava genotypes, in *Proceedings of the 2nd Scientific Meeting of the Cassava Biotechnology Network*, Vol. 1, pp. 364–374, CBN/CRIFC/AARD/CIAT.

SUDARMONOWATI, E. and HENSHAW, G.G., 1990, Cryopreservation of cassava somatic embryos and embryogenic tissue, in *Book of Abstract of International Congress for Plant Tissue and Cell Culture*, p. 140, Amsterdam, Netherlands.

TAYLOR, M., 1996a, 'Field conservation of root and tuber crop in the South Pacific', Presentation at the Consultation Meeting on the Management of Field and In Vitro Genebank, Colombia, January.

TAYLOR, M., 1996b, 'In vitro conservation of root and tuber crops in the South Pacific', Presentation at the Consultation Meeting on the Management of Field and In Vitro Genebank, Colombia, January.

TAYLOR, N.J., EDWARD, M. and HENSHAW, G.G., 1995, Production of friable embryogenic calli and suspension culture systems in two genotypes of cassava, in *Proceedings of the 2nd Scientific Meeting of the Cassava Biotechnology Network*, Vol. 1, pp. 229–240, CBN/CRIFC/ AARD/CIAT.

TAYLOR, N.J., KIERNAN, R.J., HENSHAW, G.G. and BLAKESLEY, D., 1997, Improved procedures for producing embryogenic tissues of African cassava cultivars: implications for genetic transformation, in *Proceedings of the 3rd International Scientific Meeting of Cassava Biotechnology Network*, edited by A.M. THRO and M.O. AKORODA, *African Journal of Root and Tuber Crops*, **2**(1/2), 200–203.

TERAUCHI, R., TERACHI, T. and TSUNEWAKI, K., 1989, Physical map of chloroplast DNA of aerial yam, *Dioscorea bulbifera* L., *Theoretical and Applied Genetics*, **78**, 1–10.

TERAUCHI, R., CHIKALEKE, V.A., THOTTAPPILLY, G. and HAHN, S.K., 1992, Origin and phylogeny of Guinea yams as revealed by RFLP analysis of chloroplast DNA and nuclear ribosomal DNA, *Theoretical and Applied Genetics*, **83**, 743–751.

TERI, J., WHYTE, J., EKANAYAKE, I., BIELER, P., MAHUNGU, N., ANDRADE, M., KHIZZAH, B. and MUIMBA-KANKOLONGO, A., 1997, Establish primary multiplication centres for production and distribution of cleaning planting materials, in *Project 14 Cassava Productivity in the Lowland and Mid-altitude Agroecologies of Sub-Saharan Africa Annual Report 1996*, pp. 45–46, Nigeria: IITA.

TILQUIN, J.P., 1979, Plant regeneration from stem callus of cassava, *Canadian Journal of Botany*, **57**, 1761–1763.

TONUKARI, N.J., THOTTAPPILLY, G., NG, N.Q. and MIGNOUNA, H.D., 1997, Genetic polymorphism of cassava within the Republic of Benin detected with RAPD markers, *African Crop Science Journal*, **5**(3), 219–228.

TWYFORD, C.T. and MANTELL, S.H., 1996, Production of somatic embryos and plantlets from root tip cells of the greater yam, *Plant cell, Tissue and Organ Culture*, **46**, 17–26.

TWYFORD, C.T., VIANA, A.M., JAMES, A.C. and MANTELL, S.H., 1990, Characterization of species and vegetative clones of *Dioscorea* food yams using isoelectric focusing of peroxidase and acid phosphatase isoenzymes, *Tropical Agriculture (Trinidad)*, **67**(4), 337–341.

VIANA, A.M. and MANTELL, S.H., 1989, Callus induction and plant regeneration from excised zygotic embryos of seed-propagated yams, *Dioscorea composita* Hemsi. and *D. cayenensis* Lam., *Plant Cell, Tissue and Organ Culture*, **16**, 113–122.

WITHERS, L.A., 1994, Constraints and prospects of complementary conservation methods and strategies for cassava germplasm, in *Report of the 1st Meeting of the International Network for Cassava Genetic Resources*, International Crop Network Series No. 10, pp. 135–140, Italy: IPGRI.

ZAMORA, A.B. and PAET, C.N., 1996, 'In vitro genebanking activities, Institute of Plant Breeding, College of Agriculture, University of the Philippines at Los Banos', Presentation at the Consultation Meeting on the Management of Field and In Vitro Genebank, Colombia, January.

ZOK, S., NYOCHEMBENG, L.M., TAMBONG, J. and WUTOH, J.G., 1993, Rapid seed stock multiplication of improved clones of cassava through shoot tip culture in Cameroon, in *Proceedings of the 1st International Scientific Meeting of Cassava Biotechnology Network*, edited by W.M. ROCA and A.M THRO, pp. 96–104, Colombia: CBN.

Conservation Biotechnology of Endemic and other Economically Important Plant Species of India

BINAY B. MANDAL

14.1 Introduction

India is a major centre of origin and diversity of crop plants. It occupies a special significance among the major gene-rich countries of the world owing to its immensely rich landrace diversity in major agri-horticultural crops and their wild relatives. Presently about 18 000 species of flowering plants exist in India of which nearly a third are endemic. These include 166 domesticated species of economic importance along with over 320 species of their wild relatives and around 500 species of medicinal value (Arora, 1988). The valuable plant genetic resources are, however, being lost at a very fast rate because of their replacement by high yielding varieties or due to heavy pressure on their natural habitats.

Realizing the importance of conservation of this national heritage, the National Bureau of Plant Genetic Resources (NBPGR) was established in 1976 with the national responsibility for the collection, evaluation, conservation and exchange of germplasm of various agri-horticultural crops. While germplasm conservation and exchange of plant species bearing orthodox seeds were being carried out satisfactorily, vegetatively propagated species and those bearing recalcitrant seeds were posing serious problems (Table 14.1). To minimize these problems, therefore, a need was felt to apply emerging *in vitro* techniques for the conservation and exchange of these problem species.

In India, experiments on *in vitro* conservation of endemic medicinal plants were initiated in the 1980s at several research centres such as the National Botanical Research Institute, Lucknow (Chaturvedi *et al.*, 1982) and the Department of Botany, Delhi University, Delhi (Bhojwani *et al.*, 1989). However, to make a concerted research effort in various aspects of *in vitro* conservation, NBPGR launched a special project, 'the National Facility for Plant Tissue Culture Repository' (NFPTCR) in 1986 with the financial assistance from the Department of Biotechnology, Government of India. The facility has made significant progress in the *in vitro* conservation, cryopreservation and molecular characterization of germplasm of various plant species. Today, NBPGR has become an acknowledged centre for its outstanding advancements in carrying out comprehensive work on the *in vitro* conservation of nearly two dozen crop species (about 45 plant species) of both tropical and temperate nature.

In this chapter, the major research accomplishments at NBPGR in developing and utilizing plant conservation biotechnology are described. The *in vitro* conservation activities at other Indian research centres are summarized and the implications and future prospects of these developments in the sustainable management of germplasm of endemic and/or economically important plant species of India are discussed.

14.2 *In vitro* techniques in conservation

Conservation of germplasm, in general, involves activities such as collection, propagation, characterization, evaluation, disease indexing and elimination, storage and distribution. At NBPGR, *in vitro* techniques have been used for the conservation of clonally propagated and recalcitrant seed species involving almost all the above mentioned activities. The principles of all these methods, together with practical details, are presented in Part I of this book.

For example, during the course of the *in vitro* establishment of garlic and sweet potato, mosaic like symptoms were observed in some stock plants which were then investigated for virus detection and diagnosis. Immunological studies (DAS-ELISA) in *Allium* revealed the presence of garlic mosaic virus (GMV) in several accessions of garlic and *A. ascalonicum* (Kumar *et al.*, 1990). Similarly, electron microscopy (EM, ISEM) and DAC-indirect-ELISA helped in detecting sweet potato feathery mottle virus (SPFMV) and one unidentified potyvirus in sweet potato (Kumar *et al.*, 1991). Meristem culture technique was standardized for sweet potato (Mandal and Chandel, 1990) and *Allium* spp. Meristem culture techniques, combined with virus indexing using ELISA and grafting (sweet potato) are currently being used to establish clean, healthy cultures in sweet potato and garlic in our NBPGR laboratory. (See also Martin and Postman, Chapter 5, this volume for further details.)

In the area of *in vitro* exchange and the distribution of germplasm, previously the most serious problems were the lack of infrastructure facilities and micropropagation protocols. Following the establishment of a tissue culture facility, NBPGR has been playing a vital role for the last few years in exchange and distribution of germplasm using tissue culture. More than 180 accessions of banana, indigenous obtained from the International Network for the Improvement of Banana and Plantain (INIBAP) transit centre, Leuven, Belgium, were multiplied *in vitro*, at NBPGR and established in a greenhouse and supplied in the form of suckers to various other research institutes and user scientists in the country. Similarly, cultures of nearly 75 accessions of sweet potato obtained from the International Potato Centre (CIP), Lima, Peru, were multiplied, conserved and supplied to indentors. In the case of germplasm exchange, 70 accessions of sweet potato were multiplied *in vitro* and supplied, on request, to the CIP Regional Station at New Delhi in several consignments for its exchange with CIP, Lima, Peru (see Golmirzaie *et al.*, Chapter 12, this volume), Bangladesh, Nepal, Sri Lanka and the University of Abertay Dundee, Scotland. However, the core research areas in *in vitro* conservation where NBPGR has made significant achievements are as described below.

14.3 Propagation

Clonal propagation of plant germplasm through tissue culture (see Lynch, Chapter 4, this volume) for the rapid production of plantlets is an important prerequisite for *in vitro*

Table 14.1   Species being worked upon for *in vitro* conservation at NBPGR

Crop species	Origin	Problems in field conservation	Economic importance
Banana and plantain	India/SE Asia	Virus	Major fruit crop
Ginger	India	Rhizome rot	Major spice crop
Black pepper	India	Fungal wilt	Spice/cash crop
Turmeric	India	–	Export commodity
Sweet potato	S. America	Weevils, viruses	Important root crop
Taro	India, Malaysia	–	Vegetable crop
Yams	Asia and Africa	–	Vegetable crop
Garlic and *Allium* spp.	Asia	Virus	Condiments
Rauvolfia spp.	India	Endangered	Medicinal plant
Saussurea lappa	India	Endangered	Medicinal plant
Coleus forskohlii	India	Vulnerable	Medicinal plant
Tylophora indica	India	Over exploited	Medicinal plant
Gentiana kurroo	India	Threatened	Medicinal plant
Picrorhiza kurroa	India	Threatened	Medicinal plant
Tea	China	New planting	Major beverage
Jackfruit	India	Fungal disease	Popular tree
Cocoa	S. America	–	Nutritious food
Neem	India	Minor fungus	Medicinal plant
Cardamom	India	Disease and insect	Spice crop
Almond	Asia	Disease and insect	Dry fruit
Citrus	India	Rust and blight	Major fruit crop

conservation. Many of the indigenous species of India have not been investigated regarding their amenability to micropropagation. Thus, at our laboratory, micropropagation protocols were developed for a number of crop plant species, viz. *Piper* spp. (black pepper, betel vine, etc.) (Bhat *et al.*, 1992b, 1995b), *Allium tuberosum* (Pandey *et al.*, 1992), *Alocasia indica* (Mandal, unpublished), *Dioscorea esculenta, D. hispida, D. pentaphylla*, and *D. wallichii* (edible Asian yams) (Mandal and Chandel, 1993). *In vitro* mass multiplication was also achieved in a number of endemic medicinal species which are on the verge of extinction such as *Gentiana kuroo, Picrorhiza kuroa, Coleus forskohlii, Saussurea lappa*, and *Tylophora indica* (Sharma *et al.*, 1995).

While micropropagation protocols for some of the crop species such as banana, sweet potato, ginger, several species of yams, and garlic have been reported previously, these protocols had to be refined further to be applicable to diverse species and genotypes. Most of the available micropropagation protocols were generally based on restricted genotypes and were adapted to achieve the best response from specific materials which were used for commercial purposes. Germplasm conservation programmes, on the other hand demand a protocol that ensures a sufficiently effective response over the whole range of genotypes of a crop species. Further, the objectives of micropropagation and *in vitro* conservation are quite opposite in terms of their growth requirements and it is necessary to consider this when devising media for conservation. Therefore, during the standardization of micropropagation protocols the approach at our laboratory has been to devise a simple medium capable of eliciting satisfactory responses from all the genotypes of the crop plant, obviating the need for preparing a variety of medium for each one/or a few genotypes. Many of these media comprise Murashige and Skoog (MS) medium (Murashige and Skoog, 1962) supplemented with benzyl adenine purine (BAP), kinetin (Kn), indole-3-acetic acid (IAA) and naphthalene acetic acid (NAA). For example, the medium for banana devised at our laboratory (MS + 3 mg/l BAP for multiplication and MS + 1 mg/l BAP for routine subculture) supports the multiplication and maintenance of all our accessions and the related wild species tested (Bhat and Chandel, 1993). Similarly, in sweet potato, 260 accessions are multiplied and maintained through periodical subculturing using a single medium (MS + 0.2 mg/l Kn + 0.1 mg/l IAA) and in yams, 41 accessions of six species were multiplied on a single medium (MS + 0.25 mg/l Kn + 0.25 mg/l NAA) and are currently subcultured and maintained on another medium (MS + 0.15 mg/l NAA).

14.3.1   *Somatic embryogenesis*

Regeneration through somatic embryogenesis was standardized for ginger, yam (*D. bulbifera*) and sweet potato. In all these systems long-term embryogenic cultures were established. Detailed studies on somatic embryogenesis in ginger (Kackar *et al.*, 1993) and *Dioscorea bulbifera* (Mandal, unpublished) are of special significance for cryopreservation and other biotechnological studies.

14.3.2   *Production of storage organs* in vitro

Storage organs are of special significance in *in vitro* storage and germplasm exchange (see also Golmirzaie *et al.*, Chapter 12, this volume). Protocols for development of *in vitro* rhizomes were developed in ginger (Bhat *et al.*, 1994b). Similar success (Figure 14.1) was registered for the induction of microtubers in several species of yams (*D. alata, D. bulbifera, D. pentaphylla, D. wallichii* and *D. floribunda*) and bulblets in Alliums.

14.4   Medium-term conservation

In vitro medium term conservation aims to reduce the growth rate of cultures to avoid frequent subculture. To induce slow growth low temperature incubation of cultures has been used for many species (see also Golmirzaie *et al.*, Chapter 12, this volume). Low temperature incubation in combination with media manipulation is also reported to be useful for some species. However, the potential of various other strategies of conservation for different species is not fully explored and comparative data on the feasibility of storage under non-inhibitory conditions are not available. Therefore, for a multi-crop *in vitro* repository such as at NBPGR, and in an Indian context, it was felt necessary to evolve an appropriate strategy for the safe and economic conservation of germplasm holdings.

    Various slow growth strategies, such as low temperature incubation, the use of osmotic agents, growth retardants, nutritional or hormonal manipulations, etc. were evaluated for different crop species. Low temperature incubation, in general, was effective in prolonging subculture intervals to 12–20 months in almost all crop species (Table 14.2). However, optimum temperatures varied with the crop species. For example, the optimum temperatures for Alliums and other temperate plants was 4–10°C, whereas banana and

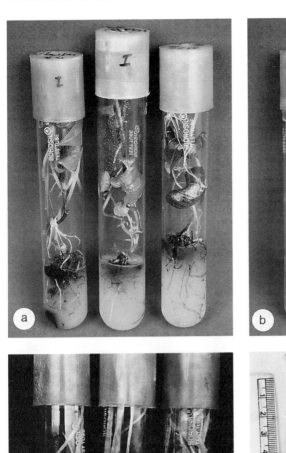

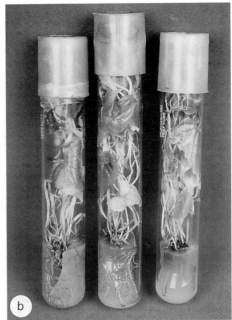

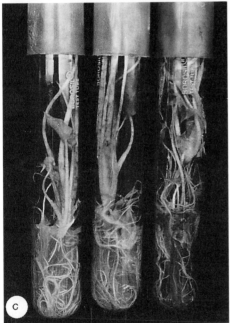

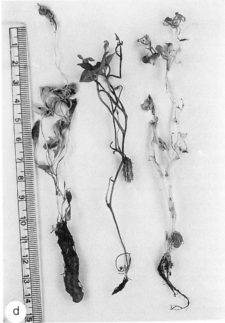

Figure 14.1 Conservation of *in vitro* plants at culture room temperature (25°C) on normal multiplication media. (a) Plantlets of *Dioscorea alata* maintained under normal growth conditions, ten months after storage; (b) plantlets of *D. floribunda* maintained under normal growth conditions, ten months after storage; (c) plantlets of *Alocasia indica* maintained under normal growth conditions (left to right: six, six and ten months after storage); (d) microtuber formation in wild edible yams (left to right: *D. pentaphylla* (basal), *D. pentaphylla* (aerial), *D. wallichii* (aerial)

Table 14.2 Status of *in vitro* medium-term conservation of germplasm at NBPGR

Species (no. of accessions)	Storage conditions	Storage period	References
Dioscorea species (41)	25°C 25°C, 0.5–1 mg/l Kn in medium	8–10 months 12–16 months	Mandal and Chandel (1996)
Ipomoea batatas (260)	25°C, 1–2% mannitol in medium 25°C, 2–2.5% sucrose in medium	12–14 months 12–14 months	Mandal and Chandel (1996)
Musa species (255)	25°C, 15°C	10–12 months 20–22 months	Bhat and Chandel (1993)
Zingiber species (140)	25°C, 25°C, 9% sucrose in medium (induced rhizome) 15°C	6–8 months 24 months 18 months	Bhat and Chandel (1996a)
Curcuma species (35)	25°C	6–8 months	Balachandran *et al.* (1990)
Piper spp. (7)	25°C	12–15 months	Bhat and Chandel (1996b)
Allium sativum and other wild *Allium* species (99)	25°C, 25°C, 9% sucrose in medium (bulblet induction) 4°C, 10°C	8–12 months 14–23 months 14–23 months	Pandey *et al.* (1992)
Rauvolfia serpentina	15°C	15 months	Sharma and Chandel (1992)
Rauvolfia species (7), *Coleus forskohlii*, (8), *Tylophora indica* (2)	25°C, 25°C 25°C	20 months 18 months 12 months	Sharma *et al.* (1995)
Gentiana kuroo (1), *Picrorhiza kurroa* (1), *Saussurea lappa* (1)	4°C, 10°C, 4°C	11 months 15 months 15 months	Sharma *et al.* (1995)

In vitro cultures maintained at 25°C were under 12–16 hour photoperiods in the culture room.

ginger could not be stored at a temperature below 15°C. Similarly, the growth of yams was greatly reduced at around 20°C. Furthermore, wide genotypic differences were recorded among genotypes of a crop species for low temperature tolerance. These results clearly showed the need for maintaining several low temperature facilities/incubators running at different temperatures for the conservation of germplasm derived from various crops.

The NBPGR laboratory is located at New Delhi where the mean year temperature is around 30°C with a minimum temperature of 100°C in January and a maximum of 45°C in June.

Under such tropical conditions the establishment and running costs of several low temperatures facilities are very high. Further, there is considerable risk of loss of germplasm due to power/equipment failure. Considering these limitations, therefore, the strategy at NBPGR was to develop protocols to conserve all the germplasm of various species at ambient culture room temperature (25°C) so that maintenance of a single temperature regime is then required. Maintenance of cultures under stress is reported to have adverse effects on many species. Thus, the strategy at the NBPGR laboratory has been to maintain cultures under normal growing conditions or under minimal stress.

14.5 Conservation at normal culture room temperature (25°C)

14.5.1 *Conservation under normal growing conditions*

This method was based on the storage of shoot cultures at a normal culture room temperature (25°C) under normal growing conditions. Studies at the NBPGR laboratory with several species such as *Musa*, yams, *Piper* spp. and *Alocasia indica* indicated that cultures of all these species could be maintained on their respective simple shoot culture media for 10–12 months at 25°C without subjecting them to growth inhibitory treatments (Figure 14.1). After initial fast growth for 3–4 months cultures entered into stationary phase and were sustained on limited nutrients available to them for the next 6–8 months. In fact, normal multiplication media, if not devised for maximum proliferation rate, could help maintain cultures of several species under non-inhibitory conditions for a considerable period of time (10–15 months) with a high percentage of survival.

14.5.2 *Conservation with media manipulation*

Medium-term conservation at 25°C was also attempted with some media manipulations to extend subculture intervals. Results of studies at the NBPGR laboratory revealed that minor modification of medium helped in maintaining cultures of several species at 25°C for 12 or more months. For example, media supplementation with a moderate dose of mannitol (1–2 per cent) could help maintain cultures of sweet potato for 12 or more months at 25°C. While genotypic variation in response to mannitol does exist, the concentration range of 1–2 per cent was suitable for maintaining 260 accessions tested without showing any serious adverse effect on the health of the cultures. A reduced dose of sucrose (2–2.5 per cent) in the medium could also maintain the cultures of sweet potato at 25°C for 12 months (Mandal and Chandel, 1996). Similarly, an increase of Kinetin from 0.25 mg/l in the normal multiplication medium to 0.5–1 mg/l could maintain cultures of yams for 12–16 months at 25°C (Mandal and Chandel, 1996).

14.5.3 *Conservation using induced storage organs*

Manipulation of media/culture conditions also helped in the induction of storage organs in several species, which in turn increased the shelf life of cultures at 25°C. In ginger, shoot cultures produced rhizomes on a high sucrose medium (9–12 per cent) and survived up to 24 months (Bhat and Chandel, 1996a). Similarly, garlic produced *in vitro* bulblets on high sucrose medium (10 per cent) and yams produced tubers on medium with NAA under reduced photoperiod/light intensity. These storage organs, when allowed to remain as such in culture vessels helped maintain both Alliums and yams *in vitro* at 25°C for up to 23 and 18 months respectively.

Table 14.3   Status of cryopreservation research at NBPGR

Species	Explant used	Freezing method	Results	References
Dioscorea alata, D. bulbifera, D. wallichii, D. floribunda	*In vitro* shoot tips	Encapsulation–dehydration, slow thawing	71% survival, up to 37% shoot regeneration with two species	Mandal *et al.* (1996)
Camellia sinensis	Embryonic axes	Rapid drying to 13.3% moisture, rapid freezing	95% survival	Chaudhury *et al.* (1991)
Artocarpus heterophyllus	Embryonic axes	Rapid drying to 14% moisture, rapid freezing	30% survival	Chandel *et al.* (1995)
Poncirus trifoliata	Embryonic axes	Rapid drying to 14% moisture, rapid freezing	68% recovery	Radhamani and Chandel (1992)
Prunus amygdalus	Embryonic axes	Rapid drying to 7% moisture, rapid freezing	100% survival, 66% regeneration	Chaudhury and Chandel (1995a)
Azadirachta indica	Seeds	4–13.5% moisture, rapid freezing	57–65% survival	Chaudhury and Chandel (1991)
Musa balbisiana	Seeds	Air drying to 13% moisture, rapid freezing	90% germination	Bhat *et al.* (1994c)
Piper nigrum	Seeds	Rapid drying to 12% moisture, rapid freezing	45% survival	Chaudhury and Chandel (1994)
Elettaria cardamomum	Seeds	Rapid drying to 8–14% moisture, rapid freezing	80% germination	Chaudhury and Chandel (1995b)

Rapid drying was carried out by placing embryonic axes on filter paper in a laminar air flow and seeds on silica gel.

14.6   Long-term conservation using cryopreservation

Cryopreservation is the only viable option available for the long-term conservation of germplasm of clonally propagated and recalcitrant seed species. India, having a large area with a humid tropical climate, has many economically important species bearing recalcitrant seeds such as mango, coconut, jackfruit, litchi, sapota, and walnut.

Further, NBPGR has established a large *in vitro* collection of vegetatively propagated species. Management of large collections, even if the subculture intervals are greatly extended, poses considerable problems (Roca *et al.*, 1989). Thus, emphasis was given to cryopreservation research at our laboratory and protocols for cryopreservation of several recalcitrant seed species and clonally propagated species have been developed. A base collection of over 100 accessions of recalcitrant seed species including tea (75 accessions) has also been established under cryopreservation.

14.6.1 *Cryopreservation of recalcitrant seed species*

Cryopreservation using zygotic embryos/embryonic axes

Successful cryopreservation using a simple method of desiccation of embryonic axes followed by rapid freezing was earlier reported for several tropical seed species. At the NBPGR laboratory, studies with embryonic axes of three predominantly recalcitrant seed species such as tea, jackfruit and cocoa revealed that partially and fully matured embryonic axes of tea and jackfruit could be desiccated to about 14 per cent moisture content and successfully cryopreserved. The degree of sensitivity of these axes to desiccation varied with the physiological maturity of the seeds (Chandel *et al.*, 1995; Chaudhury *et al.*, 1991). Similar success in cryopreservation using embryonic axes was registered in trifoliate orange (Radhamani and Chandel, 1992) and almond (Chaudhury and Chandel, 1995a).

Cryopreservation using seeds

Seed cryopreservation may be advantageous in certain recalcitrant and sub-orthodox seed species as it can avoid the inherent problems of embryo isolation and handling. Seeds of neem with/without endocarp could be desiccated and cryopreserved (Chaudhury and Chandel, 1991). Sub-orthodox seeds of pepper (Chaudhury and Chandel, 1994) and cardamom (Chaudhury and Chandel, 1995b) could also be processed and successfully cryopreserved.

Conservation of clonally propagated species using vegetative propagules is preferred because it can maintain integrity of genotype. However, wild species of various clonally propagated species can be conserved using seed as in this case conservation in the form of the gene pool may be adequate. In wild species of banana, cryopreservation using zygotic embryos has been reported earlier (Abdelnour-Esquivel *et al.*, 1992). However, at the NBPGR laboratory, cryopreservation of wild banana (*M. balbisiana*) has been simplified using seeds with a high percentage of germination (Bhat *et al.*, 1994a).

14.6.2 *Cryopreservation of clonally propagated species*

Shoot-tip/axillary bud cryopreservation is by far the best method for cryopreservation of clonally propagated species. Shoot-tips of four species of *Dioscorea* (*D. alata*, *D. bulbifera*, *D. wallichii* and *D. floribunda*) were successfully cryopreserved using the technique of encapsulation–dehydration (Figure 14.2). *In vitro* grown shoot-tips were encapsulated with 3 per cent sodium alginate, pre-cultured in 0.75 M sucrose for 48–72 hours and desiccated under laminar air flow for 4–6 hours. The treated shoot-tips could tolerate rapid freezing in liquid nitrogen with up to 75 per cent survival and ability to regenerate plants up to 37 per cent in *D. alata* and *D. wallichii* (Mandal *et al.*, 1996).

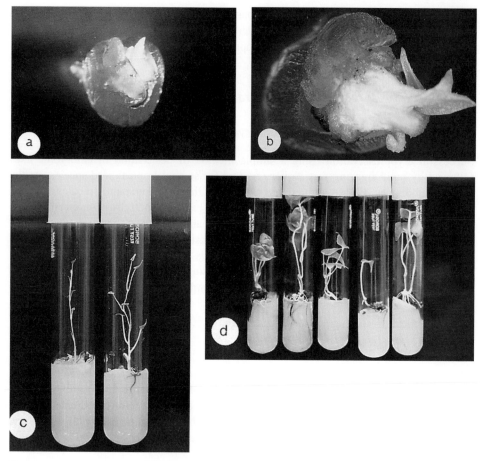

Figure 14.2 Regeneration of plantlets from encapsulated shoot apices of *Dioscorea wallichii* before (control) and after freezing in liquid nitrogen. (a, b) Shoot regeneration from frozen apices subcultured for three (a) and four (b) weeks after thawing; (c) plantlet regeneration from control (left) and frozen (right) apices; (d) fully developed plantlets from frozen apices.

14.7 Characterization, classification and monitoring

14.7.1 *Characterization and classification of germplasm*

Conservation and increased utilization of plant species stored in an *in vitro* repository requires detailed characterization, classification and documentation. Further, *in vitro* conservation of clonally propagated species has limitations on the number of accessions that can be conserved because *in vitro* conservation is relatively expensive and the management of large collections poses considerable problems. Therefore, identification of duplicates and developing core collections through characterization are of paramount importance in the *in vitro* conservation of germplasm. At our laboratory, studies on characterization and classification using various cytological, biochemical and molecular techniques were carried out on 600 accessions of more than 15 crop plant species namely, banana, ginger, *Curcuma*, yams, garlic, sweet potato, and several medicinal and aromatic plants. (See also Harris, Chapter 2, this volume for details of molecular approaches for assessing plant diversity.)

Case study: banana

Several variant forms of banana and plantains are under cultivation in India. The prevalence of large-scale synonyms and homonyms, and lack of standard documentation systems for cultivars has resulted in a confusion in cultivar identification. Thus, polymorphism in isozymes was studied in 130 accessions of banana and plantains. Seven enzyme systems were determined as ideal for identification of distinct cultivars and their classification. However, isozyme analysis could not distinguish between closely related cultivars (Bhat *et al.*, 1992a). Detailed RAPD analyses of 57 cultivars using 60 random 10-mer primers helped in selecting a set of 12 primers which was found to be sufficient for the unambiguous identification and classification of cultivars into genomic groups. Using these 12 primers a total of 150 cultivars were characterized and classified (Bhat and Jarret, 1995). The resulting groupings of the cultivars were comparable to their classification based on a set of morphological characters described by earlier workers. The reliability of the results of RAPD analysis was confirmed by RFLP analysis of nuclear DNA (Bhat *et al.*, 1994a). About 22 nuclear DNA probes were used and a very high degree of polymorphism was observed among 60 cultivars analysed. The clustering pattern of cultivars was comparable to the results obtained with RAPD data (Bhat *et al.*, 1995a).

Examples of other species

Analysis of isozyme polymorphism in 70 accessions of *Curcuma* representing 21 morphological groups revealed high genetic diversity. However, a similar study in 12 selected cultivars of ginger (using eight enzymes) revealed the prevalence of a very narrow genetic diversity among the collections. In garlic, isozyme polymorphism of 23 accessions collected from diverse agro-climatic regions of India indicated that there was less genetic diversity among the germplasm analysed. These results indicated a large number of duplicates among germplasm collections (Bhat and Chandel, 1991). Characterization of 70 accessions of taro using isozyme and RAPD analysis, 20 accessions of six species of yams and 39 accessions of sweet potato using isozyme polymorphism revealed that moderate to high numbers of polymorphs exist in these species.

14.7.2 *Monitoring genetic stability of conserved germplasm*

One of the most important aspects of the application of *in vitro* methods for germplasm conservation is the assessment of the genetic stability of conserved plants. Therefore, monitoring stability of plants conserved under *in vitro* medium-term conservation or cryopreservation is vital for any *in vitro* conservation programme. In this context, NBPGR has given careful consideration to the methods that can be applied to assess maintenance of genetic stability. (See also Harding, Chapter 7, this volume.)

The genetic stability of *in vitro* collections held at NBPGR is being periodically evaluated using morphological, cytological and biochemical (isozyme, alkaloid) analysis. For example, in sweet potato isozyme analysis was performed on *in vitro* regenerated plants to study the effect of micropropagation on maintenance of genetic stability of regenerants (Lakhanpaul *et al.*, 1990). In the course of further studies, 78 regenerants, belonging to four different accessions of sweet potato conserved *in vitro* for more than 1.6 years were analysed using morphological characters and isozyme polymorphism of five enzyme systems. All the regenerants analysed exhibited unaltered morphology and isozyme profiles indicating maintenance of genetic stability of conserved germplasm. Similarly, studies on genetic stability of tissue culture regenerated plants in *Coleus foskohlii* by biochemical analysis (forskolin content) (Sharma *et al.*, 1991) and in *Allium tuberosum* by cytological analysis (Rao *et al.*, 1992) indicated their unaltered characters. Large numbers of *Musa* cultivars maintained *in vitro* for several years were transferred to the field from time to time. These regenerants showed similar morphological characters including fruiting behaviour in field conditions. Although results of these studies indicated maintenance of genetic stability of conserved germplasm, further studies involving all the crops are necessary. Emphasis on the analysis and maintenance of genetic stability of all the crops existing in the repository is an important consideration. Further, the methods used so far are effective in determining changes in genetic products; they do not allow us to determine changes in genes. Therefore, emphasis is currently placed on the use of molecular genetic markers (RAPD, RFLP, SSR and AFLP) for the genetic analysis of our conserved germplasm. (See also Harding, Chapter 7 and Harris, Chapter 2, this volume.)

14.8 *In vitro* conservation activities at other research stations in India

Several other research stations under governmental, non-governmental (NGOs) and self-financing autonomous organizations in India are actively involved in application of *in vitro* techniques in the conservation of plant genetic resources. The crop based research institute, the Indian Institute for Spices Research (IISR), Calicut, has established and maintained cultures of several indigenous economically important spices such as black pepper, cardamom, ginger and turmeric. Similarly, the Central Tuber Crops Research Institute (CTCRI), Trivandrum, maintained some collections of sweet potato *in vitro* with periodical subculture. The Central Potato Research Institute (CPRI), Shimla, maintains large collections of potato germplasm (mostly exotic). However, emphasis on *in vitro* conservation of indigenous medicinal plants has been given by the Tropical Botanical Garden and Research Institute (TBGRI), Trivandrum, and the Central Institute for Medicinal and Aromatic Plants (CIMAP), Lucknow. Collections of various indigenous orchids are also being maintained *in vitro* at the research centre in the North East Hilly Region of the country.

Among the various Indian NGOs involved in the conservation of plant diversity, the M.S. Swaminathan Research Foundation (MSSRF), Madras, and Regional Plant Resources Centre, Bhubaneswar, are in the forefront. These centres have made significant achievements in the application of biotechnology for the conservation and characterization of indigenous plant diversity. MSSRF has continued its effort in micropropagation, conservation and reintroduction to the field of many endemic, economically useful and medicinal endangered species. These reintroduced plants are being monitored for genetic variability using RAPD markers. This research centre has also undertaken work on micropropagation and molecular genetic characterization (RAPD and RFLP) of important mangrove plants. Similarly, RPRC, Bhubaneswar, has made significant contributions in micropropagation, *in vitro* conservation, and reintroduction to the field and molecular characterization of large numbers of indigenous plant species.

14.9 Conclusions

Research in the development and application of *in vitro* techniques for the conservation of plant germplasm has significantly advanced in India. At NBPGR, micropropagation protocols have been developed and utilized for several endemic species which are on the verge of extinction. The conservation of cultures at 25°C has many practical advantages. Usual tissue culture facilities can be utilized for conservation without additional cost on infrastructure/equipment. The risk of loss of cultures due to equipment/power failure is also low as maintaining 25°C is not too difficult. Therefore, medium-term conservation through tissue culture is a technology that is readily available for many threatened endemic economically important plant species of India. Cryopreservation is now practicable for base collections of several important indigenous recalcitrant seed species formerly maintained only as field collections. Use of embryonic axes/seeds for cryopreservation of these species is easy and technically less demanding. Both cryopreservation and *in vitro* medium-term conservation technologies are available to provide added security to important clonally propagated and recalcitrant seed species of the country.

Several other research centres in India are also actively engaged in the application of *in vitro* techniques to save valuable indigenous plant species. They are using biotechnology for mass multiplication of endangered, economically important plant species, their reintroduction to the natural habitats, and molecular characterization. These are undoubtedly very important steps towards conservation of these species before they become extinct. The message of using conservation has been well received throughout the country and simple, economical protocols are now available. In the future conservation and biotechnology are sure to contribute significantly to the sustainable management of the endemic and/or economically important plant species of India.

References

ABDELNOUR-ESQUIVEL, A., MORA, A. and VILLALOBOS, V., 1992, Cryopreservation of zygoticembryos of *Musa acuminata* (AA) and *Musa balbisiana* (BB), *Cryo-Letters*, **13**, 159–164.

ARORA, R.K., 1988, The Indian gene center-priorities and prospects for collection, in PORADA, R.S., ARORA, R.K. and CHANDEL, K.P.S. (Eds), *Plant Genetic Resources Indian Perspective*, pp. 499–516, New Delhi: NBPGR.

BHAT, K.V. and CHANDEL, K.P.S., 1991, Isozyme polymorphism and genetic diversity in garlic (*Allium sativum*) germplasm, *Indian Journal of Plant Genetic Resources*, **4**(2), 50–56.

BHAT, S.R. and CHANDEL, K.P.S., 1993, *In vitro* conservation of *Musa* germplasm. Effects of mannitol and temperature on growth storage, *Journal of Hortscience*, **68**, 841–846.

BHAT, S.R. and CHANDEL, K.P.S., 1996a, Tissue culture conservation of ginger germplasm: significance of *in vitro* rhizomes, *2nd Intl. Crop Sci. Cong.*, Nov. 17–24, New Delhi, p. 335 (Abstract).

BHAT, S.R. and CHANDEL, K.P.S., 1996b, Biodiversity conservation – new biotechnological approaches, in CHOPRA, V.L., SHARMA, R.P. and SWAMINATHAN, M.S. (Eds), *Agricultural Biotechnology. 2nd Asia Pacific Conference*, Madras, Oxford and IBH Publishing Company Pvt. Ltd, New Delhi, pp. 3–12.

BHAT, K.V. and JARRET, R.L., 1995, Random amplified polymorphic DNA and genetic diversity in Indian Musa germplasm, *Genetic Resources and Crop Evolution*, **42**, 107–118.

BHAT, K.V., BHAT, S.R. and CHANDEL, K.P.S., 1992a, Survey of isozyme polymorphism for clonal identification in *Musa*. 1. Esterases acid phosphatase and catalase, *Journal of Horticultural Science*, **67**(4), 501–507.

BHAT, S.R., KACKER, A. and CHANDEL, K.P.S., 1992b, Plant regeneration from callus cultures of *Piper nigrum* L. by organogenesis, *Plant Cell Reports*, **11**, 525–528.

BHAT, K.V., JARRET, R.L. and LIU, Z., 1994a, Diversity in chloroplast and genomic DNA for germplasm characterization, clonal identification and classification of Indian *Musa* germplasm, *Euphytica*, **80**, 95–104.

BHAT, S.R., CHANDEL, K.P.S. and KACKAR, A., 1994b, *In vitro* induction of rhizomes in ginger *Zingiber officinale* Rosocoe, *Indian Journal of Experimental Biology*, **32**, 340–344.

BHAT, S.R., BHAT, K.V. and CHANDEL, K.P.S., 1994c, Studies on germination and cryopreservation of *Musa balbisiana* seed, *Seed Science and Technology*, **22**, 637–640.

BHAT, K.V., JARRET, R.L. and RANA, R.S., 1995a, DNA profiling of banana and plantain cultivars using RAPD and RFLP markers, *Electrophoresis*, **16**, 1736–1745.

BHAT, S.R., CHANDEL, K.P.S. and MALIK, S.K., 1995b, Plant regeneration from various explants of cultivated *Piper* species, *Plant Cell Reports*, **14**, 398–402.

BHOJWANI, S.S., ARUMUGAM, N., ARORA, R. and UPADHYAYA, R.P., 1989, *In vitro* conservation of some endangered plant species of India, *Indian Journal of Plant Genetic Resources*, **2**, 103–113.

CHANDEL, K.P.S., CHAUDHURY, R., RADHAMANI, J. and MALIK, S.K., 1995, Desiccation and freezing sensitivity in recalcitrant seeds of tea, cocoa and jackfruit, *Annals of Botany*, **76**, 443–450.

CHATURVEDI, H.C., SHARMA, A.K., SHARMA, M. and PRASAD, R.N., 1982, Morphogenesis, micropropagation, and germplasm preservation of some economic plants by tissue cultures, in FUGIWARA, A. (Ed.), *Proceedings of the International Congress of Plant Tissue Culture*, Maruze, Tokyo, pp. 687–688.

CHAUDHURY, R. and CHANDEL, K.P.S., 1991, Cryopreservation of desiccated seeds of neem (*Azadirachta indica* A. Juss), *Indian Journal of Plant Genetic Resources*, **4**(2), 67–72.

CHAUDHURY, R. and CHANDEL, K.P.S., 1994, Germination studies and cryopreservation of seeds of black pepper (*Piper nigrum* L.) – A recalcitrant species, *Cryo-Letters*, **15**, 145–150.

CHAUDHURY, R. and CHANDEL, K.P.S., 1995a, Cryopreservation of embryonic axes of almond (*Prunus amygdalus* Batsch.) seeds, *Cryo-Letters*, **16**, 51–56.

CHAUDHURY, R. and CHANDEL, K.P.S., 1995b, Studies on germination and cryopreservation of cardamom (*Elettaria cardamomum* maton) seeds, *Science and Technology*, **23**, 235–240.

CHAUDHURY, R., RADHAMANI, J. and CHANDEL, K.P.S., 1991, Preliminary observations on the cryopreservation of desiccated embryonic axes of tea (*Camellia sinensis* (L.) O. Kuntze) seeds for genetic conservation, *Cryo-Letters*, **12**, 31–36.

KACKAR, A., BHAT, S.R., CHANDEL, K.P.S. and MALIK, S.K., 1993, Plant regeneration via somatic embryogenesis in ginger, *Plant Cell Tissue Organ Culture*, **32**, 289–292.

KUMAR, C.A., CHANDEL, K.P.S., PANDEY, R., MANDAL, B.B., JAIN, R.K. and VARMA, A., 1990, Detection of viruses in plant germplasm under *in vitro* conservation programme, *VI Annual Convention of Indian Virological Society*, December 17–19, Abstract.

KUMAR, C.A., MANDAL, B.B., CHANDEL, K.P.S., JAIN, R.K., VARMA, A. and SHRIVASTAVA, M., 1991, Occurrence of sweet potato feathery mottle virus in germplasm of *Ipomoea batatas* L. in India, *Current Science*, **60**, 321–325.

LAKHANPAUL, S., MANDAL, B.B. and CHANDEL, K.P.S., 1990, Isozyme studies in the *in vitro* regenerated plants of *Ipomoea batatas* (L.) lanmark, *Journal Root Crops*, ISRC Special 17, 305–310.

MANDAL, B.B. and CHANDEL, K.P.S., 1990, Utilization of tissue culture technique in preservation of sweet potato germplasm, *Journal of Root Crops*, ISRC Special, **17**, 291–295.

MANDAL, B.B. and CHANDEL, K.P.S., 1993, *In vitro* conservation of edible Asian yams (*Dioscorea* spp.), ISPGR Dialogue on Plant Genetic Resources: Developing National Policy, December, 1–2, New Delhi, pp. 83–84, Abstract.

MANDAL, B.B. and CHANDEL, K.P.S., 1996, Conservation of genetic diversity in sweet potato and yams using *in vitro* strategies, in KURUP, G.T., PALANISWAMI, M.S., POTTY, V.P., PADAMAJA, G., KABEERATHUMMA, S. and PILLAI, SANTHA V. (eds.) *Tropical Tuber*

Crops: Problems, Prospects and Future Strategies, Proceedings of the International Symposium, November 6–9, 1993, Trivandrum, Oxford and IBH Publishing Company Pvt. Ltd., pp.
49–54.

MANDAL, B.B., CHANDEL, K.P.S. and DWIVEDI, S., 1996, Cyropreservation of yam (*Dioscorea*
spp.), shoot apices by encapsulation-dehydration, *Cryo-Letters*, **17**, 165–174.

MURASHIGE, T. and SKOOG, F., 1962, A revised medium for rapid growth and bioassays with
tobacco tissue cultures, *Physiologia Plantarum*, **15**, 473–497.

PANDEY, R., CHANDEL, K.P.S. and RAO, S.R., 1992, *In vitro* propagation of *Allium tuberosum*
Rottl. ex-Spring by shoot proliferation, *Plant Cell Reports*, **11**, 375–378.

RADHAMANI, J. and CHANDEL, K.P.S., 1992, Cryopreservation of embryonic axes of trifoliate
orange (*Poncirus trifoliata* (L) raf), *Plant Cell Reports*, **11**, 372–374.

RAO, S.R., PANDEY, R. and CHANDEL, K.P.S., 1992, Genetic stability studies in regenerated
plants of *Allium tuberosum* Rottl. ex-Spring – A cytological approach, *Cytologia*, **57**, 339–
347.

ROCA, W.M., CHAVEZ, R., MARIN, M.L., ARIAS, D.I., MAFLA, G. and REYES, R., 1989, *In
vitro* methods of germplasm conservation, *Genome*, **31**, 813–817.

SHARMA, N. and CHANDEL, K.P.S., 1992, Low temperature storage of *Rauvolfia serpentina*
Benth. Ex Kurz., an endangered, endemic medicinal plant, *Plant Cell Reports*, **11**, 200–203.

SHARMA, N., CHANDEL, K.P.S. and SRIVASTAVA, V.K., 1991, *In vitro* propagation of *Coleus
forskohlii* Briq., a threatened medicinal plant, *Plant Cell Reports*, **10**, 67–70.

SHARMA, N., CHANDEL, K.P.S. and PAUL, A., 1995, *In vitro* conservation of threatened plants of
medicinal importance, *Indian Journal of Plant Genetic Resources*, **8**, 107–112.

The Application of Biotechnology for the Conservation of Endangered Plants

VALERIE C. PENCE

15.1 Introduction

It is estimated that there are more than 270 000 plant species in existence and approximately 34 000, or 1 in 8, of these are considered endangered (IUCN, 1998). Competition with increasing human populations and the resulting loss of habitat are contributing to an increasing rate of plant species extinction. Some researchers have estimated that one quarter of plant species are at risk of extinction within the next generation (Raven, 1987). Approximately one-fifth of the 20 000 native species in the United States are of concern, and possibly 800 of these may be lost in the next decade unless attention is given to their conservation (Center for Plant Conservation, 1995).

There are several approaches to avoiding these losses. The preservation of species *in situ* is of primary importance for maintaining the broadest range of plant diversity. As a supplement to this, *ex situ* preservation can play a role in backing up taxa which are particularly threatened or are rare in the wild. The maintenance and propagation of species in botanical gardens and arboreta as well as in seed and spore banks has traditionally provided a valuable safeguard against loss for many rare species (Laliberté, 1997).

Modern biotechnologies offer the potential of extending these traditional *ex situ* preservation and propagation methods to an even broader range of taxa and tissues types. These techniques have been developed primarily for agricultural and horticultural species, but are increasingly being applied to collecting, propagating, preserving, and evaluating rare and endangered plant germplasm, as well.

15.2 Special aspects of endangered species work

The limited amount of plant material available is an over-riding factor in endangered species work. The ability to test protocols and even to conduct replicated experiments may be severely limited, and generally, work with related, non-endangered species is used as a guide (Campos and Pais, 1996; McComb, 1985). If only a small number of seeds are produced by a species, the use of non-seed tissues may be advised. In addition, plants may be located in remote or difficult-to-access areas and collecting trips may be expensive.

Permits are often also required for any collections that are made, as well as for transport and importation.

The resources available for work with endangered species are also often limited, compared with economically important species. In addition, there is often a disparity in the resources allocated to the conservation of plants, in comparison with animals. For example, in the United States, only 3 per cent of federal funding for the conservation of endangered species is directed towards plants, even though they comprise one-third of the total species listed as endangered or threatened in the United States (US Fish and Wildlife Service, 1995).

Despite these limitations, a significant amount of work has been done with endangered plant species using biotechnologies to solve specific problems in the areas of propagation, germplasm preservation, collection and analysis. These will be described and some examples of their application and of the particular needs of individual species will be given.

15.3   *In vitro* propagation technologies and endangered plant species

Although many endangered plant species can be propagated successfully by seed or by cuttings, there are some species that do not reproduce well by these traditional methods. One of the earliest programmes to use *in vitro* propagation methods for rare and endangered species was the Micropropagation Unit at the Royal Botanic Gardens, Kew, established in 1974. Since that time, there have been several reviews on the use of *in vitro* methods for endangered species (Fay, 1992, 1994; Fay and Gratton, 1992; Wochok, 1981), and *in vitro* propagation techniques have been used successfully for a number of rare or endangered species (see Table 15.1 and González-Benito *et al.*, Chapter 16, this volume). For specific details of the micropropagation techniques see Lynch (Chapter 4, this volume).

Certain groups of plants have received particular attention. Techniques of *in vitro* seed germination have been applied to a number of rare orchid species, while endangered cacti and succulents and insectiverous plants and lilies are other groups which have been propagated *in vitro* by several labs (Boulay, 1995; Clayton *et al.*, 1990; Dixon and Keighery, 1992; Fay, 1992; Fay and Gratton, 1992; Rubluo *et al.*, 1993; Simerda, 1990). Laboratories in Australia, Spain, India, and Hawaii, among others, have focused on propagating their endemic floras (Clemente, 1991; Dixon, 1994; Iriondo and Pérez, 1990a; Koob, 1993; Seeni, 1990), while laboratories in England, Denmark, Spain (see González-Benito *et al.*, Chapter 16, this volume) and elsewhere have also directed attention to propagating the endemic flora of islands such as St Helena, Gran Canaria, and Rodrigues, among others (Fay, 1992; González Alemán *et al.*, 1989; Krogstrup *et al.*, 1990; Ramsay, 1997). Other programmes around the world have applied *in vitro* propagation techniques to a wide variety of native and exotic endangered species.

The variety of approaches used reflects the variety and flexibility of tissue culture techniques, described in more detail elsewhere in this volume (see Lynch, Chapter 4). These approaches will be described briefly here, with some examples given in the specific context of endangered species conservation. They will be grouped into protocols which utilize seeds and those which begin with vegetative tissues.

15.3.1   In vitro *propagation using seed*

When seeds of endangered species are available, they are generally preferred for propagation, in order to maintain the maximum genetic diversity. Most endangered species

Table 15.1 Application of biotechnology to the conservation of endangered plant species

Family	Species	Status[1]	Propagation methods[2]	References
Acanthaceae	*Adhatoda beddomei*	Over-coll, med, few seeds, slow prop,	Shoot tips	Sudha and Seeni (1994)
Agavaceae	*Agave arizonica*	Prop for res or cult	Leaf – cal shoots	Powers and Backhaus (1989)
	Agave victoria-reginae	Slow prop	IV lf – som emb	Rodríguez-Garay *et al.* (1996)
Aloeaceae	*Haworthia* spp.	Self-sterile, slow prop, hort	Lf – adv sh	Rogers (1993)
	Kniphofia pauciflora	Prop for cult	Stolon – shoots	McAlister and van Staden (1996)
Alstroemeriaceae	*Leontochir ovallei*	Over-coll, med	IV germ, microprop	Lu *et al.* (1995)
Amaryllidaceae	*Crinum macowanii*	Over-coll, med	Fl stem – adv sh	Slabbert *et al.* (1995)
	Gethyllis linearis	Rare	Bulb scale – adv bulbs	Drewes and van Staden (1994)
	Leucojum aestivum	Rare	Lf – adv sh	Stanilova *et al.* (1994)
	Narcissus bugei, N. longispathus, N. nevadensis, N. tortifolius	Rare, endemic	Scale lf – adv sh	Clemente (1991)
Apiaceae	*Glehnia littoralis*	Rare, med	Rhizome – adv sh	Hui *et al.* (1996)
Apocynaceae	*Rauwolfia micrantha*	Poor germ, poor rooting, med	Sh tips, nodes	Sudha and Seeni (1996)
	Rauwolfia sepentina	Over-coll, med, poor seed viability	Nodes – microprop	Sharma and Chandel (1992)
	Wrightia tomentosa	Over-coll,	Nodes	Purohit *et al.* (1994b)
Araliaceae	*Acanthopanax koreanum*	Over-coll, med, slow prop	Stem – callus – somatic embryos	Choi *et al.* (1997)
Aristolochiaceae	*Aristolochia indica*	Over-coll, med	Sh tip and nodes Lf – adv sh	Manjula *et al.* (1997)
Asclepiadaceae	*Stapelia semota*	Rare	Ax buds	Mohamed-Yasseen *et al.* (1995)
Asteraceae	*Artemisia alba*	Rare, one population	Microprop	Ronse (1990)
	Artemisia granatensis	Few seeds available	Sh tips	Clemente *et al.* (1991)
	Atractylis arbuscula var. *schizogynophylla*	Rare, endemic	Sh – microprop	González Alemán *et al.* (1989)
	Centaurea junoniana	One pop, threatened by lava	IV germ – microprop, adv sh	Hammatt and Evans (1985)
	Leontopodium alpinum	Over-coll	Infloresc – adv sh	Zapratan (1996)
	Leptinella nana	Prop for res and reintro	Sh tips	Carson and Leung (1994)
	Olearia microdisca	Prop for reintro	Sh tips	Williams and Taji (1991)
	Psiadia coronopus	End endemic	Nodes	Krogstrup and Norgaard (1991)
	Saussurea lappa	Rare, med	Sh tip – microprop	Arora and Bhojwani (1989); Johnson *et al.* (1997)
	Senecio hadrosomus	Rare, endemic, low seed set, damaged by insects, germ rate low	Sh – microprop	Bramwell (1990)

Table 15.1 (cont'd)

Family	Species	Status[1]	Propagation methods[2]	References
Begoniaceae	*Begonia* spp.	Prop for cult	Lf – adv sh,	Bowes and Curtis (1991)
Betulaceae	*Betula uber*	30 individuals	Buds	Vijayakumar *et al.* (1990)
Boraginaceae	*Hackelia venusta*	Over-coll, hab loss	Sh tips	Edson *et al.* (1996)
Brassicaceae	*Coronopus navasii*	Rare, endemic	IV germ – seedlings – adv sh	Iriondo and Pérez (1990b)
	Draba aizoides		IV germ to microprop	Ronse (1990)
	Fibigia triquetra	Rare	IV germ – microprop	Prevalek-Kozlina *et al.* (1997)
Bromeliaceae	*Dyckia macedoi*	Rare, endemic	IV germ – microprop Lf – adv sh	Mercier and Kerbauy (1993)
	Vriesia fosteriania, V. hieroglyphica	Over-coll, prop for hort	Seedling – adv sh	Mercier and Kerbauy (1995)
Cactaceae	*Astrophytum capricorn*	Rare	IV germ – microprop	Cárdenas *et al.* (1993)
	Aztekium ritteri	Over-coll, slow growing	IV offshoots	Rodríguez-Garay and Rubluo (1992)
	Escobaria missouriensis, E. robbinsorum, Mammillaria wrightii, Pediocactus bradyi, P. despainii, P. knowltonii, P. paradinei, P. winkleri, Sclerocactus. mesae-verdae, S. spinosior, Toumeya papyracantha,	Over-coll	Shoot tips	Clayton *et al.* (1990)
	Mammilaria san-angelensis	Over-coll	IV Germ – ax and adv shoot form	Martínez-Vázquez and Rubluo (1989)
Campanulaceae	*Nesocodon mauritianus*	Rare, endemic	Sh tips, nodes – microprop	Fay and May (1990)
Caryophyllaceae	*Dianthus arenarius* subsp. *bohemicus*	Rare, endemic	Sh tips, nodes – microprop	Kovac (1995)
	Paronychia chartacea	Few seeds and low viability, hab loss, prop for reintro	IV germ – microprop	McKently and Adams (1994)
Cistaceae	*Helianthemum polygonoides*	Rare, endemic	IV germ to microprop	Iriondo *et al.* (1995)
Combretaceae	*Anogeissus rotundifolia*	Rare, endemic	IV germ – microprop	Singh and Shekhawat (1997)
Cupressaceae	*Juniperus cedrus*	Germ low, rooting difficult	Embryo – adv sh	Harry *et al.* (1995)
Dioscoreaceae	*Dioscorea balcanica, D. caucasica*	Rare, endemic	Lf and zygotic embryo – callus – somatic embryo	Chulafich *et al.* (1994)
	Trichopus zeylanicus	Few seeds, slow seed maturation, hab loss, med	IV germ – microprop	Krishnan *et al.* (1995)

Table 15.1 (cont'd)

Family	Species	Status[1]	Propagation methods[2]	References
Droseraceae	*Dionaea muscipula*	Prop for hobbyists	Lf – adv sh	Kukulczanka *et al.* (1989a)
	Dionaea inusipula, Drosera spp.	Prop for preserv	Lf – adv sh	Kukulczanka (1991)
	Drosera intermedia, D. rotundifolia	Protected species	IV germ – microprop	Ronse (1990)
Epacridaceae	*Leucopogoon obtectus*	One population	Sh tips	Bunn *et al.* (1989)
Euphorbiaceae	*Euphorbia handiensis*	Gran Canaria	Shoot culture to microprop	González Alemán *et al.* (1988)
Fabaceae	*Glycyrrhiza glabra*	Rare, med	Sh tips – microprop	Dimitrova *et al.* (1994)
	Pterocarpus marsupium	Rare	IV germ – microprop	Das and Chatterjee (1993)
	Sophora toromiro	Extinct in wild	Seedling – nodes	Iturriaga *et al.* (1994)
	Trifolium stoloniferum	Two populations	Sh tips	Singha *et al.* (1988)
Gentianaceae	*Centaurium rigualii*	Rare, endemic	Nodes	Iriondo and Pérez (1996)
	Gentiana kurroo	Over-coll, med	Sh tips, nodes	Sharma *et al.* (1993)
	Gentiana lutea	Over-coll	IV germ – microprop	Momcilovic *et al.* (1997)
	Swertia chirata	Rare	Seedling stem – callus – adv sh	Shrestha and Joshi (1992)
Globulariaceae	*Globularia ascanii*	Rare, endemic	IV germ to microprop	Cabrera-Pérez (1995)
Goodeniaceae	*Lechenaultia pulvinaris*	Rare	Sh tip – microprop	Rosetto *et al.* (1992)
Guttiferae	*Kielmayera coriacea*	Over-coll, med	IV germ – microprop	Arello and Pinto (1993)
Haemodoraceae	*Conostylis wonganensis*	Rare	Sh tip – microprop	Rosetto *et al.* (1992)
Lamiaceae	*Coleus forskohlii*	Over-coll, med	Nodes	Sharma *et al.* (1991)
	Hedeoma multiflorum	Over-coll, med	Lf – adv sh, Sh, nodes	Koroch *et al.* (1997)
Lauraceae	*Ocotea catharinensis*	Seeds poor germ, brief viability,	Zygotic emb – som emb	Moura-Costa *et al.* (1993)
	Persea indica	Seed germ low, prop for preservation	IV germ – microprop	Campos and Pais (1996)
Liliaceae	*Allium tuberosum*	Over-coll	Basal plate – adv sh	Radhamani and Chandel (1992)
	Chlorophytum borivilianum	Over-coll, med, seed germ low	Shoot bases – microprop	Purohit *et al.* (1994a)
	Fritillaria releagris	Prop for reintro and hort	Bulb scale	Kukulczanka *et al.* (1989b)
	Lilium rhodopaeum	Rare	Bulb – adv sh	Stanilova *et al.* (1994)
	Sandersonia aurantiaca	Over-coll	Seeds, embs – callus – shoots, Tubers – shoots	Finnie and van Staden (1989)
	Sowerbaea multicaulis	Rare	Sh tip – microprop	Rosetto *et al.* (1992)

Table 15.1   (cont'd)

Family	Species	Status[1]	Propagation methods[2]	References
	Stawellia dimorphantha, Wurmbea spp.	Rare	Microprop	Dixon and Keighery (1992)
	Thuranthos basuticum	Over-coll	Bulb scale	Jones *et al.* (1992)
	Trillium persistens	Rare	Stem and lf – adv shoots	Pence and Soukup (1995)
Lythraceae	*Woodfordia fruticosa*	Rare, med	Sht tips, nodes	Krishnan and Seeni (1994)
Malvaceae	*Lavatera oblongifolia*	Rare, endemic	Microprop	Iriondo and Pérez (1990b)
Marattiaceae	*Angiopteris boivinii*	Rare	IV spore germ	Fay (1992)
Meliaceae	*Turraea laciniata*	Rare, endemic	Sh tips, nodes – microprop	Krogstrup *et al.* (1990)
Myoporaceae	*Eremophila resinosa*	Rare	Sh tip – microprop	Rosetto *et al.* (1992)
Myrtaceae	*Eucalyptus graniticola*	Rare	Sh tip – microprop	Rosetto *et al.* (1992)
Nepenthaceae	*Nepenthes khasiana*	Rare, medicinal	IV germ – microprop	Latha and Seeni (1994)
	Nepenthes khasiana	Rare	Sh tips, nodes – microprop	Seeni (1990); Rathore *et al.* (1991)
	Nepenthes pervillei	Rare, prop for reintro	IV germ – microprop	Redwood and Bowling (1990)
Oleaceae	*Forsythia coreana*	Rare	Node – microprop	Lee *et al.* (1995)
Orchidaceae	*Cattleya dowiana*	Rare	IV seed germ	Marlow and Butcher (1987)
	Cypripedium calceolus	Rare	Imm seed germ	Malmgren (1992)
	Cypripedium debile, C. henryi, C. japonicum, C. tibeticum, C. montanum	Rare	IV germ, Nodes – microprop	Hoshi *et al.* (1994)
	Cypripedrum reginae	Rare	IV germ	Faletra *et al.* (1997)
	Dendrobium lindleyi	Rare	IV germ	Kaur and Sarma (1997)
	Dendrobium moniliforme	Rare	Microprop	Lim *et al.* (1993)
	Dendrobium spectatissimum	Rare	IV germ	Marlow and Butcher (1987)
	Disa uniflora, Eurychone galeandre, Laelia jongheana, Pleione formosana, Sarcochilus fitzgeraldii, S. hartmanii, Sobralia xanotholeuca	Rare	IV germ – microprop	Ronse (1990)
	Habenaria radiata	Rare	IV germ	Nagayoshi *et al.* (1996)
	Hetaeria cristata	Prop for reintro and hort	IV germ, Rhizome nodes	Yam and Weatherhead (1990)

Table 15.1   (cont'd)

Family	Species	Status[1]	Propagation methods[2]	References
	Platanthera praeclara	Rare	IV germ	From and Read (1997)
	Renanthera inschootiana	Over-coll, hab loss, prop for trade	Lf bases – adv buds	Seeni and Latha (1992)
	Spiranthes magnicamporum	Rare, prop for hort	IV germ	Anderson (1991)
	Spiranthes parksii	Less than 2000 in the wild	IV germ – microprop	Christenson (1988)
	Vanda coerulea	Over-collection	Lf bases – adv sh	Seeni (1990)
	Vanilla walkeriae	Prop for preserv	Nodes	Agrawal *et al.* (1992)
Osmundaceae	*Todea barbara*	Rare	IV spore germ	Oliphant (1989)
Pandanaceae	*Sararanga sinuosa*	Rare, endemic	IV germ to microprop	Fay (1992)
Papaveraceae	*Meconopsis paniculata*	Seed germ and survival low, hab loss	Hypocots – callus – adv sh	Sulaiman (1994)
	Mecanopsis simplicifolia	Hab loss, seedling mortality	Seeling explants – callus – adv sh	Sulaiman and Babu (1993)
Pinaceae	*Picea omorika*	Rare, grown for hort	IV germ – Sh – som emb. adv sh	Budimir and Vujicic (1992)
Piperaceae	*Peperomia reticulata*	Rare, endemic	Sh tips, nodes – microprop	Krogstrup *et al.* (1990)
Pittosporaceae	*Pittosporum balfourii*	Rare, endemic	Sh tips, nodes – microprop	Krogstrup *et al.* (1990)
Plumbaginaceae	*Limoniium calaminare, L. dufourei, L. gibertii*	Rare, endemic, preserve germplasm for breeding	IV germ – microprop	Martín and Pérez (1995)
	Limonium estevei	Few seeds available, Hort value	IV germ – microprop	Martín and Pérez (1992)
	Limonium thiniense	Rare, endemic	Herb seeds, IV germ – microprop	Lledó *et al.* (1996)
Podophyllaceae	*Podophyllum hexandrum*	Over-coll, med	Zyg emb – callus – som emb	Arumugam and Bhojwani (1990)
Polygonaceae	*Rheum emodi*	Over-coll, med	Sh tips	Lal and Ahuja (1993)
Primulaceae	*Primula scotica*	Rare, endemic	IV germ – microprop	Benson *et al.* (1998)
Proteaceae	*Grevillea scapigera*	Rare	Sh tips, nodes – microprop, also lf – adv sh	Bunn and Dixon (1992)
Ranunculaceae	*Aconitum heterophyllum*	Over-coll, med	Sh tips – microprop Lf, petiole – callus – som emb	Giri *et al.* (1993)
	Delphinium malabaricum	Low seed set, seed dormancy	Infloresc nodes	Agrawal *et al.* (1991)
Restionaceae	*Hopkinsia anoectocolea, Lepidobolus 'contorta', Loxocarya 'magna'*	Rare, low seed germ, slow prop	Embryo germ	Meney and Dixon (1995)
Rosaceae	*Cowania subintegra*	Over-grazing, poor reproduction	Sh tips – microprop	Jakobek *et al.* (1993)

Table 15.1   (cont'd)

Family	Species	Status[1]	Propagation methods[2]	References
Rutaceae	*Citrus assamensis, C. indica, C. latipes*	Rare	Sh tip – microprop	Baruah *et al.* (1996)
Rutaceae	*Citrus halinii*	Prop for germplasm preserv	IV germ, adv sh	Normah *et al.* (1997)
	Diplolaena andrewsii	Rare	Sh tip – microprop	Rosetto *et al.* (1992)
	Drummondita ericoides	Rare	Sh tip – microprop	Rosetto *et al.* (1992)
	Phebalium equestre, P. hillebrandii	Rare, endemic, prop for hort, difficult to germ	Sh tips, nodes – microprop	Jusaitis (1995)
Salicaceae	*Salix hallaisanensis*	Rare	Node – microprop	Lee *et al.* (1995)
	Salix tarraconensis	Rare, endemic	Nodes – microprop	Amo-Marco and Lledó (1996)
Scrophulariaceae	*Isoplexis canariensis*	Over-coll, med	IV germ – microprop	Arrebola *et al.* (1997)
	Penstemon haydenii	Rare	Nodes	Lindgren and McCown (1992)
	Picrorhiza kurroa	Rare, med	Sh tips, nodes – microprop	Lal *et al.* (1988); Upadhyay *et al.* (1989)
Sterculiaceae	*Sterculia urens*	Over-coll	IV germ – microprop	Purohit and Dave (1996)
	Trochetiopsis erythroxylon, T. melanoxylon	Rare, endemic, low germ rate	IV germ	Fay (1992)
Stylidiaceae	*Stylidium coroniforme*	Rare	IV germ – microprop	McComb (1985)
Theaceae	*Camellia crapnelliana, C. granthaminana, C. hongkongensis*	Rare	IV germ to microprop	Siu and Weatherhead (1995)
Thyrsopteridaceae	*Cibotium schiedei*	Rare	IV spore germ	Fay (1992)
Turneraceae	*Mathurina penduliflora*	Rare, endemic	Sh tips, nodes – microprop	Krogstrup *et al.* (1990)
Valerianaceae	*Nardostachys jatamansi*	Over-coll, med	Petiole – callus – root – adv sh	Mathur (1992)
	Valeriana wallichii	Rare, med	Sh tips, nodes – microprop	Mathur *et al.* (1988)
Zamiaceae	*Ceratozamia hildae, C. mexicana*	Rare	Megagam – adv sh, Zyg emb – som emb	Chavez *et al.* (1992)

[1] Abbreviations: germ, germination; hab, habitat; hort, horticultural value or value for cultivation; med, medicinal value; over-coll, over-collected; preserv, preservation; pop, population; prop, propagate; rare, endangered, threatened or rare; reintro, reintroduction; res, research; slow prop, slow to propagate.

[2] Procedures are given in a sequential format, with the steps in order, abbreviated and separated by hyphens. Abbreviations: adv, adventitious; cal, callus; fl, flower; imm, immature; infl, inflorescence; lf, leaf; microprop, micropropagation, using axillary bud outgrowth; som emb, somatic embryos; sh, shoot(s).

produce seeds, but in some cases they are few in number or they may be difficult to germinate. When very few seeds are available, *in vitro* germination is often used to produce sterile seedlings, which are then used to provide shoot tips and nodes as explants for micropropagation. This approach has been used for a number of species, including *Gentiana lutea, Limonium* spp. and *Nepenthes khasiana* (e.g. Latha and Seeni, 1994; Martín and Pérez, 1995; Momcilovic *et al.*, 1997), as well as for many others (Table 15.1).

When conventional procedures, such as stratification, fail to break seed dormancy or the rate of germination is very low, embryo culture may be useful. Some forms of dormancy are overcome by removing the seed coat, as with *Trochetiopsis* spp. from St Helena (Fay, 1992) and *Aster vialis*, a rare species from Oregon (Pence and Zhang, unpublished). In some cases, growth regulators have been added to stimulate germination (Meney and Dixon, 1995). Alternatively, growth regulators may be used to stimulate direct somatic embryogenesis or shoot formation from the embryo tissue, as with *Primula maguirei*, a rare species from Utah (Pence and Clark, unpublished), or to produce embryogenic callus, as with *Podophyllum hexandrum* (Arumugam and Bhojwani, 1990).

In some cases, seeds have particular requirements for germination which are not easily met by conventional germination procedures. For example, *Pholisma sonorae*, an endangered parasitic plant of the south-western United States, requires the presence of host root tissues for germination, and it has not been possible to germinate the seeds *ex situ*. Other root parasites have been successfully germinated *in vitro* (Okonkwo, 1966), and these techniques could be applied to seeds of *P. sonorae*.

Similarly, seeds of a number of rare orchid species have been asymbiotically germinated *in vitro*, such as *Vanilla alkeriae, Hetaeria cristata*, and *Cypripedium reginae*, among others (Agrawal *et al.*, 1992; Faletra *et al.*, 1997; Yam and Weatherhead, 1990). In cases such as *Spiranthes magnicamporum*, symbiotic cultures of seeds and fungus have been established *in vitro*, stimulating germination further (Anderson, 1991). Germinated orchid seeds have also been used to initiate cultures for micropropagation (Christenson, 1988).

15.3.2 *Propagation without seeds*

For some endangered species, propagation by seed is not an option. Seed viability can be low, as with *Rauvolfia micrantha* (Sudha and Seeni, 1996) and others, while in cases such as *Haworthia* spp., *Paronychia chartacea, Adhatoda beddomei*, and *Delphinium malabaricum* little or no seed is produced (Agrawal *et al.*, 1991; McKently and Adams, 1994; Rogers, 1993; Sudha and Seeni, 1994). With species such as *Artemisia granatensis* and *Limonium estevei*, so few plants were available that taking seeds was restricted (Clemente *et al.*, 1991; Martín and Pérez, 1992), while with species such as *Trillium persistens*, germination from seed is very slow (Pence and Soukup, 1995). In the majority of cases, when seeds are not available, propagation *in vitro* is accomplished by culturing shoot tips or nodes from field or greenhouse grown plants and stimulating the outgrowth of axillary shoots (Table 15.1). This is preferred, since genetic changes appear to be less likely when preformed meristems are used for propagation.

The growth habit of some species is such that the culture of apical or vegetative lateral buds would irreversibly damage the plant. In the case of monopodial orchids, such as *Phalaenopsis,* the culture of dormant buds from inflorescence nodes has been used to overcome this problem (Reisinger *et al.*, 1976). With *Delphinium malabaricum*, inflorescence nodes have also been used, since the single apical bud also grows at soil level, making it difficult to establish uncontaminated cultures (Agrawal *et al.*, 1991).

Although the culture of preformed meristems is generally preferred, because of their genetic stability, there are situations where buds are unavailable or difficult to culture or where more rapid propagation methods are desired. Organogenesis or embryogenesis has been obtained from vegetative tissues of *Meconopsis simplicifolia*, *Dionea muscipula*, *Agave victoria-reginae*, and *Haworthia* spp. (Kukulczanka *et al.*, 1989a; Rodríguez-Garay *et al.*, 1996; Rogers, 1993; Sulaiman, 1994), among others. Species in the Liliaceae and Amaryllidaceae are often propagated using bulb scales or similar tissue (Drewes and van Staden, 1994; Jones *et al.*, 1992; Kukulczanka *et al.*, 1989b; Pandy *et al.*, 1992), whereas protocorm-like bodies have been produced from leaf segments of monopodial orchids (Tanaka and Sakanishi, 1977). In the case of *Nardostachys jatamansi*, which naturally forms buds from its roots, petiole callus was used to initiate adventitious roots *in vitro*, which were then stimulated to form buds (Mathur, 1992).

15.4   Genetic diversity and genetic stability

While tissue culture is a powerful tool for multiplying individuals of a particular genetic line, it is a clonal process and, at first, may appear to contradict the goal of preserving genetic diversity. However, genetic diversity is maintained by culturing each individual available for propagation as a unique and separate line.

A related concern is that of somaclonal variation, or the introduction of genetic changes (see Harding, Chapter 7, this volume) into an otherwise clonal line. Plants obtained from preformed buds generally have a lower frequency of change than those from direct adventitious sources, while those from callus appear most likely to undergo changes (Karp, 1994). As discussed above, a number of factors are involved in developing a protocol for propagating an endangered species, and at times, buds cannot be used. In those cases, it may be necessary to regenerate adventitious shoots, but it may then be possible to propagate those shoots by axillary bud outgrowth. Another approach was taken with the micropropagation of the rare *Hackelia venusta* from the northwestern US. Cultures were grown and propagated on a minimal level of growth regulators, in order to minimize the potential for somaclonal variation (Edson *et al.*, 1996). On the other hand, somaclonal variation has been suggested as a tool for increasing the genetic diversity in species with a very narrow genetic base, such as the Easter Island endemic, *Sophora toromiro* (Jacobsen and Dohmen, 1990).

15.5   Uses of *in vitro* propagated plants of endangered species

In vitro propagation of endangered plant species is generally undertaken to increase the numbers of individuals available of extremely rare and endangered species. Some species such as the Hawaiian *Cyanea pinnatifida*, are represented by only one individual in the wild. Small parts of tissue were removed from that plant and taken to the Lyon Arboretum for culture. There they were successfully propagated and reintroduced into the wild (Lamoureaux, personal communication). Other species, such as *Sophora toromiro* are extinct in the wild, and the few plants which have been maintained *ex situ* have been used for *in vitro* work (Iturriaga *et al.*, 1994). When population numbers are low, the number of *in vitro* plants can readily surpass the number of plants in the wild (Powers and Backhaus, 1989; Tormala *et al.*, 1994).

When species have been over-collected by hobbyists or for medicine, food, or fragrance, *in vitro* propagation can provide an alternate source of plants and alleviate pressures on wild populations. Certain orchids, cacti, and wildflowers as well as a number of medicinal species have been propagated for this reason (e.g. Giri *et al.*, 1993; Pence and Soukup, 1993; Rubluo *et al.*, 1993). On the other hand, some rare species may be relatively unknown, but could have horticultural or other value if enough material is made available for breeding and development.

Tissue culture can also be used when wild grown plants are difficult to propagate for *ex situ* preservation in botanical gardens. The Center for Plant Conservation (St Louis) coordinates botanical gardens in the United States to monitor and grow endangered species *ex situ*. When traditional propagation is difficult, *in vitro* techniques are used (Christenson, 1988; Pence *et al.*, 1997). Such plants can be used as a source of seed for long-term storage, and if seed is not produced, the tissue culture lines themselves can be cryopreserved (see below). Propagated plants might also be used for *ex situ* studies on the biology of endangered plant species.

If wild populations become severely reduced or lost, propagated plants can also be used for reintroduction. This has been attempted with several species including *Artemisia granatensis, Nepenthes khasiana, Mammillaria san-angelensis, Senecio hadrosomus, Agave victoria-reginae, Bletia urbana* and several other orchid and rare Hawaiian species, including *Cyanea pinnatifida* (Bramwell, 1990; Fay, 1992; Rodríguez-Garay *et al.*, 1996; Rubluo *et al.*, 1993; Tandon *et al.*, 1990; Lamoureux, personal communication). *Rubus humulifolius*, an endangered species of Finland, known from only ten plants, was propagated *in vitro* to yield 1500 plants and was replanted in a site near its original locality (Tormala *et al.*, 1994).

15.6 Application of preservation technologies

Traditional methods for preserving plant germplasm *ex situ* have included growing plants in botanical gardens and arboreta and banking the dried seeds and spores at refrigerator (4°C) or freezer (−18° or −20°C) temperatures. The development of cryopreservation, or storage in liquid nitrogen (at −196°C), has provided a technology for even more stable, long-term storage of living tissues (see Benson, Chapter 6, this volume).

Several laboratories have applied cryopreservation protocols to the seeds of a variety of endangered species. A cooperative agreement between the National Seed Storage Lab of the US Department of Agriculture and the Center for Plant Conservation was developed to store seeds of endangered US species at the NSSL facility (Falk, 1987). Cryopreservation is being applied to the seeds of the endangered and rare flora of Western Australia at Kings Park and Botanic Garden (Touchell and Dixon, 1994), while seeds of endangered and threatened species of Ohio are cryopreserved at the Cincinnati Zoo and Botanical Garden (Pence, 1991). Several other laboratories have also developed cryogenic storage facilities for seeds of native flora (see Table 15.1).

The majority of species currently stored in liquid nitrogen are those with orthodox, or desiccation tolerant seeds. When dried, orthodox seeds generally survive liquid nitrogen exposure with little or no damage. In some cases, however, seeds may be orthodox, but short-lived unless they are carefully dried and frozen either at −20°C or in liquid nitrogen. Two examples are the short-lived seeds of *Plantago cordata* and *Salix myricoides*, listed as endangered and potentially threatened in Ohio, respectively. These have been successfully dried, cryopreserved and banked in liquid nitrogen (Pence, 1998b; Pence and Clark, 1998).

Other species have desiccation sensitive, or 'recalcitrant', seeds (see Marzalina and Krishnapillay, Chapter 17, this volume). These seeds cannot usually survive drying, and in the hydrated state they do not survive exposure to liquid nitrogen. However, cryopreservation is being increasingly applied to the excised embryos of recalcitrant seeds from tropical tree species and new approaches to recalcitrant seed cryopreservation are presently being considered (see Marzalina and Krishnapillay, Chapter 17, this volume). In general however, recalcitrant seeds do pose particular problems for long-term germplasm storage. Seeds of some large-seeded temperate trees, some wetland species and some climax species from the moist tropics fall into this category. Wetlands and moist tropical forests are two habitats that are particularly threatened, increasing the need for *ex situ* germplasm storage of species from these areas. Studies are underway at the National Seed Storage Laboratory (Ft Collins, Colorado) to determine the extent of recalcitrance in seeds of endangered species from the rainforests of Hawaii (Walters, personal communication), so that seed storage protocols can be developed for these species. A similar study is being conducted on endangered species from Ohio wetlands by the Cincinnati Zoo and Botanical Garden. In both cases, it appears that the majority of the species under study are not recalcitrant.

Cryopreservation of 'non-seed' tissues, such as immature embryos or *in vitro* cultures offers an alternative approach to be used for the preservation of recalcitrant species. These procedures centre around the techniques of slow freezing (Withers, 1985), vitrification (Sakai *et al.*, 1990), and encapsulation–dehydration (Fabre and Dereuddre, 1990). (See Benson, Chapter 6, this volume for further details on cryopreservation methodology.) Tissues most commonly used for cryogenic storage are: shoot tips from *in vitro* cultures, excised zygotic embryos and embryonic axes, somatic embryos and embryogenic or organogenic cell or callus lines (see Lynch, Chapter 4, this volume).

When endangered species are propagated by tissue culture, the tissue culture lines can be cryopreserved. Cryopreservation of *in vitro* tissue from endangered species has been accomplished using a slow freezing protocol with shoot tips of *Grevillea scapigera* and organogenic callus of *Dioscorea caucasica* and *D. balcanica* (Chulafich *et al.*, 1994; Touchell *et al.*, 1992). Other species have been cryopreserved using encapsulation–dehydration, including *Centaurium rigualii*, endemic to the Iberian peninsula (González-Benito and Pérez, 1997), *Aster vialis* and *Sisyrinchium sarmentosum*, two endangered species from Oregon (Pence and Clark, unpublished), and *Cosmos atrosanguinensis*, which is cultivated but is extinct in the wild (Wilkinson *et al.*, 1998).

In addition to tissues of seed plants, spores and gametophytes of pteridophytes and bryophytes can also be cryopreserved. Non-chlorophyllous fern spores are generally desiccation tolerant and adapt well to liquid nitrogen (LN) storage, although some are short-lived, such as those of the endangered tree fern, *Cyathea spinulosa*. These were dried and exposed to LN, with over 93 per cent recovery (Agrawal *et al.*, 1993). Chlorophyllous spores of at least some ferns, although they are generally short-lived, also can be dried and cryopreserved or cryopreserved using the encapsulation–dehydration technique (Pence, 1998d). Moss spores of several species have also survived cryopreservation (Pence, unpublished).

Some gametophytes of mosses and liverworts are naturally desiccation tolerant and when this is the case, they can be air dried and frozen directly in liquid nitrogen (Leverone and Pence, 1993). When gametophytes are sensitive to drying, pre-culture with abscisic acid (ABA) is sufficient to induce desiccation tolerance in some species of bryophytes. In other cases the encapsulation–dehydration technique has been useful in preserving both tropical and temperate byrophytes and fern gametophytes through desiccation and subsequent liquid nitrogen exposure (Pence, 1998a, 1998c). Slow freezing has also been used to cryopreserve protoplasts of *Marchantia* (Takeuchi *et al.*, 1980), while pre-culture with mannitol or ABA and proline has been shown to provide protection of moss tissues through slow freezing protocols (Christianson, 1998; Grimsley and Withers, 1983). These techniques should be readily transferable to rare or endangered bryophyte and pteridophyte species.

Slow growth is another option for preserving *in vitro* cultures of endangered species for medium-term storage of several months to several years. Good survival has been reported for shoots of *Centaurium rigualii* for three years, *Picrorhiza kurroa* for ten months, *Saussurea lappa* for 12 months, and of *Coronopus navasii, Lavatera oblongifolia,* and *Centauriuim rigualii* for six months, when stored at 5°C (Arora and Bhojwani, 1989; Iriondo and Pérez, 1991, 1996; Upadhyay *et al.*, 1989). Similarly, *Drosera* spp. and *Dionaea muscipula* have been maintained for up to ten months at 0–6°C (Kukulczanka, 1991). *Rauvolfia serpentina* remained healthy after 15 months at 15°C, although lower temperatures were deleterious (Sharma and Chandel, 1992). A number of rare Hawaiian species are maintained at the Lyon Arboretum on minimal medium *in vitro*, and only transferred at six month intervals (Lamoureux, personal communication).

15.7   Application of *in vitro* collection biotechnology

In vitro collection, or IVC, is the initiation of tissue cultures in the field. It can be used to broaden germplasm collection to include species for which seeds are not available and for which cuttings may be difficult to maintain or transport (see Hummer, Chapter 3 and Lynch, Chapter 4, this volume).

IVC is a very flexible technique and can be adapted to a variety of situations. Either partial or full sterilization of the tissue is made on site and the tissue is transferred to containers of medium for transport back to the lab. In some cases, such as with *Cocos nucifera*, tissues have been collected with minimal treatments in the field (Assy Bah *et al.*, 1987). Once they were transported back to the lab, they were resterilized and dissected further for culture. In other cases, sterilization and dissection have been completed in the field and the growth and development of the cultures initiated at that point (Pence, 1996). IVC has been used to collect a variety of plant tissues, including orchid seeds (Warren, 1983), embryos (Assy Bah *et al.*, 1987), apical or nodal buds (e.g. Altman *et al.*, 1990; Ruredzo, 1991; Yidana *et al.*, 1987), and leaf and stem tissue (Pence, 1996).

Different strategies have been used to minimize contamination in IVC cultures. In some cases, a portable glove box was used to reduce contamination from ambient sources (Sossou *et al.*, 1987), whereas in other cases the work has been done quickly in the open air. Internal contamination can be a more serious problem than ambient contamination, since many plants harbour endophytic fungi and bacteria. However, the use of fungicides and antibiotics in the medium can reduce this contamination to a workable level (Pence, 1996).

The initiation of *in vitro* cultures in the field can facilitate the transport of the tissues. Because of the small size of the explants, more material can be transported, compared to the transport of whole plants or cuttings. In addition, the cultures are initiated with fresh material which can begin the process of growth *in vitro* immediately, compared with whole plant materials which may undergo some deterioration in transport before they are planted *ex situ*. Finally, the transport of clean plant materials *in vitro* generally facilitates their movement through international border inspections.

IVC could have broad applicability in collecting materials from endangered plant species when seeds or spores are not available or to supplement seed collection when seeds are of low viability or are difficult to germinate. Wild collected seeds of *Drosera rotundifolia*, a potentially threatened species in the State of Ohio, proved difficult to germinate in early studies. Thus, as a supplement to seed collection, leaf tissues were collected by IVC and plants were successfully regenerated from these (Pence and Bezold, unpublished). Seed collection is prohibited from several other Ohio species, because of their rarity. IVC will be used as an alternative, to collect vegetative tissue for propagation and preservation, with minimal disturbance to the plants in the field.

IVC can be used as a source of material for both the propagation and preservation of endangered plant germplasm. For example, leaf and bud tissue collected by IVC from *Brunfelsia densifolia*, a rare Puerto Rican tree growing at the Fairchild Tropical Garden in Florida, was transported to the Cincinnati Zoo and Botanical Garden where plants were regenerated from the cultures and tissue was cryopreserved (Pence, 1990).

The limitations of IVC are those of tissue culture in general, since conditions for growing some species *in vitro* have not yet been developed. However, the number and variety of species which have been successfully propagated *in vitro* continues to grow, and this will, in turn, be reflected in the widening applicability of IVC techniques for the collection of rare or endangered plant germplasm.

15.8 Application of genetic analysis and molecular techniques to endangered plant species conservation

The development of molecular techniques (see Harris, Chapter 2, this volume) has opened the door to a number of areas relevant to monitoring the stability (see Harding, Chapter 7, this volume) and diversity of endangered plant germplasm held *in situ* and *ex situ*.

Molecular techniques, particularly RAPD analyses, are being used to monitor the genetic diversity of populations of rare species, as well as to define species themselves, all of which impact on endangered species management. A high level of diversity was found for the ten known populations of *Banksia cuneata*, an endangered species of south-western Australia (Maguire and Sedgley, 1997). RAPD analysis also indicated that the one known individual of *Eucalyptus graniticola* was a hybrid of two more common species, rather than an endangered relict species (Rosetto *et al.*, 1997). Thus, rather than reinforcing the population *in situ*, the species was backed up *ex situ*. At the Chicago Botanic Garden, RAPDs are being used to compare populations of the rare *Cirsium pitcheri*. DNA from Indiana and Wisconsin populations is being analysed and compared with DNA from 130-year-old herbarium specimens of the extirpated Illinois population (Havens, personal communication).

Methods for detecting somaclonal variation in tissue culture propagated species have traditionally included phenotypic evaluation, isozyme analysis, chromosome counts, and DNA cytofluorometry (Hakman *et al.*, 1984; Iriondo and Pérez, 1996; Martín and Pérez, 1992; Sudha and Seeni, 1994). For example, isozyme analysis and chromosome counts were made to evaluate plants micropropagated from *Centaurium rigualii*, an endangered species of Spain, and no variations were found using these methods (Iriondo and Pérez,

1996). The use of RAPDs can provide an even more precise tool for the detection of somaclonal variation in micropropagated endangered plants (Martin and Pérez, 1994).

In preserving plant germplasm, DNA can be banked as a back-up or supplement to the storage of living tissue, or may be used when living tissues cannot be stored. Techniques for isolating DNA from dried tissue, as well as dried-frozen material have been developed (Adams *et al.*, 1992), making collection and banking from wild, remote species a possibility. Libraries of DNA from rare or endangered species are being set up to store this information for future use (Mattick *et al.*, 1992) and recommendations and guidelines for DNA banking have been made (Adams and Adams, 1992).

Molecular technologies offer the possibility of collecting ancient DNA from extinct species, as well. Within the past decade, the isolation of DNA fragments from dried herbarium specimens (Pyle and Adams, 1989; Havens, personal communication) as well as from fossil materials (Rogers and Bendich, 1985; Suyama *et al.*, 1996) has been demonstrated. A fragment of DNA, identified as part of the RuBisCo gene, was obtained from a fossil leaf compression of the extinct *Magnolia latahensis* (Golenberg *et al.*, 1990). This was compared with gene sequences from extant Magnoliaceae and other species in order to determine the genetic relationship of the extinct and extant species. A portion of this same gene was also obtained from fossil *Taxodium* (Soltis *et al.*, 1992).

15.9 Summary

In vitro techniques have found applications in a number of areas impacting on the conservation and management of rare and endangered plant species. This will be expanded upon in the subsequent chapter by González-Benito *et al.* of this volume. *In vitro* propagation methods have been used for endangered species when traditional methods using seeds, off-shoots or cuttings have not been adequate. Cryopreservation technologies have improved storage of endangered plant germplasm, by increasing the longevity of preserved seeds and spores and allowing the long-term storage of other tissues, such as shoot tips, embryos, gametophytes, callus and cell cultures, when seed preservation is not possible. *In vitro* collection methods expand the collection of living tissues in the field, beyond seeds and cuttings, and should allow collection of endangered species with minimal disturbance. Finally, molecular technologies have provided new tools for analysing the genetic diversity of endangered species populations as well as the genetic fidelity of *in vitro* propagated plants, and for looking into the past at species which have become extinct. Taken together, these biotechnologies can help define the needs of particular endangered species and provide techniques to supplement traditional methods of endangered species management.

References

ADAMS, R.P. and ADAMS, J.E., 1992, Task group reports, DNA Bank-Net Workshop, in *Conservation of Plant Genes*, edited by R.P. ADAMS and J.E. ADAMS, pp. 325–340, San Diego: Academic Press.

ADAMS, R.P., DO, N. and CHU, G., 1992, Preservation of DNA in plant specimens from tropical species by desiccation, in *Conservation of Plant Genes*, edited by R.P. ADAMS and J.E. ADAMS, pp. 135–152, San Diego: Academic Press.

AGRAWAL, D.C., PAWAR, S.S., MORWAL, G.C. and MASCARENHAS, A.F., 1991, *In vitro* micropropagation of *Delphinium malabaricum* (Huth) Munz. – a rare species, *Annals of Botany*, **68**, 243–245.

AGRAWAL, D.C., MORWAL, G.C. and MASCARENHAS, A.F., 1992, *In vitro* propagation and slow growth storage of shoot cultures of *Vanilla walkeriae* Wight – an endangered orchid, *Lindleyana*, **7**, 95–99.

AGRAWAL, D.C., PAWAR, S.S. and MASCARENHAS, A.F., 1993, Cryopreservation of spores of *Cyathea spinulosa* Wall. ex. Hook. f. – an endangered tree fern, *Journal of Plant Physiology*, **142**, 124–126.

ALTMAN, D.W., FRYXELL, P.A., KOCH, S.D. and HOWELL, C.R., 1990, *Gossypium* germplasm conservation augmented by tissue culture techniques for field collecting, *Economic Botany*, **44**, 106–113.

AMO-MARCO, J.B. and LLEDÓ, M.D., 1996, *In vitro* propagation of *Salix tarraconensis* Pau ex Font Quer, an endemic and threatened plant, *In Vitro Cellular and Developmental Biology – Plant*, **32**, 42–46.

ANDERSON, A.B., 1991, Symbiotic and asymbiotic germination and growth of *Spiranthes magnicamporum* (Orchidaceae), *Lindleyana*, **6**, 183–186.

ARELLO, E.F. and PINTO, J.E.B.P., 1993, *In vitro* propagation of *Kielmeyera coriacea, Pesquisa Agropecuaria Brasileira*, **28**, 25–31.

ARORA, R. and BHOJWANI, S.S., 1989, *In vitro* propagation and low temperature storage of *Saussurea lappa* C.B. Clarke – an endangered, medicinal plant, *Plant Cell Reports*, **8**, 44–47.

ARREBOLA, M.L., SOCORRO, O. and VERPOORTE, R., 1997, Micropropagation of *Isoplexis canariensis* (L.) G. Don, *Plant Cell Tissue and Organ Culture*, **49**, 117–119.

ARUMUGAM, N. and BHOJWANI, S.S., 1990, Somatic embryogenesis in tissue cultures of *Podophyllum hexandrum, Canadian Journal of Botany*, **68**, 487–491.

ASSY BAH, B., DURAND-GASSELIN, T. and PANNETIER, C., 1987, Use of zygotic embryo culture to collect germplasm of coconut (*Cocos nucifera* L.), *FAO/IBPGR Plant Genetic Resources Newsletter*, **71**, 4–10.

BARUAH, A., NAGARAJU, V. and PARTHASARATHY, V.A., 1996, Micropropagation of three endangered *Citrus* species. 1. Shoot proliferation in vitro, *Annals of Plant Physiology*, **10**, 124–128.

BENSON, E.E., DANAHER, J.E., PIMBLEY, I., WAKE, J. and DALEY, S., 1998, A micropropagation system for *Primula scotica*: a rare and at risk Scottish species endemic to Orkney, *In Vitro Cellular and Developmental Biology – Plant*, **34**, 61A.

BOWES, B.G. and CURTIS, E.W., 1991. Conservation of the British National *Begonia* collection by micropropagation, *New Phytologist*, **119**, 169–181.

BOULAY, J., 1995, Carnivorous plants, micropropagation assays, *Bulletin des Academie & Societe Lorraines des Sciences*, **34**, 151–159.

BRAMWELL, D., 1990, The role of *in vitro* cultivation in the conservation of endangered species, in HERNÁNDEZ BERMEJO, J.D., CLEMENTE, M. and HEYWOOD, V. (Eds), *Conservation Techniques in Botanic Gardens*, pp. 3–15, Koenigstein: Koeltz Scientific Books.

BUDIMIR, S. and VUJICIC, R., 1992, Benzyladenine induction of buds and somatic embryogenesis in *Picea omorika* (Pancic) Purk, *Plant Cell Tissue and Organ Culture*, **31**, 89–94.

BUNN, E. and DIXON, K.W., 1992, *In vitro* propagation of the rare and endangered *Grevillea scapigera* (Proteaceae), *HortScience*, **27**, 261–262.

BUNN, E., DIXON, K.W. and LANGLEY, M.A., 1989, *In vitro* propagation of *Leucopogon obtectus* Benth. (Epacridaceae), *Plant Cell Tissue and Organ Culture*, **19**, 77–84.

CABRERA-PÉREZ, M.A., 1995, Explant establishment in the micropropagation of *Globularia ascanii*, a threatened species from Gran Canaria, *Botanic Gardens Micropropagation News*, **1**, 111–113.

CAMPOS, P.S. and PAIS, M.S.S., 1996, *In vitro* micropropagation of the Macarronesian evergreen tree *Persea indica* (L.) K. Spreng, *In Vitro Cellular and Developmental Biology – Plant*, **32**, 184–189.

CÁRDENAS, E., del CARMEN OJEDA, M., TORRES, T.E. and SÁENZ, E.O., 1993, Micropropagation of *Astrophytum capricorne* an endangered cactus from NE Mexico, *Botanic Gardens Micropropagation News*, **1**, 75–76.

CARSON J.A. and LEUNG, D.W.M., 1994, *In vitro* flowering and propagation of *Leptinella nana* L, *New Zealand Journal of Botany*, **32**, 79–83.

Center for Plant Conservation, 1995, National Collection of Endangered Plants, Center for Plant Conservation, Missouri Botanical Garden, St Louis, MO.

CHAVEZ, V.M., LITZ, R.E. and NORSTOG, K., 1992, *In vitro* morphogenesis of *Ceratozamia hildae* and *C. mexicana* from megagametophytes and zygotic embryos, *Plant Cell Tissue and Organ Culture*, **30**, 93–98.

CHOI, Y.-E., KIM, J.-W. and SOH, W.-Y., 1997, Somatic embryogenesis and plant regeneration from suspension cultures of *Acanthopanax koreanum* Nakai, *Plant Cell Reports*, **17**, 84–88.

CHRISTENSON, E.A., 1988, Conservation of *Spiranthes parksii*: a beginning, *Orchid Review*, **96**, 148–149.

CHRISTIANSON, M.L., 1998, A simple protocol for cryopreservation of mosses, *The Bryologist*, **101**, 32–35.

CHULAFICH, L., GRUBISHICH, D., VUIICHICH, R., VOLKOVA, L.A. and POPOV, A.S., 1994, Somatic embryo production *in vitro* in *Dioscorea caucasica* Lipsky and *Dioscorea balcanica* Kosanin and cryopreservation of their organogenic callus tissues, *Russian Journal of Plant Physiology*, **41**, 821–826.

CLAYTON, P.W., HUBSTENBERGER, J.F., PHILLIPS, G.C. and BUTLER-NANCE, S.A., 1990, Micropropagation of members of the Cactaceae subtribe Cactinae, *Journal of the American Society for Horticultural Science*, **115**, 337–343.

CLEMENTE, M., 1991, The micropropagation unit at the Cordoba Botanic Garden, Spain, *Botanic Gardens Micropropagation News*, **1**, 30–33.

CLEMENTE, M., CONTRERAS, P., SUSÍN, J. and PLIEGO-ALFARO, F., 1991, Micropropagation of *Artemisia granatensis, HortScience*, **26**, 420.

DAS, T. and CHATTERJEE, A., 1993, *In vitro* studies of *Pterocarpus marsupium* – an endangered tree, *Indian Journal of Plant Physiology*, **36**, 269–272.

DIMITROVA, D., VARBANOVA, K., PEEVA, I., ANGELOVA, S. and CUTEVA, Y., 1994, A study on in vitro cultivation of *Glycyrrhiza glabra, Plant Genetic Resources Newsletter*, **100**, 12–13.

DIXON, K.W., 1994, Towards integrated conservation of Australian endangered plants: the Western Australia model, *Biodiversity and Conservation*, **3**, 148–159.

DIXON, B. and KEIGHERY, G., 1992, WA lilies. Notes on the propagation and cultivation of Western Australian plants from the Liliiflorae families, *Australian Plants*, **17**, 7–19.

DREWES, F.E. and van STADEN, J., 1994, *In vitro* propagation of *Gethyllis linearis* L. Bol., a rare indigenous bulb. *South African Journal of Botany*, **60**, 295–296.

EDSON, J.L., LEEGE-BRUSVEN, A.D., EVERETT, R.L. and WENNY, D.L., 1996, Minimizing growth regulators in shoot culture of an endangered plant, *Hackelia venusta* (Boraginaceae), *In Vitro Cellular and Developmental Biology – Plant*, **32**, 267–271.

FABRE, J. and DEREUDDRE, J., 1990, Encapsulation–dehydration: a new approach to cryopreservation of *Solanum* shoot tips, *Cryo-Letters*, **11**, 413–426.

FALETRA, P., DOVHOLUK, A., KING, T. and SOKOLSKI, K., 1997, Saving *Cypripedium reginae, Orchids*, 1997, 139–143.

FALK, D., 1987, Exploring seed storage of endangered plants, *Plant Conservation*, **2**, 7.

FAY, M.F., 1992, Conservation of rare and endangered plants using in vitro methods, *In Vitro Cellular and Developmental Biology – Plant*, **28**, 1–4.

FAY, M.F., 1994, In what situations is *in vitro* culture appropriate to plant conservation? *Biodiversity and Conservation*, **3**, 176–183.

FAY, M.F. and GRATTON, J., 1992, Tissue culture of cacti and other succulents – a literature review and a report of micropropagation at Kew, *Bradleya*, **10**, 33–48.

FAY, M.F. and MAY, N.W.M., 1990, The *in vitro* propagation of *Nesocodon mauritianus, Botanic Gardens Micropropagation News*, **1**, 6–8.

FINNIE, J.F. and van STADEN, J., 1989, *In vitro* propagation of *Sandersonia* and *Gloriosa, Plant Cell Tissue and Organ Culture*, **19**, 151–158.

FROM, M. and READ, P.E., 1997, *Platanthera praeclara:* a threatened temperate prairie orchid species, *In Vitro Cellular and Developmental Biology*, **33**, 17A.

GIRI, A., AHUJA, P.S. and KUMAR, P.V.A., 1993, Somatic embryogenesis and plant regeneration from callus cultures of *Aconitum heterophyllum* Wall, *Plant Cell Tissue and Organ Culture*, **32**, 213–218.

GOLENBERG, E.M., HENDERSON D. and ZURAWSKI, G., 1990, Chloroplast DNA sequence from a Miocene *Magnolia* species, *Nature*, **344**, 656–658.

GONZÁLEZ ALEMÁN, C., ORTEGA GONZÁLEZ, C.I. and RUBIO HERNÁNDEZ, A.M., 1988, Propagación 'in vitro' de endemismos canarios en peligro de extinción: *Euphorbia handiensis* Burchd, *Botanica Macaronesica*, **16**, 15–28.

GONZÁLEZ ALEMÁN, C., RUBIO HERNÁNDEZ, A.M. and ORTEGA GONZÁLEZ, C.I., 1989, Propagación 'in vitro' de endemismos canarios en peligro de extinción: *Atractylis arbuscula* Svent. et Michaelis, *Botanica Macaronesica*, **17**, 47–56.

GONZÁLEZ-BENITO, M.E. and PÉREZ, C., 1997, Cryopreservation of nodal explants of an endangered plant species (*Centaurium rigualii* Esteve) using the encapsulation–dehydration method, *Biodiversity and Conservation*, **6**, 583–590.

GRIMSLEY, N.H. and WITHERS, L.A., 1983, Cryopreservation of the moss, *Physcomitrella patens, Cryo-Letters*, **4**, 251–258.

HAKMAN, I., von ARNOLD, S. and BENGTSSON, A., 1984, Cytofluorometric measurement of a nuclear DNA in adventitious buds and shoots of *Picea abies* regenerated *in vitro*, *Physiologia Plantarum*, **60**, 321–325.

HAMMATT, N. and EVANS, P.K., 1985, The *in vitro* propagation of an endangered species: *Centaurea junoniana* Svent. (Compositae), *Journal of Horticultural Science*, **60**, 93–97.

HARRY, I.S., PULIDO, C.M. and THORPE, T.A., 1995, Plantlet regeneration from mature embryos of *Juniperus cedrus, Plant Cell Tissue and Organ Culture*, **41**, 75–78.

HOSHI, Y., KONDO, K. and HAMATANI, S., 1994, *In vitro* seed germination of four Asiatic taxa of *Cypripedium* and notes on the nodal micropropagation of American *Cypripedium montanum, Lindleyana*, **9**, 93–97.

HUI, H., JIANG, N. and LIU, Q.X., 1996, On the culture of test-tube plantlet from rhizomes of threatened plant *Glehnia littoralis* Fr. Schmidt ex Miq, *Journal of Plant Resources and Environment*, **5**, 57–58.

IRIONDO, J.M. and PÉREZ, C., 1990a, Application of *in vitro* culture techniques to the conservation of Iberian endemic plant species, *Botanic Gardens Micropropagation News*, **1**, 4–6.

IRIONDO, J.M. and PÉREZ, C., 1990b, Micropropagation of an endangered plant species: *Coronopus navasii* (Brassicaceae), *Plant Cell Reports*, **8**, 745–748.

IRIONDO, J.M. and PÉREZ, C., 1991, *In vitro* storage of three endangered species from S.E. Spain, *Botanic Gardens Micropropagation News*, **1**, 46–48.

IRIONDO, J.M. and PÉREZ, C., 1996, Micropropagation and *in vitro* storage of *Centaurium rigualii* Esteve (Gentianaceae), *Israel Journal of Plant Science*, **44**, 115–123.

IRIONDO, J.M., PRIETO, C. and PÉREZ-GARCÍA, F., 1995, *In vitro* regeneration of *Helianthemum polygonoides* Peinado *et al.*, an endangered salt meadow species, *Botanic Gardens and Micropropagation News*, **2**, 2–5.

ITURRIAGA, L., JORDAN, M., ROVERARO, C. and GOREUX, A., 1994, *In vitro* culture of *Sophora toromiro* (Papilionaceae), an endangered species, *Plant Cell Tissue and Organ Culture*, **37**, 201–204.

IUCN, 1998, 1997, *IUCN Red List of Threatened Plants*, Cambridge: IUCN Publications Services Unit.

JACOBSEN, H.-J. and DOHMEN, G., 1990, Modern plant biotechnology as a tool for the reestablishment of genetic variability in *Sophora toromiro, Courier forschungs Institut Senckenberg*, **125**, 233–237.

JAKOBEK, J.L., STUTZ, J.C., BESS, V.H. and BACKHAUS, R.A., 1993, Micropropagation of *Cowania subintegra* and *Cowania stansburiana, Plant Cell Tissue and Organ Culture*, **35**, 297–299.

JOHNSON, T.S., NARAYAN, S.B. and NARAYANA, D.B.A., 1997, Rapid in vitro propagation of *Saussurea lappa*, an endangered medicinal plant, through multiple shoot cultures, *In Vitro Cellular and Developmental Biology – Plant*, **33**, 128–130.

JONES, N.B., MITCHELL, J., BAYLEY, A.D. and van STADEN, J., 1992, *In vitro* propagation of *Thuranthos basuticum, South African Journal of Botany*, **58**, 403–404.

JUSAITIS, M., 1995, *In vitro* propagation of *Phebalium equestre* and *Phebalium hillebrandii* (Rutaceae), *In Vitro Cellular and Developmental Biology – Plant*, **31**, 140–143.

KARP, A., 1994, Origins, causes and uses of variation in plant tissue cultures, in *Plant Cell and Tissue Culture*, edited by I.K. VASIL and T.A. THORPE, pp. 139–151, Dordrecht, Netherlands: Kluwer Academic Publishers.

KAUR, S. and SARMA, C.M., 1997, Selection of best medium for in vitro propagation of *Dendrobium lindleyi* Steud, *Advances in Plant Sciences*, **10**, 1–5.

KOOB, G.A., 1993, Tissue culture at Harold L. Lyon Arboretum, *Plant Conservation* (Center for Plant Conservation), **7**(2), 2–3.

KOROCH, A.R., JULIANI, H.R., Jr, JULIANI, H.R. and TRIPPI, V.S., 1997, Micropropagation and acclimatization of *Hedeoma multiflorum, Plant Cell Tissue and Organ Culture*, **48**, 213–217.

KOVAC, J., 1995, Micropropagation of *Dianthus arenarius* subsp. *bohemicus* – an endangered endemic from the Czech Republic, *Botanic Gardens Micropropagation News*, **1**, 106–108.

KRISHNAN, P.N. and SEENI, S., 1994, Rapid micropropagation of *Woodfordia fruticosa* (L.) Kurz (Lythraceae), a rare medicinal plant, *Plant Cell Reports*, **14**, 55–58

KRISHNAN, P.N., SUDHA, C.G. and SEENI, S., 1995, Rapid propagation through shoot tip culture of *Trichopus zeylanicus* Gaertn., a rare ethnomedicinal plant, *Plant Cell Reports*, **14**, 708–711.

KROGSTRUP, P. and NORGAARD, J.V., 1991, Micropropagation of *Psiadia coronopus* (Lam.) Benth, a threatened endemic species from the island of Rodrigues, *Plant Cell Tissue and Organ Culture*, **27**, 227–230.

KROGSTRUP, P., NORGAARD, J.V. and HAMANN, O., 1990, Micropropagation of threatened endemic and indigenous plant species from the island of Rodrigues, *Botanic Gardens Micropropagation News*, **1**, 8–11.

KUKULCZANKA, K., 1991, Micropropagation and *in vitro* germplasm storage of Droseraceae, *Botanic Gardens Micropropagation News*, **1**, 38–42.

KUKULCZANKA, K., CZASTKA, B. and ARCZEWSKA, A., 1989a, Regeneration from leaves of *Dionea muscipula* Ellis cultured *in vitro*, *Acta Horticulturae*, **251**, 155–160.

KUKULCZANKA, K., KROMER, K. and CZASTKA, B., 1989b, Propagation of *Fritillaria meleagris* L. through tissue culture, *Acta Horticulturae*, **251**, 147–153.

LAL, N. and AHUJA, P.S., 1993, Assessment of liquid culture procedures for *in vitro* propagation of *Rheum emodi, Plant Cell Tissue and Organ Culture*, **34**, 223–226.

LAL, N., AHUJA, P.S., KUKREJA, A.K. and PANDEY, B., 1988, Clonal propagation of *Picrorhiza kurroa* Royle ex Benth. by shoot tip culture, *Plant Cell Reports*, **7**, 202–205.

LALIBERTÉ, B., 1997, Botanic garden seed banks/genebanks worldwide, their facilities, collections and network, *Botanic Gardens Conservation News*, **2**(9), 18–23.

LATHA, P.G. and SEENI, S., 1994, Multiplication of the endangered Indian pitcher plant (*Nepenthes khasiana*) through enhanced axillary branching *in vitro*, *Plant Cell Tissue and Organ Culture*, **38**, 69–71.

LEE, B.C., KIM, S.C. and KWON, H.M., 1995, *In vitro* propagation of a rare species, *Forsythia coreana* for. *aureoreticulata* and *Salix hallaisanensis*, *Research Report of the Forest Genetics Research Institute Kyonggido*, **31**, 129–133.

LEVERONE, L. and PENCE, V.C., 1993, Recovery of moss and liverwort gametophyte *in vitro* following desiccation and cryopreservation, *In Vitro Cellular and Developmental Biology*, **29**, 88A.

LIM, H.C., PARK, N.B., CHOI, D.C., JIN, S.G., PARK, K.H. and CHOI, B.J., 1993, Studies on micropropagation *in vitro* and hardening culture in *Dendrobium moniliforme, RDA Journal of Agricultural Science, Biotechnology*, **35**, 221–225.

LINDGREN, D.T. and McCOWN, B., 1992, Multiplication of four *Penstemon* species *in vitro*, *HortScience*, **27**, 182.

LLEDÓ, M.D., CRESPO, M.B. and AMO-MARCO, J.B., 1996, Micropropagation of *Limonium thiniense* Erben (Plumbaginaceae) using herbarium material, *Botanic Gardens Micropropagation News*, **2**, 18–21.

Lu, C., Ruan, Y. and Bridgen, M., 1995, Micropropagation procedures for *Leontochir ovallei, Plant Cell Tissue and Organ Culture*, **42**, 219–221.

Maguire, T.L. and Sedgley, M., 1997, Genetic diversity in *Banksia* and *Dryandra* (Proteaceae) with emphasis on *Banksia cuneata*, a rare and endangered species, *Heredity*, **79**, 394–401.

Malmgren, S., 1992, Large scale asymbiotic propagation of *Cypripedium calceolus* – Plant physiology from a surgeon's point of view, *Botanic Gardens Micropropagation News*, **1**, 59–63.

Manjula, S., Thomas, A., Daniel, B. and Nair, G.M., 1997, *In vitro* plant regeneration of *Aristolochia indica* through axillary shoot multiplication and organogenesis, *Plant Cell Tissue and Organ Culture*, **51**, 145–148.

Marlow, S.A. and Butcher, D., 1987, Propagation of two rare orchid species from seed, *Orchid Review*, **95**, 331–334.

Martín, C. and Pérez, C., 1992, Multiplication *in vitro* of *Limonium estevei* Fdez. Casas, *Annals of Botany*, **70**, 165–167.

Martín, C. and Pérez, C., 1994, The use of RAPD to determine the genetic variability of micropropagated plant from endangered species. Application to the Spanish endemism, *Limonium estevei, Phyton*, **56**, 65–72.

Martín, C. and Pérez, C., 1995, Micropropagation of five endemic species of *Limonium* from the Iberian peninsula, *Journal of Horticultural Science*, **70**, 97–103.

Martínez-Vázquez, O. and Rubluo, A., 1989, *In-vitro* mass propagation of the near-extinct *Mammillaria san-angelensis* Sánchez-Mejorada, *Journal of Horticultural Science*, **64**, 99–105.

Mathur, J., 1992, *In vitro* morphogenesis in *Nardostachys jatamansi* DC: shoot regeneration from callus derived roots, *Annals of Botany*, **70**, 419–422.

Mathur, J., Ahuja, P.S., Mathur, A., Kukreja, A.K. and Shah, N.C., 1988, *In vitro* propagation of *Valeriana wallichii, Planta Medica*, **54**, 82–83.

Mattick J.S., Ablett, E.M. and Edmonson, D.L., 1992, The gene library – preservation and analysis of genetic diversity in Australasia, in *Conservation of Plant Genes*, edited by. R.P. Adams and J.E. Adams, pp. 15–35, San Diego: Academic Press.

McAlister, B.G. and van Staden, J., 1996, *In vitro* propagation of *Kniphofia pauciflora* Bak. for conservation purposes, *South African Journal of Botany*, **62**, 219–221.

McComb, J.A., 1985, Micropropagation of the rare species *Stylidium coroniforme* and other *Stylidium* species, *Plant Cell Tissue and Organ Culture*, **4**, 151–158.

McKently, A.H. and Adams, J.B., 1994, *In vitro* propagation of *Paronychia chartacea, HortScience*, **29**, 921.

Meney, K.A. and Dixon, K.W., 1995, *In vitro* propagation of Western Australian rushes (Restionaceae and related families) by embryo culture. Part 1. *In vitro* embryo growth, *Plant Cell Tissue and Organ Culture*, **41**, 107–113.

Mercier, H. and Kerbauy, G.B., 1993, Micropropagation of *Dyckia macedoi* – an endangered endemic Brazilian bromeliad, *Botanic Gardens Micropropagation News*, **1**, 70–72.

Mercier, H. and Kerbauy, G.B., 1995, The importance of tissue culture technique for conservation of endangered Brazilian bromeliads from Atlantic rain forest canopy, *Selbyana*, **16**, 147–149.

Mohamed-Yasseen, Y., Barringer, S.A. and Splittstoesser, W.E., 1995, Micropropagation of the endangered succulent, *Stapelia semota*, by axillary proliferation, *Cactus and Succulent Journal*, **67**, 366–368.

Momcilovic, I., Grubisic, D. and Neskovic, M., 1997, Micropropagation of four *Gentiana* species (*G. lutea, G. cruciata, G. purpurea* and *G. acaulis*), *Plant Cell Tissue and Organ Culture*, **49**, 141–144.

Moura-Costa, P.H., Viana, A.M. and Mantell, S.H., 1993, *In vitro* plantlet regeneration of *Ocotea catharinensis*, an endangered Brazilian hardwood forest tree, *Plant Cell Tissue and Organ Culture*, **35**, 279–286.

Nagayoshi, T., Hatanaka, T. and Suzuki, T., 1996, Seed propagation of *Habenaria radiata, Nature and Human Activities*, **1**, 67–81.

NORMAH, M.N., HAMIDAH, S. and GHANI, F.D., 1997, Micropropagation of *Citrus halimii* – an endangered species of South-east Asia, *Plant Cell Tissue and Organ Culture*, **50**, 225–227.

OKONKWO, S.N.C., 1966, Studies on *Striga senegalensis* Benth. III. *In vitro* culture of seedlings. Establishment of cultures, *American Journal of Botany*, **53**, 679–687.

OLIPHANT, J.L., 1989, *In vitro* cultivation of *Todea barbara* – from spore to sporophyte, *International Plant Propagators Society Proceedings*, **38**, 324–325.

PANDY, R., CHANDEL, K.P.S. and RAMA RAO, S., 1992, *In vitro* propagation of *Allium tuberosum* Rottl. ex. Spreng. by shoot proliferation, *Plant Cell Reports*, **11**, 204–206, 211.

PENCE, V.C., 1990, *In vitro* collection, regeneration, and cryopreservation of *Brunfelsia densifolia*, *VIIth International Congress on Plant Tissue and Cell Culture*, p. 377, Amsterdam, Netherlands: IAPTC.

PENCE, V.C., 1991, Cryopreservation of seeds of Ohio native plants and related species, *Seed Science and Technology*, **19**, 235–251.

PENCE, V.C., 1996, *In vitro* collection (IVC), in *In Vitro Conservation of Plant Genetic Resources*, edited by NORMAH, M.N., NARIMAH, M.K. and CLYDE, M.M., pp. 181–190, Plant Biotechnology Laboratory, Universiti Kebangsaan Malaysia, Kuala Lumpur, Malaysia.

PENCE, V.C., 1998a, Cryopreservation of bryophytes: the effects of ABA and encapsulation dehydration, *The Bryologist*, **101**, 279–218.

PENCE, V.C., 1998b, Cryostorage of seeds of *Salix* spp. (submitted).

PENCE, V.C., 1998c, A comparison of two methods for the cryopreservation of in vitro grown fern gametophytes (submitted).

PENCE, V.C., 1998d, Cryopreservation of chlorophyllous and nonchlorophyllous fern spores (in preparation).

PENCE, V.C. and CLARK, J., 1998, Desiccation and cryopreservation of seeds of the rare wetland species, *Plantago cordata* Lam. (submitted).

PENCE, V.C. and SOUKUP, V.G., 1993, Factors affecting the initiation of mini-rhizomes from *Trillium erectum* and *T. grandiflorum* tissues *in vitro, Plant Cell Tissue and Organ Culture*, **35**, 229–235.

PENCE, V.C. and SOUKUP, V.G., 1995, Propagation of the rare *Trillium persistens in vitro, Botanic Gardens Micropropagation News*, **1**, 109–110.

PENCE, V.C., CLARK, J., ZHANG, H. and AVASARALA, S., 1997, A CREW–CPC collaboration for the propagation of endangered plants, *In Vitro Cellular and Developmental Biology*, **33**, 16A.

POWERS, D.E. and BACKHAUS, R.A., 1989, *In vitro* propagation of *Agave arizonica* Gentry & Weber, *Plant Cell Tissue and Organ Culture*, **16**, 57–60.

PREVALEK-KOZLINA, B., KOSTOVIC-VRANJES, V. and SLADE, D., 1997, *In vitro* propagation of *Fibigia triquetra* (DC) Boiss., a rare stenoendemic species, *Plant Cell Tissue and Organ Culture*, **51**, 141–143.

PUROHIT, S.D. and DAVE, A., 1996, Micropropagation of *Sterculia urens* Roxb. – an endangered tree species, *Plant Cell Reports*, **15**, 704–706.

PUROHIT, S.D., DAVE, A. and KUKDA, G., 1994a, Micropropagation of safed musli (*Chlorophytum borivilianum*), a rare Indian medicinal herb, *Plant Cell Tissue and Organ Culture*, **39**, 93–96.

PUROHIT, S.D., KUKDA, G., SHARMA, P. and TAK, K., 1994b, *In vitro* propagation of an adult tree *Wrightia tomentosa* through enhanced axillary branching, *Plant Science*, **103**, 67–72.

PYLE, M.M. and ADAMS, R.P., 1989, *In situ* preservation of DNA in plant specimens, *Taxon*, **38**, 576–581.

RADHAMANI, J. and CHANDEL, K.P.S., 1992, *In vitro* propagation of *Allium tuberosum* Rottl. ex. Spreng. by shoot proliferation, *Plant Cell Reports*, **11**, 204–206, 211.

RAMSAY, M.M., 1997, An integrated approach to endangered plant conservation at the Royal Botanic Gardens, Kew: the role of *in vitro* techniques, *In Vitro Cellular and Developmental Biology*, **33**, 16A.

RATHORE, T.S., TANDON, P. and SHEKHAWAT, N.S., 1991, *In vitro* regeneration of pitcher plant (*Nepenthes khasiana* Hook. f.) – A rare insectivorous plant of India, *Journal of Plant Physiology*, **139**, 246–248.

RAVEN, P.H., 1987. The scope of the plant conservation problem world-wide, in *Botanic Gardens and the World Conservation Strategy*, edited by D. BRAMWELL, O. HAMANN, V. HEYWOOD and H. SYNGE, pp. 19–29, London: Academic Press.

REDWOOD, G.N. and BOWLING, J.C., 1990, Micropropagation of *Nepenthes* species, *Botanic Gardens Micropropagation News*, **1**, 19–20.

REISINGER, D.M., BALL, T.A. and ARDITTI, J., 1976, Clonal propagation of *Phalaenopsis* by means of flower stalk node cultures, *Orchid Review*, **84**, 45–52.

RODRÍGUEZ-GARAY, B. and RUBLUO, A., 1992, *In vitro* morphogenetic responses of the endangered cactus *Aztekum ritteri* (Boedeker), *Cactus and Succulent Journal*, **64**, 116–119.

RODRÍGUEZ-GARAY, B., GUTIÉRREZ-MORA, A. and ACOSTA-DUEÑAS, B., 1996, Somatic embryogenesis of *Agave victoria-reginae* Moore, *Plant Cell Tissue and Organ Culture*, **46**, 85–87.

ROGERS, S.M.D., 1993, Optimization of plant regeneration and rooting from leaf explants of five rare *Haworthia, Scientia Horticulturae*, **56**, 157–161.

ROGERS, S.O. and BENDICH, A.J., 1985, Extraction of DNA from milligram amounts of fresh, herbarium and mummified plant tissues, *Plant Molecular Biology*, **5**, 69–76.

RONSE, A., 1990, *In vitro* culture at the National Botanic Garden of Belgium, *Botanic Gardens Micropropagation News*, **1**, 14–15.

ROSETTO, M., DIXON, K.W. and BUNN, E., 1992, Aeration: a simple method to control vitrification and improve *in vitro* culture of rare Australian plants, *In Vitro Cellular and Developmental Biology – Plant*, **28**, 192–196.

ROSETTO, M., LUCAROTTI, F., HOPPER, S.D. and DIXON, K.W., 1997, DNA fingerprinting of *Eucalyptus graniticola*: a critically endangered relict species or a rare hybrid? *Heredity*, **79**, 310–318.

RUBLUO, A., CHÁVEZ, V., MARTÍNEZ, A.P. and MARTÍNEZ-VÁZQUEZ, O., 1993, Strategies for the recovery of endangered orchids and cacti through *in vitro* culture, *Biological Conservation*, **63**, 163–169.

RUREDZO, T.J., 1991, A minimum facility method for *in vitro* collection of *Digitaria eriantha* ssp. *pentzii* and *Cynodon dactylon, Tropical Grasslands*, **25**, 56–63.

SAKAI, A., KOBAYASHI, S. and OIYAMA, I., 1990, Cryopreservation of nucellar cells of navel orange (*Citrus sinensis* Osb. var. *brasiliensis* Tanaka) by vitrification, *Plant Cell Reports*, **9**, 30–33.

SEENI, S., 1990, Micropropagation of some rare plants at the Tropical Botanic Garden and Research Institute, Trivandrum, India, *Botanic Gardens and Micropropagation News*, **1**, 16–18.

SEENI, S. and LATHA, P.G., 1992, Foliar regeneration of the endangered Red Vanda, *Renanthera imschootiana* Rolfe (Orchidaceae), *Plant Cell Tissue and Organ Culture*, **29**, 167–172.

SHARMA, N. and CHANDEL, K.P.S., 1992, Low-temperature storage of *Rauvolfia serpentina* Benth. ex Kurz.: an endangered, endemic medicinal plant, *Plant Cell Reports*, **11**, 200–203.

SHARMA, N., CHANDEL, K.P.S. and SRIVASTAVA, V.K., 1991, *In vitro* propagation of *Coleus forskohlii* Briq., a threatened medicinal plant, *Plant Cell Reports*, **10**, 67–70.

SHARMA, N., CHANDEL, K.P.S. and PAUL, A., 1993, *In vitro* propagation of *Gentiana kurroo* – an indigenous threatened plant of medicinal importance, *Plant Cell Tissue and Organ Culture*, **34**, 307–309.

SHRESTHA, J.N. and JOSHI, S.D., 1992, Tissue culture technique for medicinally important herbs – *Orchis incarnata* and *Swertia chirata, Banko Janakari*, **3**, 25–26.

SIMERDA, B., 1990, Effective ways of propagating endangered cacti, *British Cactus and Succulent Journal*, **8**, 9–12.

SINGH, R.P. and SHEKHAWAT, N.S., 1997, Micropropagation of *Anogeissus rotundifolia* Blatt. and Hallb. – an endemic and rare tree of the Thar Desert, *Journal of Sustainable Forestry*, **4**, 159–170.

SINGHA, S., BAKER, B.S. and BHATIA, S.K., 1988, Tissue culture propagation of running buffalo clover (*Trifolium stoloniferum* Muhl. ex A. Eaton), *Plant Cell Tissue and Organ Culture*, **15**, 79–84.

SIU, L.P. and WEATHERHEAD, M.A., 1995, Tissue culture of the Camellias of Hong Kong, *Botanic Gardens Micropropagation News*, **2**, 6–8.

SLABBERT, M.M., de BRUYN, M.H., FERREIRA, D.I. and PRETORIUS, J., 1995, Adventitious *in vitro* plantlet formation from immature floral stems of *Crinum macowanii, Plant Cell Tissue and Organ Culture*, **43**, 51–57.

SOLTIS, P.S., SOLTIS, D.E. and SMILEY, C.J., 1992, An *rbc*L sequence from a Miocene *Taxodium* (bald cypress), *Proceedings of the National Academy of Sciences USA*, **89**, 449–451.

SOSSOU, J., KARUNARATNE, S. and KOVOOR, A., 1987, Collecting palm: *in vitro* explanting in the field, *FAO-IBPGR Plant Genetic Resources Newsletter*, **69**, 7–18.

STANILOVA, M.I., ILCHEVA, V.P. and ZAGORSKA, N.A., 1994, Morphogenetic potential and in-vitro micropropagation of endangered plant species *Leucojum aestivum* L. and *Lilium rhodopaeum* Delip., *Plant Cell Reports*, **13**, 451–453.

SUDHA, C.G. and SEENI, S., 1994, *In vitro* multiplication and field establishment of *Adhatoda beddomei* C.B. Clarke, a rare medicinal plant, *Plant Cell Reports*, **13**, 203–207.

SUDHA, C.G. and SEENI, S., 1996, *In vitro* propagation of *Rauwolfia micrantha*, a rare medicinal plant, *Plant Cell Tissue and Organ Culture*, **44**, 243–248.

SULAIMAN, I.M., 1994, Regeneration of plantlets through organogenesis in the Himalayan yellow poppy, *Meconopsis paniculata, Plant Cell Tissue and Organ Culture*, **36**, 377–380.

SULAIMAN, I.M. and BABU, C.R., 1993, *In vitro* regeneration through organogenesis of *Meconopsis simplicifolia* – an endangered ornamental species, *Plant Cell Tissue and Organ Culture*, **34**, 295–298.

SUYAMA, Y., KAWAMURO, K., KINOSHITA, I., YOSHIMURA, K., TSUMURA, Y. and TAKAHARA, H., 1996, DNA sequence from a fossil pollen of *Abies* spp. from Pleistocene peat, *Genes and Genetic Systems*, **71**, 145–149.

TAKEUCHI, M., MATSUSHIMA, H. and SUGAWARA, Y., 1980, Long-term freeze preservation of protoplasts of carrot and *Marchantia, Cryo-Letters*, **1**, 519–524.

TANAKA, M. and SAKANISHI, Y., 1977, Clonal propagation of *Phalaenopsis* by leaf tissue culture, *American Orchid Society Bulletin*, **46**, 733–737.

TANDON, P., RATHORE, T.S. and DANG, J.C., 1990, Mass multiplication and conservation of some threatened plant species of Northeast India through tissue culture, *Abstracts VIIth International Congress on Plant Tissue and Cell Culture*, p. 136, Amsterdam, Netherlands: IAPTC.

TORMALA, T., RAATIKAINEN, M., PUSKA, R. and VALOVIRTA, I., 1994, Biotechnology in conserving endangered plant species: a case from Finland, *Aquilo Ser Botanica*, **33**, 135–140.

TOUCHELL, D.H. and DIXON, K.W., 1994, Cryopreservation for seedbanking of Australian species, *Annals of Botany*, **74**, 541–546.

TOUCHELL, D.H., DIXON, K.W. and TAN, B., 1992, Cryopreservation of shoot-tips of *Grevillea scapigera* (Proteaceae): a rare and endangered plant from Western Australia, *Australian Journal of Botany*, **40**, 305–310.

UPADHYAY, R., ARUMUGAM, N. and BHOJWANI, S.S., 1989, In vitro propagation of *Picrorhiza kurroa* Royle ex Benth. – an endangered species of medicinal importance, *Phytomorphology*, **39**, 235–242.

US Fish and Wildlife Service, 1995, *Endangered and Threatened Wildlife and Plants*, US Government Printing Office, Washington, DC.

VIJAYAKUMAR, N.K., FERET, P.P. and SHARIK, T.L., 1990, In vitro propagation of the endangered Virginia roundleaf birch (*Betula uber* [Ashe] Fern.) using dormant buds, *Forest Science*, **36**, 842–846.

WARREN, R., 1983, Tissue culture, *Orchid Review*, **91**, 306–308.

WILKINSON, T., WETTEN, A. and FAY, M.F., 1998, Cryopreservation of *Cosmos atrosanguineus* shoot tips by a modified encapsulation/dehydration method (submitted).

WILLIAMS, R.R. and TAJI, A.M., 1991, Effect of temperature, gel concentration and cytokinins on vitrification of *Olearia microdisca* (J.M. Black) in vitro shoot cultures, *Plant Cell Tissue and Organ Culture*, **26**, 1–6.

WITHERS, L.A., 1985, Cryopreservation of cultured cells and protoplasts, in *Cryopreservation of Plant Cells and Organs*, edited by K.K. KARTHA, pp. 243–267, Boca Raton: CRC Press.

WOCHOK, Z.S., 1981, The role of tissue culture in preserving threatened and endangered plant species, *Biological Conservation*, **20**, 83–89.

YAM, T.W. and WEATHERHEAD, M.A., 1990, Early growth of *Hetaeria cristata* seedlings and plantlet initiation from rhizome nodes, *Lindleyana*, **5**, 199–203.

YIDANA, J.A., WITHERS, L.A. and IVINS, J.D., 1987, Development of a simple method for collecting and propagating cocoa germplasm *in vitro*, *Acta Horticulturae*, **212**, 95–98.

ZAPRATAN, M., 1996, Conservation of *Leontopodium alpinum* using *in vitro* techniques in Romania, *Botanic Gardens and Micropropagation News*, **2**, 26–28.

Conservation of the Rare and Endangered Plants Endemic to Spain

M.E. GONZÁLEZ-BENITO, C. MARTÍN, J.M. IRIONDO AND C. PÉREZ

16.1 Introduction

Spain is the European country with the richest and most exclusive flora. When the Canary Islands are included, there are approximately 1300 endemic plant species of which many have a high biological interest. Unfortunately a high number of these species are on the verge of extinction. Recent surveys of these endangered plant species of continental Spain and the Balearic and Canary Islands (Gómez-Campo, 1987, 1996; Sainz et al., 1994) denote around 600 endemic species which, according to the categories followed by IUCN, are threatened to some degree. The current situation could develop, in the near future, not only with the loss of very important germplasm useful to breeding programmes, but also with reduced biodiversity which would undermine ecosystem stability.

Although some of these threatened species are protected by local or international laws, measures for the conservation of endangered plant species in their natural habitats are only very slowly being introduced. Therefore, it is highly desirable to develop appropriate techniques for *ex situ* conservation in order to enable a quick and efficient response to the most critical situations.

There are two different strategies in which *in vitro* culture techniques (see Lynch, Chapter 4, this volume) can be used in the conservation of endangered species, (see also Pence, Chapter 15, this volume). The first one is the use of micropropagation techniques in order to rapidly increase the number of individuals in species with reproductive problems and/or with extremely reduced populations. Plant material thus obtained can be of great value for research, living collections, to reduce pressure of botanical collection on the natural population and, if considered appropriate, for plant introduction programmes. The second strategy is the development of *in vitro* storage techniques (see Benson, Chapter 6, this volume) which are particularly useful when the conservation of seeds is not possible. These techniques allow *ex situ* conservation at a very low maintenance cost.

Using these techniques, a large amount of plant material can be produced in a short period of time from a minimum of starting plant material, and hence, with minimal impact on the native population. Another advantage of *in vitro* culture techniques is that they can be an alternative to seed banks with highly restricted species, where the simple act of gathering seeds in the natural population may affect its survival. However, the convenience of these techniques is controversial, particularly with regard to the possible genetic variability produced throughout the process. In these circumstances, monitoring regenerated plantlets through somaclonal variation detection techniques may be of interest to eliminate undesirable variants.

The choice of an adequate explant, the micropropagation cycle and the type and concentration of growth regulators may all contribute to reduce the appearance of somaclonal variants, but it will not completely eliminate the probability of inducing them. On the other hand, the limited availability of starting plant material, especially prevalent with threatened plant species, further complicates the process of micropropagation, leaving little margin for the previous considerations. In this chapter we will describe several studies on the preservation of Spanish endemic plant species using biotechnology.

16.2 The application of micropropagation and *in vitro* conservation to endangered plants endemic to Spain

In the past years there has been intense activity with regard to the development of micropropagation protocols for Spanish endemics. Table 16.1 includes some of the Spanish endangered endemics on which micropropagation methods have been published. Actually, botanical gardens, universities and other research institutions have worked on a much longer list of Spanish endangered endemics, but the results have not always been published. Information on much of this unpublished activity has been gathered by Iriondo *et al.* (1994).

It is remarkable that 39 per cent of the species listed in Table 16.1 belong to the genus *Limonium*. More than half of these species (56 per cent) belong to two families: *Plumbaginaceae* and *Asteraceae*. These high percentages are easily understood when it is considered that nearly half of all the Iberian endemics belong to just 27 genera, one of which is the genus *Limonium*. In the Iberian Peninsula there are 86 endemic, mostly endangered or rare, species of *Limonium* (Erben, 1993). In contrast, only 8 per cent of the species are monocotyledons.

Research on *in vitro* storage of Spanish endemics is much more scarce. Minimum growth techniques have been used with *Centaurium rigualii, Coronopus navasii, Lavatera oblongifolia, Limonium calaminare, L. catalaunicum, L. dichotomum, L. dufourii, L. estevei* and *L. gibertii* with relative success (Iriondo and Pérez, 1991a; Martín, 1993). In general, reduction of temperature was the most effective way of decreasing growth. Cultures stored at 5°C in MS medium (Murashige and Skoog, 1962) alone or supplemented with 4.44 µM BAP + 0.54 µM NAA for periods ranging from four to six months maintained a good survival rate.

In some cases, as in the *in vitro* storage protocol for *L. estevei*, the maintenance of the cultures in darkness together with the reduction of temperature (5°C) allowed for a longer period of conservation with good recovery.

On the other hand, in *Coronopus navasii* a higher survival rate was achieved when cultures were stored under a 16 hour light photoperiod (Figure 16.1). Cryopreservation techniques have also been used in the conservation of *in vitro* propagated explants of Spanish endemics, and they will be discussed in the next section.

These techniques constitute an important tool for the conservation of the rich endemic flora of Spain, where an effective *in situ* conservation policy is not always possible for all species. At present, a good representation of endangered plants is being obtained through

Table 16.1 Spanish endangered endemics on which micropropagation methods have been developed for conservation purposes

Family	Species	Reference
Amaryllidaceae	*Narcissus longispathus* Pugsley	Clemente (1991)
	Narcissus nevadensis Pugsley	Clemente (1991)
	Narcissus tortifolius Fdez. Casas	Clemente (1991)
Asteraceae	*Artemisia granatensis* Boiss.	Clemente *et al.* (1991)
	Atractylis arbuscula Svent. & Mich.	González *et al.* (1989)
	Centaurea balearica Rodr.	Estades and Medrano (1990)
	Centaurea carratracensis Lange	Clemente (1991)
	Pericallis hadrosoma (Svent.) B. Nord.	Ortega and González (1990)
	Senecio hermosae Pitard	Ortega and González (1985)
Brassicaceae	*Coronopus navasii* Pau	Iriondo and Pérez (1990)
	Vella lucentina M.B. Crespo	Lledó *et al.* (1995)
Cistaceae	*Cistus heterophyllus* Desf. subsp. *carthaginensis* (Pau) M.B. Crespo & Mateo	Arregui *et al.* (1993)
	Helianthemum polygonoides Pein., Mtnez-Parras, Alc. & Esp.	Iriondo *et al.* (1995)
Euphorbiaceae	*Euphorbia handiensis* Burchd.	González *et al.* (1988)
Gentianaceae	*Centaurium rigualii* Esteve	Iriondo and Pérez (1996a)
Globulariaceae	*Globularia ascanii* Bramwell & Kunkel	Bramwell (1990); Cabrera (1995)
Lamiaceae	*Thymus richardii* Pers.	Estades and Medrano (1990)
Leguminosae	*Lotus berthelotii* Masferrer	Ortega (1982)
	Lotus kunkelii (Esteve) Bramwell & Davis	Bramwell (1990)
Malvaceae	*Lavatera oblongifolia* Boiss.	Iriondo and Pérez (1992)
Plumbaginaceae	*Limonium arborescens* (Brouss.) O. Kuntze	Fay (1993)
	Limonium calaminare Pignatti	Martín and Pérez (1995)
	Limonium dufourii (Girard) O. Kuntze	Lledó *et al.* (1993); Martín and Pérez (1995)
	Limonium estevei Fdez. Casas	Martín and Pérez (1992)
	Limonium fruticans (Webb) O. Kuntze	Fay (1993)
	Limonium gibertii (Sennen) Sennen	Martín and Pérez (1995)
	Limonium imbricatum (Webb & Berthel.) Hubbard	Fay (1993)
	Limonium parvibracteatum Pignatti	Lledó *et al.* (1993)
	Limonium redivivum (Svent.) Kunkel & Sund.	Fay (1993)
	Limonium rumicifolium (Svent) Kunkel & Sund.	Fay (1993)
	Limonium spectabile (Svent) Kunkel & Sund	Fay (1993)
	Limonium rigualii M.B. Crespo & Erben	Lledó *et al.* (1993)
	Limonium santapolense Erben	Lledó *et al.* (1993)
	Limonium thiniense Erben	Lledó *et al.* (1996)
Scrophulariaceae	*Antirrhinum microphyllum* Roth.	González-Benito *et al.* (1996)
Thymeleaceae	*Daphne rodriguezii* Teixidor	Estades and Medrano (1990)

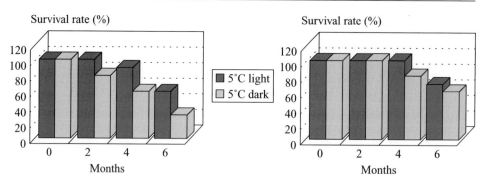

Figure 16.1 Survival rate after different periods of shoots of *Coronopus navasii* stored *in vitro* in different incubation conditions

micropropagation. These plants are being used for scientific studies and for display in botanical gardens (e.g. Jardín Botánico de Córdoba).

16.3   Cryopreservation and the conservation of endangered endemic Spanish plants

Seeds of many Spanish endemics have been stored dry at −20°C in the Seed Bank run by the Department of Plant Biology at the Universidad Politécnica de Madrid. There is, however, an increasing interest in cryopreservation (see Benson, Chapter 6, this volume for full details of the technique) as some physiological or even genetic damage can occur in seeds stored for long periods under such conditions. Such deterioration could be of great concern when the species are somehow threatened due to the scarcity of their populations, their low seed production or their habitat degradation. The development of cryopreservation protocols of vegetative explants is also of interest for those species whose micropropagation has already been carried out.

16.3.1   *Cryopreservation of orthodox seeds*

Studies largely concern species of the families Asteraceae, Betulaceae, Brassicaceae, Cariophylaceae, Cistaceae, Leguminosae, Plumbaginaceae and Scrophulariaceae (Table 16.2). Cryopreservation was carried out by direct immersion of seeds (contained in polypropylene cryovials) in liquid nitrogen, where they were maintained for one or 30 days. Thawing took place at room temperature.

Most seeds studied either had a low initial moisture content or this was reduced by desiccation with silica gel (3.9 to 8.4 per cent fresh weight basis). Only seeds of *Gypsophila struthium* had a slightly higher moisture content (10.8 per cent). Desiccation did not improve seed survival after storage in liquid nitrogen in those species where this factor was studied (Iriondo *et al.*, 1992). Storage for one or 30 days in liquid nitrogen did not significantly affect germination percentages in most species. However, cryopreservation seemed to improve germination in *Antirrhinum microphyllum*. This increase could be explained by a scarification effect of the freezing and thawing process. Removal of

Table 16.2 Results of seed cryopreservation studies carried out with Spanish endemics

Family	Species	MC[a]	GT	C[b]	Germination (%) LN (1 day)[c]	LN (30 days)[c]	Reference
Asteraceae	*Onopordum*	10.7	C	9[d]	6[d]	8[d]	Iriondo *et al.* (1992)
	nervosum Boiss	6.5	B	72	75	87	González-Benito *et al.* (1998a)
		5.0	C	9[d]	3[d]	12[d]	Iriondo *et al.* (1992)
	O. nogalesii Svent	7.0	B	6[e]	1[e]	7[e]	González-Benito *et al.* (1998b)
		7.0	B	4	4	3	González-Benito *et al.* (1998b)
	Centaurea	6.8	E	65[d]	73[d]	67[d]	González-Benito
	hyssopifolia Vahl	6.8	E		64[df]	62[df]	*et al.* (1998a)
		6.8	E	48	47	38	González-Benito
		6.8	E		52[f]	52[f]	*et al.* (1998a)
Betulaceae	*Betula celtiberica*	10.6	C	3[d]	4[d]	3[d]	Iriondo *et al.* (1992)
	Rothm. et Vasc.	6.1	C	1[d]	3[d]	7[d]	Iriondo *et al.* (1992)
Brassicaceae	*Coronopus navasii* (Cav.) DC.	ND	D	30	34	40	González-Benito *et al.* (1998b)
	Diplotaxis virgata (Cav.) DC.	7.7	D	90	93	88	González-Benito *et al.* (1998b)
	Iberis pectinata Boiss	7.2	A	97	97	98	González-Benito *et al.* (1998b)
	Vella pseudocytisus L.	7.3	D	97	98	100	González-Benito *et al.* (1998b)
Cariophylaceae	*Gypsophila struthium* Loefl.	10.8	B	100	93	96	González-Benito *et al.* (1998b)
Cistaceae	*Cistus osbeckiifolius* Webb ex Christ.	4.3	A	57[e]	74[e]	56[e]	González-Benito *et al.* (1998b)
		4.3	A	30	25	24	González-Benito *et al.* (1998b)
	Halimium	6.4	C	6[d]	1[d]	1[d]	Iriondo *et al.* (1992)
	atriplicifolium (Lam.) Spach	5.3	C	4[d]	2[d]	0[d]	Iriondo *et al.* (1992)
	Helianthemum	3.9	D	42[e]	42[e]	55[e]	González-Benito *et al.* (1998b)
	polygonoides Peinado *et al.*	3.9	D	7	6	4	González-Benito *et al.* (1998b)
	H. squamatum (L.) Dum. Cours.	ND	D	32[e]	29[e]	39[e]	González-Benito *et al.* (1998b)
		ND	D	21	18	17	González-Benito *et al.* (1998b)
Leguminosae	*C. atlantica* Browicz	11.0	C	4[d]	2[d]	0[d]	Iriondo *et al.* (1992)
		7.8	C	2[d]	1[d]	0[d]	Iriondo *et al.* (1992)
	Onobrychis *eriophora* Desv.	7.7	C	48[de]	40[de]	–	González-Benito *et al.* (1994)
		5.3	C	60[de]	70[de]	–	González-Benito *et al.* (1994)
		6.0	C	42[d]	51[d]	54[d]	Iriondo *et al.* (1992)
		5.2	C	55[d]	46[d]	31[d]	Iriondo *et al.* (1992)

Table 16.2   (cont'd)

Family	Species	MC[a]	GT	C[b]	LN (1 day)[c]	LN (30 days)[c]	Reference
					Germination (%)		
	O. peduncularis (Cav.) D.C. ssp. *matritensis* (Boiss. et Reuter) Maire	ND	A	83	78	73	González-Benito *et al.* (1998b)
Plumbaginaceae	*Limonium dichotomum* (Cav.) Kuntze	8.4	E	83	82	83	González-Benito *et al.* (1998a)
		8.4	E		84[f]	83[f]	González-Benito *et al.* (1998a)
Scrophulariaceae	*Antirrhinum majus* L. ssp. *barrelieri* Boreau	ND	A	93	92	91	González-Benito *et al.* (1998a)
	A. microphyllum Rothm	ND	A	38	52	55	González-Benito *et al.* (1998b)

[a] MC, moisture content (fresh weight basis).
[b] C, control (unfrozen) seeds.
[c] Storage for 1 or 30 days in liquid nitrogen.
[d] Humidification by placing seeds in a saturated atmosphere.
[e] Dormancy breaking treatment carried out.
[f] Fast thawing (water bath at 40°C).
ND, not determined.
GT, germination temperature: A, 15°C; B, 20°C; C, 25°C; D, 25°C (day) 16 h/15°C (night) 8 h;
E, 10 days at 15°C, then 25/15°C.

hard-seededness has been observed in legume seeds, especially with the increase of the number of alternating temperature cycles (Boyce, 1987; Pritchard *et al.*, 1988).

Some species require dormancy breaking treatments before sowing to achieve high germination percentages. The possible interaction of this treatment with cryopreservation has been studied in three Cistaceae and one Asteraceae species (González-Benito *et al.*, 1998b). Seeds of *Cistus osbeckiifolius*, *Helianthemum polygonoides* and *H. squamatum* were placed in boiling water and soaked for up to 24 hours as the water cooled down to room temperature. *Onopordum nogalesii* seeds were immersed in a solution of giberellic acid (1000 mgl^{-1}) for 24 hours at 20°C. For cryopreserved seeds, dormancy breaking treatments were performed after thawing. The results indicated that the dormancy breaking treatment applied increased germination in control and frozen seeds in the first three species (Table 16.2). In *C. osbeckiifolius* and *H. polygonoides* a significant cryopreservation × dormancy-breaking interaction was found ($p < 0.05$). However, cryopreserved and dormancy treated seeds never showed lower germination than the dormancy treated non-cryopreserved seeds.

Other factors studied include thawing speed and humidification of seeds before sowing. Seeds of two endemic species (*Centaurea hyssopifolia* and *Limonium dichotomum*) were cryopreserved by immersion in liquid nitrogen for one or 30 days. Seeds were then thawed either at room temperature (slow thawing) or in a 40°C water bath (fast thawing). Half of the seeds were subjected to humidification before sowing by placing them in a sealed container above water at room temperature for three days. The results showed that for both species there was not a significant effect of cryopreservation or the speed of thawing on germination (Table 16.2).

Humidification increased germination in *C. hyssopifolia* and did not have any effect in *L. dichotomum*. This could have been due to the fact that the first species has seeds with hard sclerenchyma layers. No interaction between humidification and freezing in liquid nitrogen was observed in these two species.

16.3.2 *Cryopreservation of vegetatively propagated germplasm*

Cryopreservation of nodal explants of *Centaurium rigualii* Esteve (Gentianaceae), an endemic species from the south-east Iberian Peninsula, has been studied using two different techniques: vitrification and encapsulation–dehydration (González-Benito and Pérez, 1994; González-Benito *et al.*, 1997; see also Benson, Chapter 6, this volume, for methodological details). In the first study Sakai *et al.* (1990) vitrification solution was used. No survival was achieved unless explants had been previously cultured on semi-solid medium containing 2 per cent (v/v) dimethyl sulphoxide or glycerol for two days, when 15 per cent survival was observed.

A protocol to improve survival of *C. rigualii* nodal explants after direct immersion in liquid nitrogen was developed using the encapsulation–dehydration technique to protect against ice crystal formation (González-Benito *et al.*, 1997). Each node was included in an alginate bead, cultured in liquid medium with 0.75 M sucrose, desiccated with silica gel to around 20 per cent moisture content and subjected to immersion in liquid nitrogen. The preculture of nodes in 0.1 M sucrose (standard concentration) for three days improved survival after freezing and fast thawing (32 per cent) compared to non-precultured nodes (5 per cent). When node preculture was carried out in a higher sucrose concentration (0.3 M) for one day, survival after freezing was increased to 70 per cent on the eighth week of culture.

This technique has also been employed with nodal segments of another Spanish endemic species: *Antirrhinum microphyllum* (González-Benito *et al.*, 1998c). Different desiccation times and preculture media were studied. A survival rate of 85 per cent was obtained after cryopreservation when beads had been desiccated for four or five hours (21 and 20 per cent moisture content, respectively). However, the number of frozen nodal explants that showed bud elongation as a response after four hours desiccation was higher when the preculture had been carried out with 0.3 M sucrose (70 per cent vs 53 per cent), although this difference was not found after five hours of desiccation (68 per cent).

Preculture of explants on medium with a high sucrose concentration to improve survival after freezing has been reported in many encapsulation–dehydration protocols. In some cases no survival was observed after immersion in liquid nitrogen without preculture on 0.3 M sucrose (Matsumoto *et al.*, 1995). Similarly, survival was improved for cryopreserved nodal segments of *Dianthus hybridus* after a two day preculture on 0.6 M sucrose (Fukai *et al.*, 1994).

16.4 Stability assessments

The use of *in vitro* culture techniques bears the risk of somaclonal variation occurring (Larkin and Scowcroft, 1981; see also Harding, Chapter 7, this volume). It is therefore necessary to assess the genetic stability of the cultures, particularly when the aim of the *in vitro* techniques is genetic conservation. Although a great number of Spanish endemic species have been micropropagated (see Table 16.1), the occurrence of genetic changes has only been assessed in a few of them. Several techniques can be used to detect morphological variations and genetic changes. The latter can include studies on chromosome structure and number, or the use of molecular markers (isozymes or DNA markers). A combination of several techniques is recommended for the evaluation of the regenerated plants, especially when one of them is based on morphological characters.

In *Lavatera oblongifolia* (Iriondo and Pérez, 1996b), shoots regenerated from calli showed relevant morphological variants. In some cases, shoots looked extremely thick and were surrounded by a large number of leaves arranged in an abnormal phyllotaxis. However, most of the regenerated plantlets that lived through the acclimatization process did not present markedly abnormal morphological features. Leaf morphology of regenerated *Lavatera* plantlets was further examined through computer-assisted image analysis techniques. No single variants were found in leaf shape, and no significant differences were found in petiole length among non-acclimatized vitroplants, acclimatized vitroplants and plants obtained from cuttings, a situation commonly described in plants of other species regenerated from calli, e.g. *Saintpaulia ionantha* (Cassells and Plunkett, 1986). However, qualitative differences in leaf hairiness between non-acclimatized plants and the rest of the groups were found. Stellate hairs of non-acclimatized plants were shorter in height, narrower, and more transparent. No differences among regenerants were detected when stability was assessed with the following six isozyme systems: malate dehydrogenase (MDH), glutamate oxaloacetate transaminase (GOT), cytochrome oxidase (Cy-O), acid phosphatase (Ac-P), catechol oxidase (CO) and peroxidase (PER). Nevertheless, four different types of esterase (EST) zymograms were obtained from leaves of acclimatized regenerants derived from a single clone. The use of the esterase isozyme system has been discussed for this type of study as it is highly variable among tissues and artefacts easily appear (Richardson *et al.*, 1986).

L. oblongifolia has a chromosome number of $2n = 42$ ($x = 7$, hexaploid) (Luque and Devesa, 1986). No changes in the ploidy level were found among regenerants, although chromosome counts with $2n = 36\text{--}38$ suggest possible hypo-aneuploids.

In *Coronopus navasii*, neither morphological nor isozyme variation was detected among regenerants from a single clone. However, a chromosome duplication was detected in one of the acclimatized plants ($2n = 64$ vs $2n = 32$) (Iriondo and Pérez, 1991b).

Seven isozyme systems (PER, EST, GOT, Ac-P, MDH, Cy-O and CO) were studied in *Centaurium rigualii* in the ninth subculture of a clonal line of shoot cultures (Iriondo and Pérez, 1996a). All isozyme systems but esterase showed recurring banding patterns. In this system, one of the somaclones diverged from the rest of the sample. However, when the somaclones were resampled at the twelfth subculture no differences were found in their esterase banding patterns. In these plants, caryological observations showed no departures from the natural chromosome number ($2n = 20$).

Molecular markers based on DNA offer the advantage of revealing changes occurring directly in DNA sequences. Some of these techniques, such as RAPD (random amplified polymorphic DNA, Figure 16.2), have the additional advantage of requiring a small quantity of DNA, which is an important factor when working with threatened species. Another benefit of this method is that no previous knowledge of the genome studied is required.

This method was employed to evaluate the possible somaclonal variation occurring in plants of *Limonium estevei* obtained through micropropagation (Martín and Pérez, 1994).

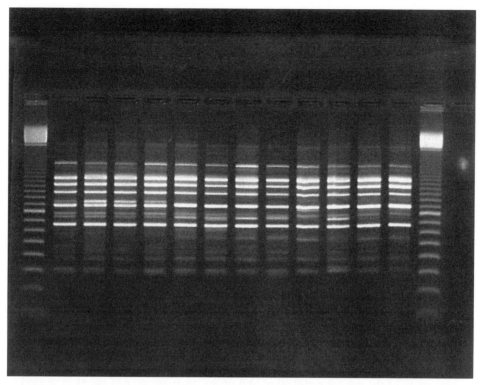

Figure 16.2 RAPD band patterns of *Betula pendula* subsp. *fontqueri* used to detect somaclonal variation of *in vitro* plantlets. Band patterns were obtained with primer OPO-16 (Operon Technology Co.). Each test track represents a single *in vitro* plant and every fourth track is assigned to a different set of clonal lines (i.e. three clones with four representative plants). Outer tracks are markers

Six plants from each of five clonal lines subcultured every month for two years were tested with three different primers. The band patterns revealed differences between clonal lines, but intraclonal variation was not detected.

The same technique was used in the assessment of genetic stability of plants of *Centaurium rigualii* which were maintained in *in vitro* conditions for six years with monthly subcultures (Martín *et al.*, 1996). All the plants came from the same original clonal line. In this case, markers from 13 different primers were considered to establish the differences of the samples through an UPGMA (unweighted pair group method arithmetic average) analysis. From 31 plants tested, 15 different patterns were detected.

These examples show that somaclonal variation can occur and thus justify the need to monitor the stability of *in vitro* cultured plants at different stages of the process, especially when genetic conservation is the aim of these techniques (see also Harding, Chapter 7, this volume).

16.5   Plant diversity assessment for endangered species conservation

Many researchers (Avise and Hamrick, 1996; Ellstrand and Elam, 1993; Falk and Holsinger, 1991; Hamrick, 1983) consider that the knowledge of the species genetic composition is essential for any comprehensive conservation plan. The study of the genetic composition and, therefore, the species diversity is important to interpret its demography, reproductive biology and evolution (see Harris, Chapter 2, and Pence, Chapter 15, this volume).

The knowledge of the population genetic variability is especially necessary when the conservation strategy is focused on *ex situ* conservation of plant material, which should be representative of the genetic variation in the natural population (Frankel and Soulé, 1981; Simberloff, 1988). This is even more relevant if introduction or reintroduction measures are to be considered.

There are only a few studies about the genetic composition of wild species, and the number is even lower when considering endangered or rare species. In recent years this kind of work has proliferated with the employment of molecular markers, firstly with the use of isozymes, and more recently with DNA markers (RFLP, DNA-fingerprinting, RAPD).

As previously mentioned in Section 16.4, one of the main advantages that the RAPD technique offers is that it only requires a small amount of DNA and, therefore, a small amount of plant material, which is in many cases a limiting factor when working with endangered plants.

In the species in which the establishment of *in vitro* cultures is the chosen conservation strategy, assessment of the genetic diversity of the initial plant material guarantees the representation of the diversity of the natural population in the sample. At the same time, it can be used as a reference to check the occurrence of somaclonal variation after the *in vitro* culture period. This kind of strategy has been used in the conservation project for *Antirrhinum microphyllum* undertaken in our laboratory. Using RAPD markers the genetic study of the natural populations and of the micropropagated plants is studied to check whether the material propagated is representative of the original population and its stability after culture.

Some other studies on the genetic diversity of natural populations of endemic Spanish plants using RAPD markers have been published: *Erodium paularense* (Martín *et al.*, 1997), *Limonium cavanillesii* (Palacios and González-Candelas, 1997a), *Limonium dufourii* (Palacios and González-Candelas, 1997b). The information obtained from these works may be very useful for further studies or for the application of conservation strategies (see also Harris, Chapter 2, this volume).

16.6   Conclusions

Throughout this chapter, we have shown some of the research where biotechnology has been applied in the conservation of Spanish endemics. It should be borne in mind that this type of work should be integrated in a broader perspective and not be an aim in itself (see Benson, Chapter 1, this volume). The final goal should be the species conservation. These studies are now being extended in some of the species already mentioned (e.g. *Erodium paularense* and *Antirrhinum microphyllum*), comprising demography, population models, reproductive biology, etc. All this knowledge will allow the appropriate application of biotechnology to species preservation.

The two main questions to be considered in *ex situ* conservation plans of an endangered plant species are: 'What to conserve?' and 'How to conserve?'.

The selection of the species to preserve is based in many cases on economical factors. The priority of some species above others will often take into account the imminence and type of threat. In addition, what to conserve within a species has not always been so carefully considered or has not been based on scientific knowledge. It is precisely in this situation where molecular techniques can play an important role. Firstly, these methods can show the genetic diversity of the species populations. Secondly, they can be used to check whether the 'conserved population' represents the diversity of the original population. With this knowledge some other practical problems can also be solved, such as whether to preserve populations separately or in a single sample.

Economic factors also dictate the procedure of conservation. Seed banking is probably the easiest and cheapest way of preserving a wide range of genotypes of a certain species. When this option is not possible (e.g. as is the case for vegetatively propagated species and recalcitrant seeds), one should compare conventional methods (field collections) with using biotechnological approaches. If the species has recalcitrant seeds (i.e. they cannot be desiccated without viability loss), cryopreservation of embryos could be employed (Pence, 1995). In this or other cases, conservation of vegetative plant material could be considered through *in vitro* conservation or cryopreservation. The most appropriate technique for the conservation of a species can be selected after a careful analysis of the nature of the propagules, the techniques that could be used and the resources available (both material and human).

When *in vitro* conservation techniques are used, or when cryopreservation implies the use of *in vitro* culture at some point, special care should be taken to check the possible appearance of somaclonal variation. The genetic composition of the stored sample should be compared after different conservation periods with that of the sample at the starting point (see Harding, Chapter 7, this volume). It should also be considered that not only could genetic changes occur during this process, but that there also could be an accidental selection towards genotypes better fit for the *in vitro* technique employed. However, in some extreme cases where the number of individuals is very low, micropropagation is clearly justified. The emergency and seriousness of the situation weighed against the probability of somaclonal variation could make the latter irrelevant.

The final question to be answered could be: What for? The preserved sample could be used for research work or to display in botanical gardens. The convenience of the use of micropropagation in plant reintroduction programmes should be carefully studied. A great number of genetically cloned specimens obtained through this technique from a few individuals of the native population may not be of great value. If planted alongside the original population, genetic impoverishment could possibly occur. On the other hand, if planted apart from the original population, its narrow genetic base would decrease the chances of adaptation to the habitat and survival in extremely mutable conditions. Studies on the genetic diversity of a given population, knowledge of the physiology of reproduction of the species and the estimation of the number of clone lines needed to adequately represent the native population's genetic diversity are all of great value when preserving endangered species.

We believe that further research is needed in order to determine correctly the value of these techniques as a method to avoid the extinction of threatened species. The experience obtained from the application of these techniques to different species, each one with its own characteristics, will be of great significance in this respect.

Acknowledgements

The authors thank Dr Julian S. Taylor for the manuscript revision.

References

ARREGUI, J.M., JUAREZ, J., LAGUNA, E. and NAVARRO, L., 1993, Micropropagación de *Cistus heterophyllus*. Un ejemplo de la aplicación del cultivo de tejidos a la conservación de especies amenazadas, *Vida Silvestre*, **74**, 24–29.

AVISE, J.C. and HAMRICK, J.L. (Eds), 1996, *Conservation Genetics: Case Histories from Nature*, New York: Chapman and Hall.

BOYCE, K.G., 1987, Cryopreservation of seeds of lucerne, medic and subclover cultivars in liquid nitrogen, *Seed Science & Technology*, **15**, 466–468.

BRAMWELL, D., 1990, The role of *in vitro* cultivation in the conservation of endangered species, in HERNÁNDEZ-BERMEJO, J.E., CLEMENTE, M. and HEYWOOD, V. (Eds), *Conservation Techniques in Botanic Gardens*, pp. 3–15, Koenigstein: Koeltz Scientific Books.

CABRERA, M.A., 1995, Explant establishment in the micropropagation of *Globularia ascanii*, a threatened species from Gran Canaria, *Botanic Gardens Micropropagation News*, **1**, 8.

CASSELLS AND PLUNKETT, 1986, Habit differences in African violets produced from leaf cuttings and *in vitro* from leaf discs and recycled axenic leaves, in WITHERS, L.A. and ALDERSON, P.G. (Eds), *Plant Tissue Culture and its Agricultural Applications*, pp. 105–111, London: Butterworths.

CLEMENTE, M., 1991, The micropropagation unit at the Cordoba Botanic Garden, Spain, *Botanic Gardens Micropropagation News*, **1**, 30–32.

CLEMENTE, M., CONTRERAS, P., SUSIN, J. and PLIEGO-ALFARO, F., 1991, Micropropagation of *Artemisia granatensis*, *HortScience*, **26**, 420–420.

ELLSTRAND, N.C. and ELAM, D.R., 1993, Population genetic consequences of small population size. Implications for plant conservation, *Annual Review of Ecology and Systematics*, **24**, 217–242.

ERBEN, M. 1993, *Limonium* Mill, in CASTROVIEJO, S., AEDO, C., CIRUJANO, S., LAÍNZ, M., MONSERRAT, P., MORALES, R., *et al.* (Eds), *Flora Iberica*, Vol. 3, pp. 2–143, Madrid: Real Jardín Botánico, CSIC.

ESTADES, J. and MEDRANO, H., 1990, Contribución a la conservación de plantas endémicas de Baleares mediante cultivo *in vitro*, in HERNÁNDEZ-BERMEJO, J.E., CLEMENTE, M. and HEYWOOD, V. (Eds), *Conservation Techniques in Botanic Gardens*, pp. 121–123, Koenigstein: Koeltz Scientific Books.

FALK, D.A. and HOLSINGER K., 1991, *Genetics and Conservation of Rare Plants*, New York: Oxford University Press.

FAY, M., 1993, Micropropagation of plants from the Canary Islands, *Bol. Mus. Mun. Funchal*, **2**, 85–88.

FRANKEL, O.H. and SOULÉ, M.E., 1981, *Conservation and Evolution*, Cambridge: Cambridge University Press.

FUKAI, S., TOGASHI, M. and GOI, M., 1994, Cryopreservation of *in vitro*-grown *Dianthus* by encapsulation-dehydration, *Technical Bulletin of Faculty of Agriculture, Kagawa University*, **46**, 101–107.

GÓMEZ-CAMPO, C. (Ed.), 1987, *Libro Rojo de especies vegetales amenazadas de España Peninsular e Islas Baleares*, Madrid: ICONA.

GÓMEZ-CAMPO C. (Ed.), 1996, *Libro Rojo de Especies Vegetales Amenazadas de las Islas Canarias*, Viceconsejería de Medio Ambiente, Gobierno de Canarias.

GONZÁLEZ, C., ORTEGA, C. and RUBIO, A., 1988, Propagación *in vitro* de endemismos canarios en peligro de extinción: *Euphorbia handiensis* Burchd, *Botanica Macaronesica*, **16**, 15–28.

GONZÁLEZ, C., RUBIO, A. and ORTEGA, C., 1989, Propagación *in vitro* de endemismos canarios en peligro de extinción: *Atractylis arbuscula* Svent. et Michaelis, *Botanica Macaronesica*, **17**, 47–56.

GONZÁLEZ-BENITO, M.E. and PÉREZ, C., 1994, Studies on the cryopreservation of nodal explants of *Centaurium rigualii* Esteve, an endemic threatened species, through vitrification, *Botanic Gardens Micropropagation News*, **1**, 82–84.

GONZÁLEZ-BENITO, M.E., CAZÉ-FILHO, J. and PÉREZ, C., 1994, Cryopreservation of seeds of several legume species, *Plant Varieties & Seeds*, **7**, 23–27.

GONZÁLEZ-BENITO, M.E., TAPIA, J., RODRÍGUEZ, N. and IRIONDO, J.M., 1996, Micropropagation of commercial and wild genotypes of snapdragon (*Antirrhinum spp.*), *Journal of Horticultural Science*, **71**, 11–15.

GONZÁLEZ-BENITO, M.E., VIVIANI, A.B. and PÉREZ, C., 1997, Cryopreservation of nodal explants of an endangered plant species (*Centaurium rigualii* Esteve) using the encapsulation-dehydration method, *Biodiversity and Conservation*, **6**, 583–590.

GONZÁLEZ-BENITO, M.E., FERNANDEZ-LLORENTE, F. and PEREZ-GARCIA, F., 1998a, Interaction between cryopreservation, rewarming rate and seed humidification on the germination of two Spanish endemic species. Annals of Botany, **82**, 683–686.

GONZÁLEZ-BENITO, M.E., IRIONDO, J.M. and PÉREZ-GARCÍA, F., 1998b, Seed cryopreservation: an alternative method for the conservation of Spanish endemics, *Seed Science & Technology*, **26**, 257–262.

GONZÁLEZ-BENITO, M.E., NUÑEZ-MORENO, Y. and MARTÍN, C., 1998c, A protocol to cryopreserve nodal explants of *Antirrhinum microphyllum* by encapsulation-dehydration, *Cryo-Letters*, **19**, 225–230.

HAMRICK, J.L., 1983, The distribution of genetic variation within and among natural plant populations, in SCHONEWALD-COX, C.M., CHAMBERS, S.M., MACBRYDE, B. and THOMAS, W.L. (Eds), *Genetics and Conservation*, pp. 335–348, Menlo Park, CA: Benjamin-Cummings.

IRIONDO, J.M. and PÉREZ, C., 1990, Micropropagation of a threatened plant species: *Coronopus navasii* (Brassicaceae), *Plant Cell Reports*, **8**, 745–748.

IRIONDO, J.M. and PÉREZ, C., 1991a, *In vitro* storage of three endangered species from S.E. Spain, *Botanic Gardens Micropropagation News*, **1**, 46–48.

IRIONDO, J.M. and PÉREZ, C., 1991b, 'Variaciones cromosómicas en plántulas de *Coronopus navasii* regeneradas *in vitro*', presentation at IX Reunión de la SEFV, II Congreso Hispano-Luso de Fisiología Vegetal, Madrid, September.

IRIONDO, J.M. and PÉREZ, C., 1992, *In vitro* plant regeneration of *Lavatera oblongifolia* (*Malvaceae*), an endangered species, *Botanic Gardens Micropropagation News*, **1**, 54–57.

IRIONDO, J.M. and PÉREZ, C., 1996a, Micropropagation and *in vitro* storage of *Centaurium rigualii* Esteve (Gentianaceae), *Israel Journal of Plant Sciences*, **44**, 115–123.

IRIONDO, J.M. and PÉREZ, C., 1996b, Somaclonal variation in *Lavatera*, in BAJAJ, Y.P.S. (Ed.), *Somaclonal Variation in Crop Improvement*, Vol II, pp. 280–295, Berlin: Springer-Verlag.

IRIONDO, J.M., PÉREZ, C. and PÉREZ GARCÍA, F., 1992, Effect of seed storage in liquid nitrogen on germination of several crop and wild species, *Seed Science & Technology*, **20**, 165–171.

IRIONDO, J.M., DE HOND, L. and GÓMEZ-CAMPO, C., 1994, Current research on the biology of threatened plant species of the Mediterranean Basin and Macaronesia: a database, *Bocconea*, **4**, 1–385.

IRIONDO, J.M., PRIETO, C. and PÉREZ-GARCIA, F., 1995, *In vitro* regeneration of *Helianthemum polygonoides* Peinado *et al.*, an endangered salt meadow species, *Botanic Gardens Micropropagation News*, **2**, 2–5.

LARKIN, P.J. and SCOWCROFT, W.R., 1981, Somaclonal variation – a novel source of variability from cell cultures for plant improvement, *Theoretical and Applied Genetics*, **60**, 197–214.

LLEDÓ, M.D., CRESPO, M.B. and del AMO-MARCO, J.B., 1993, Preliminary remarks on micropropagation of threatened *Limonium* species (Plumbaginaceae), *Botanic Gardens Micropropagation News*, **1**, 72–74.

LLEDÓ, M.D., CRESPO, M.B. and del AMO-MARCO, J.B., 1995, *In vitro* multiplication of *Vella lucentiana* M.B. CRESPO (Brassicaceae), a threatened Spanish endemic species, *In Vitro Cell. Dev. Biol.*, **31**, 199–201.

LLEDÓ, M.D., CRESPO, M.B. and del AMO-MARCO, J.B., 1996, Micropropagation of *Limonium thiniense* Erben (Plumbaginaceae) using herbarium material, *Botanic Gardens Micropropagation News*, **2**, http://www.rbgkew.org.uk/science/micropropagation/bgmn2-2-1.html.

Luque and Devesa, 1986, Contribución al estudio citotaxonómico del género *Lavatera* (Malvaceae) en España, *Lagascalia*, **14**, 227–239.

Martín, C., 1993, 'Micropropagación y conservación *in vitro* de seis especies del género *Limonium* endémicas de la Península Ibérica', unpublished PhD thesis, Universidad Politécnica de Madrid.

Martín, C. and Pérez, C., 1992, Multiplication *in vitro* of *Limonium estevei* Fdez. Casas, *Annals of Botany*, **70**, 165–167.

Martín, C. and Pérez, C., 1994, The use of RAPD to determine the genetic variability of micropropagated plants from endangered species. Application to the Spanish endemism *Limonium estevei*, *PHYTON*, **56**, 65–72.

Martín, C. and Pérez, C., 1995, Micropropagation of five endemic species of *Limonium* from the Iberian Peninsula, *Journal of Horticultural Science*, **70**, 97–103.

Martín, C., Torres, E. and Iriondo, J.M., and Pérez, C., 1996, Utilización de marcadores moleculares para la detección de variación somaclonal, *Proceedings of the IV Coloquio Internacional de Biotecnología de las Plantas*, pp. 84–86, Santa Clara, Cuba.

Martín, C., González-Benito, M.E. and Iriondo, J.M., 1997, Genetic diversity within and among populations of a threatened species: *Erodium paularense* Fern. Gonz & Izco, *Molecular Ecology*, **6**, 813–820.

Matsumoto, T., Sakai, A., Takahashi, C. and Yamada, K., 1995, Cryopreservation of *in vitro*-grown apical meristems of wasabi (*Wasabia japonica*) by encapsulation-vitrification method, *Cryo-Letters*, **16**, 189–196.

Murashige, T. and Skoog, F., 1962, A revised medium for rapid bioassays with tobacco tissue cultures, *Physiologia Plantarum*, **15**, 473–497.

Ortega, C., 1982, Micropropagación de *Lotus berthelotii* Masf. (Leguminosae), un endemismo canario en peligro de extinción, *Botanica Macaronesica*, **10**, 19–25.

Ortega, C. and González, C., 1985, Contribución a la conservación *ex situ* de especies canarias en peligro: propagación *in vitro* de *Senecio hermosae* Pitard, *Botanica Macaronesica*, **14**, 59–72.

Ortega, C. and González, C., 1990, *Senecio hadrosomus*: cultivos *in vitro* y reintroducción en su hábitat natural, in Hernández-Bermejo, J.E., Clemente, M. and Heywood, V. (Eds), *Conservation Techniques in Botanic Gardens*, pp. 161–162, Koenigstein: Koeltz Scientific Books.

Palacios, C. and González-Candelas, F., 1997a, Lack of genetic variability in the rare and endangered *Limonium cavanillesii* (Plumbaginaceae) using RAPD markers, *Molecular Ecology*, **6**, 671–675.

Palacios, C. and González-Candelas, F., 1997b, Analysis of population genetic structure and variability using RAPD markers in the endemic and endangered *Limonium dufourii* (Plumbaginaceae), *Molecular Ecology*, **6**, 1107–1121.

Pence, V.C., 1995, Cryopreservation of recalcitrant seeds, in Bajaj, Y.P.S. (Ed.), *Biotechnology in Agriculture and Forestry, Vol. 32, Cryopreservation of Plant Germplasm*, pp. 29–50, Berlin: Springer Verlag.

Pritchard, H.W., Manger, K.R. and Prendergast, F.G., 1988, Changes in *Trifolium arvense* seed quality following alternating temperature treatment using liquid nitrogen, *Annals of Botany*, **62**, 1–11.

Richardson, B.J., Baverstock, P.R. and Adams, H., 1986, *Allozyme Electrophoresis: A Handbook for Animal Systematics and Population Structure*, Sidney: Academic Press.

Sainz, H., Moreno, J.C., Domínguez, F., Galicia, D. and Moreno, L., 1994, Conservación de la flora española amenazada, *Ecosistemas*, **9/10**, 58–65.

Sakai, A., Kobayashi, S. and Oiyama, I., 1990, Cryopreservation of nucellar cells of navel orange (*Citrus sinensis* Obs. var *brasiliensis* Tanaka) by vitrification, *Plant Cell Reports*, **9**, 30–33.

Simberloff, D., 1988, The contribution of population and community biology to conservation science, *Annual Review of Ecology and Systematics*, **19**, 473–512.

Recalcitrant Seed Biotechnology Applications to Rain Forest Conservation

M. MARZALINA AND B. KRISHNAPILLAY

17.1 Introduction

Within the equatorial region, rain forest covers more than 40 per cent of the land, providing a rich sanctuary for both flora and fauna. Rain forests are priceless reservoirs of plant germplasm and play a vital role in maintaining global environment stability (see also Viana *et al.*, Chapter 18, this volume). The composition, structure and ecology of natural forests, with their large array of plant and animal species is very complex and still remains to be fully understood. Malaysia, as one of the top 12 countries in the megadiversity league, consists of 0.2 per cent of the world's land area but is estimated to harbour more than 2800 species of trees (Saw, 1992) or 6 per cent of the world's flowering plant species. A majority of the rain forest species, especially timber trees, produce recalcitrant seeds. The importance and difficulty of conserving recalcitrant seeded species are well recognized. In recent years, awareness over the loss of plant diversity has captured worldwide attention. Though several conservation activities of plant genetic resources have been performed, they are very labour intensive and complex and encompass various tasks such as germplasm collection, storage and evaluation. Germplasm conservation has become necessary for future sustainable harvesting systems and as a means of maintaining species diversity to prevent genetic erosion. Presently, whilst recalcitrant species can be conserved *in situ* in national parks and forest reserves, some have been taken out of their natural habitat and conserved *ex situ* in aboreta and botanic gardens. In the field of biotechnology, many other complementary methods for *ex situ* conservation have been explored, including cryopreservation (see Benson, Chapter 6, this volume). According to Ahuja (1991), the success of biotechnological approaches is largely dependent upon three factors:

- The ability of germplasm to survive storage treatments (i.e. dehydration/desiccation, low and ultra low temperature stress).
- *In vitro* morphogenetic competence.
- The development of widely applicable, routine and economically viable conservation methods.

Biotechnology therefore, has the potential for application in crop improvement and genetic resource management programmes. Among other reasons for conservation is the need for providing a continuous supply of seeds and seedlings for the various planting programmes. The current trend the world over is to establish forest plantations in order to reduce the logging impact on natural forests with greater emphasis towards the conservation of the existing biodiversity. In fact, one of the major constraints with tropical forest management is the availability of good quality seeds. For the establishment of successful forest plantations there is a need for quality planting materials that are selected from elite mother trees in order to produce seedlings that have improved survival and produced greater economic returns.

17.2   Seed characteristics

The majority of tropical tree species have very complex life cycles. Many of these species produce recalcitrant seeds. These species have a long juvenile phase and may come into first flowering and fruiting only after 15–20 years of growth or more. In addition, fruiting and seed sets in many tropical species is not an annual occurrence. When it does occur, the seed production period continues for a short period, which is termed as a mass flowering season that normally occur once in 3–7 years. It is also not possible to predict their flowering times owing to their erratic flowering patterns. Consequently, it causes problems to secure large quantities of fruits on a regular basis. Generally, the seeds that are produced do not undergo dormancy, but instead they are metabolically primed for immediate germination, as soon as the seeds mature on mother plants. According to studies done by Hong and Ellis (1990) and Finch-Savage (1992), desiccation tolerance in recalcitrant seeds increases during seed development on the mother plant; however, unlike orthodox seeds, maturation drying to low moisture contents does not occur. The moisture content range of fresh recalcitrant seeds can be from as low as 36 per cent (*Hevea brasiliensis*) to as high as 90 per cent (*Sechium edule*) (Hong and Ellis, 1996). Hence, the seeds shed from trees in a fully matured stage undergo rapid post-collection deterioration if the correct environments for continued growth are not provided. The rate of deterioration is dependent on the environment condition around the seeds. This phenomenon leads to reduced rates of germination and seed growth, decreased ability to germinate under stressful conditions and increased probability of abnormal seedlings and longer field emergence (Krishnapillay and Marzalina, 1998). Tropical forest seeds can be broadly divided into three major groups, based on their sensitivity to desiccation and to low temperatures, as follows.

17.2.1   *Orthodox seeds*

This group includes all the seeds which desiccate naturally on the mother plants. The seeds can be dried to lower moisture contents (less than 10 per cent) without any serious deleterious effects. In fact, the lower the seed moisture content and storage temperature, the longer they survive.

17.2.2   *Recalcitrant seeds*

This group of seeds easily die if they are dried below certain moisture limits (12–30 per cent). They are also killed when exposed to low temperatures (less than 16°C). Even at optimal moist conditions, survival of seeds in this group is limited from a few days to a few weeks. If they are not collected at maturity and sowed immediately, they will die. The period between collection and sowing is very short. These seeds therefore are difficult to store and do not conform to the rules applicable to orthodox seeds. To date raising planting material from recalcitrant seeds is best done in nurseries but this involves high costs for maintenance and space.

17.2.3 *Intermediate seeds*

This is yet another category that has been recently defined (Ellis *et al.*, 1990). The seeds in this category have storage characters intermediate between orthodox and recalcitrant. Intermediate seeds can be dried to seed moisture levels almost similar to that of orthodox seeds without their viability being affected. However, the dry seeds are easily injured when exposed to low temperatures and viability drops rapidly while in storage. Therefore these seeds can be stored under conditions used for orthodox seeds but only for a short period.

17.3 Seed collection and handling

For conservation purposes, recalcitrant seeds should be collected from healthy trees with good shape and form. Several techniques of seed collection, namely ground collection, shaking seed bearing branches, free climbing (normally done by natives) or climbing the trees using equipment can be done. Records from Seed Technology Laboratory FRIM show that seeds meant for conservation are best collected freshly from the tree crown by climbing. Criteria for collection of good seeds should include mature seeds, uniformity in colour and size, and healthy. Most recalcitrant seeds respire intensively because of their high moisture content, and hence require good ventilation. If large quantities are closely packed, suffocation, physiological breakdown, fungal growth and overheating will occur, resulting in rapid death of the seeds. On the other hand, the seeds will also deteriorate rapidly if moisture content is reduced too much or too rapidly. This is likely to happen during transporting in open vehicles due to air movements. Plastic bags should be used with their top left open or small holes perforated in the sides. Hessian or jute bags that have been loosely woven are also suitable for transport. Temperatures below 16°C or above 32°C should be avoided for such seeds. Seeds should be kept shaded from direct sun at all times during transport to the conservation centres. Collection journeys should not be more than three days. If it is unavoidable, efforts must be made to dispatch the seeds to their destination by the second day of collection. Any germinated seeds during the collection should be separated from the batch as these can be planted right away at the nursery and later be useful as a source of shoot tips for cryopreservation. It is important to keep the materials moist during transit. At FRIM, a prototype mobile seed–seedling chamber has been invented for the transporting of tropical seeds, especially the recalcitrant type over long journeys, i.e. more than three days with minimal deterioration to the seeds. This vehicle is incorporated with sensors that control the chamber environment, including temperature (20±5°C), humidity (80–95 per cent), and light (if needed).

17.4 Seed storage

In the past, many types of storage methods have been proposed for recalcitrant seeded species, but without exception their use has been limited or unsuccessful. Thus to conserve recalcitrant seeds, availability of seeds should be given priority as this will determine the period of storage as viability is reduced during storage.

17.4.1 Short-term and mid-term storage methods

These can be applied to seeds for which viability is maintained for less than 12 months. This duration may be sufficient in cases where short-term conservation is required to 'hold' the germplasm for short periods during which land preparation is required or for the advent of suitable weather for planting. The strategy is to desiccate recalcitrant seeds to just above a critical moisture content, treat them with an effective fungicide and then store them in semi-sealed packaging which allows adequate gaseous exchange, but restricts moisture loss. Several common conventional storage methods include: imbibed storage using sawdust, ground charcoal, pearlite and vermiculite; storage in airtight containers or partial vacuum; regular ventilation or maintaining oxygen levels above 10 per cent; and incorporating germination inhibitors into the storage system (Khare *et al.*, 1987; Sakhibun, 1981; Sasaki, 1980; Song *et al.*, 1984). For many other recalcitrant seed species this approach has not shown any significant results. The lowest safe moisture content of many of the recalcitrant species falls within the range of 20–60 per cent (Hor, 1996; Tompsett, 1992; Yap, 1981, 1986).

However, fungicide treatment followed by partial desiccation and storage under ambient or low temperatures have yielded some fruitful results for some crop seeds such as *Hevea* and *Cacao*. Hor (1984) found that after treating *Cacao* seeds with 0.2 per cent Benlate/Thiram mixture, partially desiccating the seeds by air drying and storing them loosely packed in polythene bags at 21–24°C, he was able to prolong the viability of the seeds from one week to about 24 weeks with a 50 per cent germination. Similarly, soaking *Hevea* seeds in 0.3 per cent Benlate, air drying and storing them loosely packed in perforated polythene bags at ambient temperatures, was able to prolong seed viability from three months to one year with about 50 per cent germinability (Normah, 1987). Although this approach has potential for recalcitrant seed storage, it provides only a short- to medium-term storage strategy.

17.4.2 Long-term storage

It is evident that with the present limitations, the maximum period of storage that can be achieved using conventional storage methods is about a year or so for recalcitrant seeds. At FRIM, three methods have been devised as protocols for long-term storage: seedling chamber conservation; storage of seedlings on the forest floor; and cryopreservation.

Seedling chamber

Marzalina *et al.* (1994) reported the possibility of using a seedling chamber to slow down the germination and growth of seedlings of recalcitrant seeds. Through this procedure, seed viability in a slow growth condition can be maintained for periods longer than a year. Freshly collected seeds are surface treated with a 0.1 per cent Benlate/Thiram fungicide and allowed to germinate under ambient conditions in containers kept moist using wet tissue papers. Once the radicle has emerged, they are inspected for insect attack or fungal infection. Infection-free germinated seeds are loosely packed in polythene bags or plastic boxes containing sterilized sand or lined moist tissue paper and stored in a specially constructed seedling chamber. This chamber has a temperature, humidity and light control and an alarm system. The temperature is maintained at 16°C, the relative humidity at 80 per cent and the photoperiod maintained for four hours each day. The germinated seeds placed in these chambers develop slowly and barely attain the height of 20–25 cm. To date, 20 recalcitrant seeded species have been tested and the storage period obtained was between four and 24 months (Krishnapillay and Marzalina, 1998). When these seedlings are removed and planted in polybags, they require weaning in at least 70 per cent shade for a period of 2–3 weeks before they can be placed under direct sunlight. Post-storage survival percentages are between 60 and 80 per cent when they are planted out. This method of storage is advantageous when there is an excessive seed fall and also it can be used to tied over germplasm during long periods of unfavourable weather and delayed replanting programmes. However, the disadvantage is that a large facility is required and seedlings have to be hardened before field planting (Hor, 1996).

Conservation on the forest floor

The other approach being adopted at FRIM is the storage of seedlings on the forest floor under natural subdued light; this method is not new and has been practiced by nature over time. In this approach, areas of the forest floor are cleared and freshly collected seeds are sown in these areas; they then germinate and develop into seedlings. Owing to the sub-dued light conditions, the seedlings develop very slowly and can remain within manageable heights for up to a few years. Over a period of five years, seedlings of *Hopea odorata* do not grow above 10 cm under these conditions. Once these seedlings are potted into polybags and placed under nursery conditions, they rapidly expand and begin to increase in height. Seventeen species have been tested using this method and the results, to date, are very encouraging. This approach could be adopted to keep stock of planting materials with minimal cost and space. Although this method has good potential, many problems need to be resolved, including protection against field pests and diseases, and improvement of transplanting techniques.

17.5 Cryopreservation

Cryopreservation protocol development has now reached the stage applicable for conservation of recalcitrant seeds (for details of general methodology see Benson, Chapter 6, this volume).

17.5.1 *Cryopreservation of whole seeds*

Orthodox or semi-recalcitrant seeds of tropical species tested can be cryopreserved without much problem as their lowest safe moisture content can be reduced to less than 20 per cent (Table 17.1) (Hor, 1996; Krishnapillay and Marzalina, 1993; Marzalina *et al.*, 1997; Normah and Marzalina, 1996). However, when truly recalcitrant seeds are dried and directly plunged into liquid nitrogen, there is no recovery. As mentioned earlier, many tropical recalcitrant seeds are commonly large in size, highly variable in water content, and often attached to appendages adapted for dispersal and thus they will not tolerate a reduction in water content below a relatively high level without loss of viability. Therefore,

Table 17.1 Summary of non-recalcitrant species that have been successfully cryopreserved

Species	Part cryopreserved	Temperature/method	MC	Viability
Forest				
Alstonia angustiloba	Whole seed	−20°C, deep freezer	5%	81%
Bambusa arundinacea	Whole seed	−196°C, desiccation	10.7%	73%
Dendrocalamus brandisii	Whole seed	−196°C, desiccation	7.3%	48%
Dendrocalamus membranaceus	Whole seed	−196°C, desiccation	8.5%	53%
Dialium platysepalum	Whole seed	−20°C, deep freezer	10.6%	37%
Dipterocarpus alatus	Whole seed	−196°C, desiccation	10%	68%
Dipterocarpus intricartus	Whole seed	−196°C, desiccation	6–8%	70%
Dyera costulata	Whole seed	−196°C, direct plunge	6.7%	37%
Hopea odorata	Embryonic axis	−40°C, −196°C, desiccation, slow cool	6%	15%
Melia azaderach	Whole seed	−20°C, deep freezer	12.4%	43%
Pterocarpus indicus	Whole seed	−196°C, desiccation	8%	80%
Swietenia macrophylla	Whole seed	−196°C, desiccation	5–6%	63%
Swietenia macrophylla	Embryonic axis	−196°C, encap; desiccation	4–5%	63%
Thyrsostachys siamensis	Whole seed	−196°C, desiccation	7.8%	86%
Fruit tree				
Annona squamosa	Whole seeds	−196°C, direct plunge	7.1%	99%
Averhoa carambola	Whole seeds	−196°C, direct plunge	7%	93%
Baccaurea polyneura	Embryonic axes	−196°C, desiccation	11.3%	25%
Baccaurea polyneura	Whole seed	−196°C, desiccation	5–8%	15%
Carica papaya	Whole seed	−196°C, desiccation	10–11.5%	66%
Citrus aurantifolia	Embryonic axis	−196°C, desiccation	9–11%	100%
Citrus halimii	Embryonic axis	−196°C, desiccation	16.6%	100%
Citrus mitis	Embryonic axis	−196°C, desiccation	14%	80%
Maikara zapota	Whole seed	−30°C, desiccation	11%	18%
Manilkara sapota	Whole seeds	−196°C, direct plunge	10.5%	100%
Psidium quajava	Whole seeds	−196°C, direct plunge	10.4%	96%
Plantation crops				
Acacia mangium	Whole seed	−196°C, direct plunge	7.6%	76%
Albizia falcataria	Whole seed	−20°C, deep freezer	6.6%	100%
Coffea liberica	Whole seed	−196°C, desiccation	15–20%	86%
Tectona grandis	Whole seed	−20°C, deep freezer	7–9%	97%
Ornamentals				
Adenanthera pavonina	Whole seed	−20°C, deep freezer	9.5%	97%
Cassia nodosa	Whole seed	−196°C, direct plunge	9.5%	17%

Table 17.1   (cont'd)

Species	Part cryopreserved	Temperature/method	MC	Viability
Cassia spectabilis	Whole seed	−196°C, direct plunge	9.5%	20%
Casuarina sumatrana	Whole seed	−196°C, direct plunge	11.2%	23%
Lagestroemia speciosa	Whole seed	−196°C, direct plunge	9.8%	87%
Lagestroemia floribunda	Whole seed	−20°C, deep freezer	12.8%	24%
Livistona chinensis	Embryo	−196°C, desiccation	20%	55%
Ptychosperma macarthurii	Embryo	−196°C, desiccation	19%	65%
Veitchia merilli	Embryo	−196°C, desiccation	17%	80%

(From Hor, 1996; Marzalina *et al.*, 1997; Normah and Marzalina, 1996).

rapid and systematic approaches are required to initially screen diverse tropical tree seed germplasm for their ability to survive pre-growth, cryoprotectants and dehydration treatments as well as exposure to liquid nitrogen (Benson *et al.*, 1996).

17.5.2   *Cryopreservation of excised embryos*

Advances in cryopreservation now offer many possibilities (through the application of controlled freezing, cryoprotective dehydration, vitrification and encapsulation–dehydration; see Benson, Chapter 6 this volume) for conserving the embryos of recalcitrant seeds. Excised embryos are used for storage because they are small, more resistant to desiccation and relatively uniform in size and moisture content (Fu *et al.*, 1993; Hiew, 1991; Hor *et al.*, 1990). Results indicate that when embryos are subjected to cryopreservation, there is some recovery (see Table 17.2). Most of the embryos are dissected aseptically and left to partially desiccate in the laminar flow for 4–6 hours, during which time the moisture content can be reduced to 4–33.5 per cent. Krishnapillay and Engelmann (1996) suggested that a more precise and reproducible desiccation could be achieved by placing plant material in a stream of compressed air or in an airtight container with silica gel. Optimal survival rates are generally noted when embryos are dehydrated down to 10–20 per cent water content (fresh weight basis). The embryos can then be placed in cryovials for storage. Direct immersion of these vials in liquid nitrogen has been practiced but some methods also involve a slow-cooled technique in which the temperature is lowered at a rate of −1°C per minute to −40°C followed by direct plunging into liquid nitrogen. Techniques for rapid thawing using a 40°C water bath or slow thawing in a laminar flow cabinet do appear to have an effect on the survival of the embryos. The recovery of the embryos can be low (5–15 per cent) using these procedures and viable seedlings grow very slowly. The recovery of the embryos can be culture-media dependent. In some cases modified re-growth patterns such as callusing or incomplete development are observed. Other serious problems include the formation of abnormal seedlings and microbial contamination (Normah and Marzalina, 1996); the latter appears to be due to seed-borne fungi. Modified recovery conditions, notably the hormonal balance of the culture medium, can significantly improve the survival rate of cryopreserved material (Krishnapillay and Engelmann, 1996).

Table 17.2 Summary of recalcitrant species that have been successfully cryopreserved

Species	Part cryopreserved	Temperature/method	MC	Viability
Arthocarpus heterophyllus	Embryonic axes	−196°C, cryoprotectant, slow cool	15%	60%
Calamus manan	Embryonic axes	−196°C, direct plunge	7.8%	60%
Elaeis guineensis	Embryonic axes	−196°C, encap; desiccation	10%	60%
Elaeis guineensis	Embryonic axes	−196°C, direct plunge	11.7%	50%
Hevea brasiliensis	Embryonic axis	−196°C, direct plunge	16%	87%
Nephelium lappaceum	Embryonic axes	−40°C, −196°C, desiccation, slow cool	25%	40%
Shorea leprosula	Embryonic axes	−40°C, −196°C, desiccation, slow cool	4–5%	12%
Shorea macrophylla	Embryonic axes	−40°C, −196°C, desiccation, slow cool	10%	5%
Shorea ovalis	Embryonic axes	−40°C, −196°C, desiccation, slow cool	8%	7%
Shorea parvifolia	Embryonic axes	−40°C, −196°C, desiccation, slow cool	5–7%	10%

(From Hor, 1996; Marzalina *et al.*, 1997; Normah and Marzalina, 1996).

Cryopreservation using encapsulation–dehydration (see Benson, Chapter 6, this volume) has been successfully applied to many temperate and some tropical species (Engelmann *et al.*, 1995; Fabre and Dereuddre, 1990; Hatanaka *et al.*, 1994). Thus, encapsulated zygotic embryos are pre-grown for various periods of time in liquid medium with high sucrose concentrations; they are then partially dehydrated in a laminar flow cabinet or using silica gel and subsequently directly immersed into liquid nitrogen. For recovery, samples are usually placed directly under standard culture conditions. Marzalina (1995) found that dehydrating encapsulated embryo axes of *Swietenia macrophylla* treated with 0.5 M sucrose for two days, followed by eight hours desiccation before cryopreservation gave 63 per cent survival levels. She also found that the survival percentage of encapsulated embryos fell with increasing molarity of sucrose solution and the period of drying. Thus, this technique is largely dependent upon the optimization of pre-growth dehydration and desiccation treatments. Once excised, the zygotic embryos of tropical tree seeds undergo phenolic oxidation evidenced by rapid browning or blackening of tissue and surrounding medium. This is also exacerbated by surface sterilization procedures. Studies by Benson *et al.* (1996) indicated that encapsulated embryos of *Hopea weightiana* and *Vatica cinerea* subjected to sucrose dehydration and subsequent air desiccation have made it possible to achieve lower moisture contents which make the embryos more amenable to cryopreservation. Using these techniques, they found that embryos did not suffer the deleterious oxidative phenomenon (tissue browning). Thus, this method offers considerable advantages for the preparation of 'stress-sensitive' germplasm for cryopreservation. Normah and Marzalina (1996) did suggest the use of activated charcoal to be incorporated into the recovery medium in order to absorb the toxic phenolic compounds. As mentioned by Benson *et al.* (1996) increasing importance is being given to the role of stress in recalcitrant seed systems and novel approaches to ameliorating oxidative damage in *in vitro* cultures should also be explored for tropical germplasm.

17.5.3 *Cryopreservation of shoot tips*

Cryopreservation of tropical tree shoot tips or apical meristems normally involves using alginate encapsulation techniques mentioned above. Most studies have been performed on tropical orthodox and intermediate species. Thus, axillary buds of *Morus indica* (Bapat *et al.*, 1987), adventitious buds of *M. alba* (Machii, 1992), and axillary buds of *Betula platyphylla* var. *japonica* (Kinoshita and Saito, 1992) have been encapsulated in alginate beads and stored for 34–80 days at 4°C and have shown no loss of viability. Maruyama *et al.* (1997) have stored encapsulated shoot tips of *Cedrela odorata*, *Guazuma crinita* and *Jacaranda mimosaefolia* above freezing temperatures at 12°C, 20°C and 25°C and obtained 80, 90 and 70 per cent viability respectively. At the Seed Technology Section at FRIM, work on encapsulation and desiccation techniques is being attempted on excised embryos of *Shorea leprosula,* but to date no significant results have been obtained.

Vitrification (see Benson, Chapter 6, this volume) is another technique that can be employed to cryopreserve shoot tips and apices. Such a procedure eliminates the need for controlled freezing and enables cells and meristems to be cryopreserved by direct transfer into liquid nitrogen. To date, most studies have been carried out on temperate species and monocots (Matsumoto *et al.*, 1995; Niino *et al.*, 1992; Sakai, 1993; Thinh, 1997). As vitrification involves placing samples for pretreatment in extremely concentrated cryoprotective solutions and ultra-rapidly freezing them, the intracellular solutes vitrify and form an amorphous glassy structure; the formation of detrimental intracellular ice crystals is avoided. Applications of this technique to shoot tips of apple (*Malus* spp.) and pear (*Pyrus* spp.) from cold-hardened plantlets resulted in 80 per cent and 70 per cent shoot tip survival respectively after 40 days after re-planting (Niino *et al.*, 1992). Little work has been reported for tropical recalcitrant species and some are sensitive to the toxic cryoprotectants used in the vitrification mixtures (Benson *et al.*, 1996).

17.5.4 *Cryopreservation of somatic embryos*

Both encapsulation–dehydration and vitrification techniques can be employed to cryopreserve somatic embryos. Little work has been performed on tropical recalcitrant seeds (e.g. as an alternative means of conserving tree germplasm) at the moment and this may not be due to the failure of the cryopreservation techniques, but to the problems relating to somatic embryo induction by tissue cultures and the regeneration of these tissues to full plants. Once this can be achieved, no doubt somatic embryos will provide a suitable germplasm source in terms of their response to freezing. They have the advantages of being small in size and being useful for cloning. Micropropagation of tropical trees has been reported to be very difficult as these plants are problematic to culture; thus to date, encapsulated shoot tips offer the most effective means of cryo-conservation (Normah and Marzalina, 1996).

17.6 Conclusions

The need for cryopreservation becomes evident when one examines conventional preservation systems and their limitation. Reports show that many species decline in viability with time under such storage conditions (Stanwood and Bass, 1981). The potential use of biotechnology for the conservation of tropical tree germplasm is considerable (Benson *et al.*, 1996; Krishnapillay and Engelmann, 1996; Zakri *et al.*, 1991). However, for many tropical recalcitrant species, there are several prerequisites, which have to be fulfilled before *in vitro* conservation becomes possible. With proper reduction of moisture content and controlled desiccation, seed-derived material can be stored in liquid nitrogen. Therefore elucidating the mechanisms of seed recalcitrance will greatly benefit the implementation of conservation strategies. Significant progress can be achieved if more collaborative links between research institutes from tropical and temperate countries working in the same area are established.

References

AHUJA, M.R., 1991, Application of biotechnology to the preservation of forest tree germplasm, in AHUJA, M.R. (Ed.) *Woody Plant Biotechnology,* pp. 307–313, New York: Plenum Press.

BAPAT, V.A., MATHRE, M. and RAO, P.S., 1987, Propagation of *Morus indica* L. (Mulberry) by encapsulated shoot buds, *Plant Cell Reports*, **6**, 393–395.

BENSON, E.E., KRISHNAPILLAY, B. and MARZALINA, M., 1996, The potential of biotechnology in the *in vitro* conservation of Malaysian forest germplasm: an integrated approach, in NORHARA, H., BACON, P.S. and KHOO, K.C. (Eds), *Proceedings of the 3rd Conference on Forestry and Forest Products Research*, 3–4 October 1995, FRIM, Vol. 1, pp. 76–90.

ELLIS, R.H., HONG, T.D. and ROBERTS, E.H., 1990, An intermediate category of seed behaviour? 1. Coffee, *J. Exp. Bot.*, **41**, 1167–1174.

ENGELMANN, F., BENSON, E.E., CHABRILANGE, M.T., GONZALES ARNAO, S., MICHAUAX-FERRIERE, N., PAULET, F., GLASZMANN, J.C. and CHARRIER, A., 1995, Cryopreservation of several tropical plant species using encapsulation/dehydration of apices, in TERZI (Ed.), *Current Issues in Plant Molecular and Cellular Biology*, pp. 315–320, Lancaster: Kluwer Academic.

FABRE, J. and DEREUDDRE, J., 1990, Encapsulation-dehydration: a new approach to cryopreservation of *Solanum* shoot tips, *Cryo. Letts.*, **17**, 413–426.

FINCH-SAVAGE, W.E., 1992, Seed development in the recalcitrant species *Quercus robur* L.: germinability and desiccation tolerance, *Seed Sci. Res.*, **2**, 17–22.

FU, J.R., XIA, Q.H. and TANG, L.F., 1993, Effects of desiccation on excised embryonic axis of three recalcitrant seeds and studies on cryopreservation, *Seed Sci. Tech.*, **21**, 85–95.

HATANAKA, T., YASUDA, T., YAMAGUCHI, T. and SAKAI, A., 1994, Direct regrowth of encapsulated somatic embryos of coffee (*Coffea canephora*) after cooling in liquid nitrogen, *Cryo. Letts.*, **15**, 47–42.

HIEW, Y.H., 1991, Towards the development of a protocol for cryopreservation of embryos of rambutan (*Nephelium lappaceum* L.), M. Agric. Sc. Thesis, Universitit Pertanian, Malaysia.

HONG, T.D. and ELLIS, R.H., 1990, A comparison of maturation drying, germination and desiccation tolerance between developing seeds of *Acer pseudoplatanus* L. and *Acer platanoides* L., *New Phytologist*, **116**, 589–596.

HONG, T.D. and ELLIS, R.H., 1996, A protocol to determine seed storage behavior, *IPGRI Tech. Bull.*, No. 1.

HOR, Y.L., 1984, Storage of cocoa (*Theobroma cacao*) seeds and changes associated whith their deterioration, unpublished PhD thesis, Universiti Pertanian Malaysia.

HOR, Y.L., 1996, Storage of tropical tree seeds with special reference to cryopreservation, in YAPA, A.C. (Ed.), *Proceedings of the International Symposium: On Recent Advances in*

Tropical Tree Seed Technology and Planting Stock Production, 12–14 June 1995, AFTSC Thailand. pp. 63–69.

HOR, Y.L., STANWOOD, P.C. and CHIN, H.F., 1990, Effects of dehydration on freezing characteristics and survival in liquid nitrogen of three recalcitrant seeds, *Pertanika,* **13**, 309–314.

KHARE, P.K., YADAR, V.K. and MISHRA, G.P., 1987, Collection, germination and storage of *Shorea robusta* seeds, in KARMA, S.K. and AYLING, R.D. (Eds), *Proceedings of the IUFRO Symposium on Tree Seed Problems in Africa,* Harare, Zimbabwe, SUAS, pp. 154–158.

KINOSHITA, I. and SAITO, A., 1992, Regeneration of Japanese white birch plants from encapsulated axillary buds, in KUWAHARA, M. and SHIMADA, M. (Eds), *5th International Conference on Biotechnology in Pulp and Paper Industry,* Kyoto, Japan, Uni Publishers, Tokyo, pp. 493–496.

KRISHNAPILLAY, B. and ENGELMANN, F., 1996, Alternative methods for the storage of recalcitrant and intermediate seeds: slow growth and cryopreservation, in OUEDRAOGO, A.S., POULSEN, K. and STUBSGAARD, F. (Eds), *Intermediate/recalcitrant tropical forest tree seeds, Proceedings of a Workshop on Improved Methods for Handling and Storage of Intermediate/ Recalcitrant Tropical Forest Tree Seeds,* 8–10 June 1995, IPGRI, pp. 34–39.

KRISHNAPILLAY, B. and MARZALINA, M., 1993, Cryopreservation of some tropical forest tree seeds, in *Proceedings of the 5th National Biotechnology Seminar,* 13–14 December 1993, Port Dickson.

KRISHNAPILLAY, B. and MARZALINA, M., 1998, Seed biology and handling, in APPANAH, S. and COSSALTER, C. (Eds), *Dipterocarps: State of Knowledge and Needs for Future Research,* CIFOR/FRIM Publication (in press).

MACHII, H., 1992, *In vitro* growth of encapsulated adventitious buds in mulberry, *Morus alba* L. *Japan J. Breed,* **42**, 445–559.

MARUYAMA, E., KINOSHITA, I., ISHII, K., OHBA, K. and SAITO, A., 1997, Germplasm conservation of the tropical forest trees, *Cedrela odorata* L., *Guazuma crinita* Mart., and *Jacaranda mimosaefolia* D. Don., by shoot tip encapsulation in calcium-alginate and storage at 12–25°C, *Plant Cell Reports,* **16**, 393–396.

MARZALINA, M., 1995, Penyimpanan biji benih mahogani (*Swietenia macrophylla* King.), unpublished PhD thesis, Universiti Kebangsaan Malaysia.

MARZALINA, M., KRISHNAPILLAY, B., HARIS, M. and SITI ASHA, A.B., 1994, A possible new technique for seedling storage of recalcitrant species, in *Proceedings Bio-Refor: Meeting of Experts in the Asian Pacific region,* Kangar, Perlis, pp. 135–136.

MARZALINA, M., KRISHNAPILLAY, B. and NASHATUL ZAIMAH, N.A., 1997, *In vitro* conservation of tropical rainforest germplasm via cryopreservation, in *Proceedings of the 4th Conference on Forestry and Forest Products Research,* 2–4 October 1997, FRIM.

MATSUMOTO, T., SAKAI, A., TAKAHASHI, C. and YAMADA, K., 1995, Cryopreservation of *in vitro*-grown apical meristems of wasabi (*Wasabia japonica*) by encapsulation-vitrification method, *Cryo. Letts,* **16**, 189–196.

NIINO, T., SAKAI, A., YAKUWA, H. and NOJIRI, K., 1992, Cryopreservation of *in vitro*-grown shoot tips of apple and pear by vitrification, *Plant Cell, Tissue and Organ Culture,* **28**, 261–266.

NORMAH, M.N., 1987, Effects of temperature on rubber (*Hevea brasiliensis*) seed storage, unpublished PhD thesis, Universiti Pertanian Malaysia.

NORMAH, M.N. and MARZALINA, M., 1996, Achievements and prospects of *in vitro* conservation for tree germplasm, in NORMAH, M.N. (Ed.), *In vitro Conservation of Plant Genetic Resources,* UKM, PP. 253–261.

SAKAI, A., 1993, Cryogenic strategies for survival of plant cultured cells and meristems cooled to –196°C, *JICA GRP,* No. 6.

SAKHIBUN, M.S., 1981, Storage and viability of *Hevea* seeds, unpublished MSc thesis, Universiti Pertanian Malaysia.

SASAKI, S., 1980, Storage and germination of dipterocarp seeds, *Malayan Forester,* **43**, 290–308.

SAW, L.G., 1992, Forest resources and ecosystem conservation in Malaysia, FAO, Italy, *Forest Genetic Resources Info.,* **20**, 32–37.

SONG, X., CHEN, Q., WANG, D. and YANG, J., 1984, A study on the principal storage conditions of *Hopea hainanensis* seeds, *Scientia Silvae Sinicae*, **20**, 225–236.

STANWOOD, P.C. and BASS, L.N., 1981, Seed germplasm perservation using liquid nitrogen, *Seed Sci. Tech.*, **9**, 423–437.

THINH, N.T., 1997, Cryopreservation of germplasm of vegetatively propagated tropical monocots by vitrification, unpublished PhD thesis, Kobe University.

TOMPSETT, P.B., 1992, A review of the literature on storage of dipterocarp seeds, *Seed Sci. Tech.*, **20**(2), 251–267.

YAP, S.K., 1981, Collection, germination and storage of dipterocarp seeds, *Malaysian Forester*, **44**, 281–300.

YAP, S.K., 1986, Effect of dehydration on the germination of dipterocarp fruits, in *Proceedings of the IUFRO Symposium: Seed Problems Under Stressful Conditions,* FFRI, Vienna, Vol. 12, pp. 168–181.

ZAKRI, A.H., NORMAH, M.N. and ABDUL KARIM, A.G., 1991, Conservation of plant genetic resources through *in vitro* methods, in *Proceedings of the MNCPGR/CSC International Workshop on Tissue Culture for the Conservation of Biodiversity and Plant Genetic Resources*, 28–31 May 1990, FRIM.

18

Applications of Biotechnology for the Conservation and Sustainable Exploitation of Plants from Brazilian Rain Forests

ANA MARIA VIANA, MARIA CRISTINA MAZZA AND SINCLAIR MANTELL

18.1 Introduction

Brazil is located between latitudes 5° North and 32° South, and its altitude ranges from sea level to more than 3000 metres. It covers several different ecological conditions varying from equatorial to temperate, which ensure the presence of a rich and diversified flora and fauna. The Brazilian tropical and subtropical forests are the original sources of several economically important products and biomolecules of industrial significance which can be utilized by the fine chemistry sector as models for the synthesis of new molecules (Myers, 1983, 1988). The increasing deforestation and degradation of natural habitats, not only in the Amazon Region but also in the south-east and south of Brazil, has provoked substantial erosion of genetic diversity and many tree species are now on the verge of extinction (CIMA, 1991; Fundação SOS Mata Atlântica, 1992).

The development of appropriate strategies for genetic improvement, management and for *in situ* and *ex situ* conservation of native tree species requires integrated information on breeding systems, gene flow patterns, population genetic structure, reproductive biology and ecology (see also Marzalina and Krishnapillay, Chapter 17, this volume). Biotechnological tools present a broad range of possibilities for the conservation, utilization and protection of forestry resources (Haynes, 1994). The objective of this chapter is to evaluate the potential of biotechnology in the context of conserving Brazilian trees which are native to the Amazon, Atlantic and Araucaria Forests, and particular emphasis will be given to the application of techniques involving molecular genetic markers, plant cell and tissue culture, growth limitation and cryopreservation.

18.2 Background

18.2.1 *The Brazilian forests*

The most important forest formations in Brazil, which cover ca. 70 per cent of its territory, are the Dense Ombrophylous Forest (the Amazon and the Atlantic Forests) and the Mixed Ombrophylous Forest (the Subtropical Evergreen Seasonal Conifer Forest – also called the Araucaria Forest). The Amazon tropical humid forest is the largest and covers 42 per cent of the national territory (ca. 260 million hectares).

Geographical localization

Most of the territory occupied by the Amazon Forest is located between the Equator and latitude 10°S and longitudes 55°W to 70°W. The Atlantic Forest is located on the Eastern coast of Brazil between latitudes 5° and 30°S and longitudes 30° to 50°W, extending from Rio Grande Norte to Rio Grande do Sul. The Araucaria Forest occurs in southern and south-eastern Brazil, mainly in the States of Paraná, Santa Catarina and Rio Grande do Sul, extending from latitudes 19° to 31°S and longitudes 41° to 54°W.

Ecological and economical importance

The Amazon Forest and the eastern coastal forests of Brazil represent several centres of endemism of plant and animal species. Studies carried out on the distribution of woody plant families in tropical South America have shown the existence of as many as 26 endemism centres, 14 of them being in the Amazon and four in the Atlantic (Mata Atlântica) Forest (Mori, 1989; Prance, 1973, 1982, 1985). Wood production is the most important economical activity in the Amazon Forest. Of the main tree species currently extracted, only five are responsible for more than 74 per cent of the exportations from the Amazon region. These species are: *Virola surinamensis* (Myristicaceae), *Swietenia macrophylla* and *Carapa guianensis* (Meliaceae), *Cordia goeldiana* (Boraginaceae) and *Diplotropis* spp. Other species such as *Torresea acreana* (Leguminosae-Papilionideae), *Cedrela odorata* (Meliaceae), *Vouacapoua americana* (Leguminosae-Caesalpinioideae), *Euxylophora paraensis* (Rutaceae), *Calophyllum brasiliense* (Guttiferae), *Ocotea cymbarum* (Lauraceae), *Bowdichia nitida* (Leguminosae-Faboideae), *Aspidosperma* spp. (Apocynaceae) and *Bertholletia excelsa* (Lecythidaceae) are also important for wood production in certain areas of the Amazon (Lisboa *et al.*, 1991). In the last few years, the markets for the Amazonian non-wood resources, including edible fruits, oils, latex, fibres and medicines, have increased. Besides the more traditional non-wood species, such as *Hevea brasiliensis* (rubber), *Theobroma cacao* (Sterculiaceae) (cocoa), *Bertholettia excelsa* (Brazilian nuts), *Paulinia cupana* (Sapindaceae) (guarana), *Aniba roseodora* (Lauraceae) (linalol oil) and *Bixa orellana* (urucum, natural food pigment), the following native fruit tree species have become economically important: *Euterpe oleracea* (Palmae) (açai), *Theobroma grandiflorum* (Sterculiaceae) (cupuaçu), *Bactris gasipaes* (Palmae) (pupunha palm), and *Myrciaria dubia* (Myrtaceae) (camu-camu). In addition, several medicinal species have gained increased importance with the recent commercialization of homeopathic practices.

   The Araucaria Forest supports a complex ecosystem composed of many species, some of which are endemic (Klein, 1963). It is typically dominated by *Araucaria angustifolia* (Araucariaceae), which comprises more than 40 per cent of the forest trees in natural lands (Galvão *et al.*, 1989; Longhi, 1980; Oliveira and Rotta, 1982). The genus *Araucaria* has been regarded as one of the most primitive genera in the Coniferales (Nikles, 1980; Stockey, 1982) and *Araucaria angustifolia* is the only species in the genus and one of the three native coniferous trees in Brazil. In several succession stages, it forms multiple associations with two genera of the Lauraceae family – *Ocotea* and *Nectandra*, with *Ocotea porosa* being the most frequent. *Ilex paraguariensis* St. Hil. (Aquifoliaceae), *Podocarpus lambertii* Klotz and *Podocarpus sellovii* (Podocarpaceae) (Carvalho, 1994; Maack, 1981) are also associated with this type of forest.

The Atlantic Forest represents one of the highest levels of plant biodiversity in Brazil, with ca. 10 000 plant species and the percentage of endemism of trees, ca. 53.5 per cent, reflects its botanical uniqueness (Fonseca, 1985; Fundação SOS Mata Atlântica, 1992; Mori *et al.*, 1981; Viana *et al.*, 1997). Before the drastic decrease in the availability of raw forest materials in the 1960s, the Araucaria Forest was the most significant in Brazil for the wood industry and the main economically important species explored were *Araucaria angustifolia* and *Ocotea porosa* (Lauraceae), and from the Atlantic Forest *Ocotea catharinensis* and *Ocotea odorifera* (Lauraceae), and *Aspidosperma olivaceum* (Apocynaceae). Other valuable timber producing species which have been heavily exploited in the past few decades are: *Cordia trichotoma* and *Patagonula americana* (Boraginaceae), *Cedrela fissilis* and *Cabralea canjerana* (Meliaceae), *Apuleia leiocarpa, Parapiptadenia rigida* and *Myrocapus frondosus* (Leguminosae) and *Balfourodendron riedelianum* (Rutaceae) (Reitz *et al.*, 1978, 1979). *Caesalpinia echinata* Lamark (Leguminosae-Caesalpinioideae) used to be the most economically important species from the Atlantic Forest in the north-east of Brazil. Other economically important native species which are sources of non-wood products are *Euterpe edulis* (Palmae) (palmito) and *Ilex paraguariensis* (Aquifoliaceae) (erva mate). It is important to emphasize that the Atlantic Forest is an important genetic pool of many potentially important and still underexploited native fruit tree species from the Myrtaceae family such as *Feijoa sellowiana, Campomanesia xanthocarpa, Eugenia brasiliensis, Eugenia uniflora* and *Myrcia cauliflora,* the Annonaceae (*Duguetia lanceolata*) and the Leguminosae (*Inga* spp.) (Silva, 1996).

Medicinal importance

The Amazon, Araucaria and Atlantic Forests are sources of essential oils and many other biologically active compounds useful for the comestic and pharmaceutical industries. Table 18.1 shows the main tree families and genera which have been studied in Brazil over the last five years from the pharmacological and/or phytochemical perspectives, totalling ca. 190 species. The most important medicinal species from the Amazon Forest are *Aniba roseodora* (linalol, anti-epilepsy) and *Dicypellium caryophyllatum* (Lauraceae) (eugenol), *Bertholletia excelsa* (oil, vitamins), *Carapa guianensis* (medicinal oil, antiseptic, anti-fever, wound regenerative), *Paulinia cupana* (guarana, tonic, stimulant), *Pilocarpus microphyllus* (Rutaceae) (pilocarpine) and *Psychotria ipecacuanha* (Rubiaceae) (alkaloid, analgesic). In the Araucaria and Atlantic Forests, apart from *Ocotea odorifera,* important for the production of safrol and metileugenol, the main species studied for their pharmacological properties and phytochemical constituents belong to the genera *Maytenus* (particularly *Maytenus ilicifolia,* diuretic, antiseptic), *Bauhinia* (particularly *Bauhinia forficata,* diuretic, hypoglycaemic), *Copaifera* (in special *Copaifera langsdorffii,* essential oil, antiseptic), *Tabebuia* (particularly *Tabebuia avellanedae,* antitumour, analgesic) and *Erythrina* (particularly *Erythrina mulungu,* sedative, tranquillizer). These trees together with the Amazonian trees *Bertholletia excelsa, Carapa guianensis* and *Paulinia cupana* are already being used in commercial homeopathy in Brazil (Teske and Trentini, 1994).

Table 18.1 Families and genera of Brazilian trees of medicinal importance

Family	Genus
Agavaceae	*Cordiline*
Annonaceae	*Annona, Rollinia*
Apocynaceae	*Aspidosperma, Peschiera, Rauvolfia*
Aquifoliaceae	*Ilex*
Bignoniaceae	*Jacaranda, Tabebuia*
Boraginaceae	*Cordia*
Burseracea	*Protium*
Celastraceae	*Maytenus*
Chloranthaceae	*Hedyosmum*
Compositae	*Gochnatia, Vernonia*
Erythroxylaceae	*Erythroxylum*
Euphorbiaceae	*Croton*
Guttiferae	*Calophyllum*
Lauraceae	*Aniba, Nectandra, Ocotea, Persea*
Lecythidaceae	*Cariniana, Bertholettia*
Leguminosae	*Acacia, Apuleia, Bauhinia, Caesalpinia, Cassia, Copaifera, Erythrina, Hymenaea, Parapiptadenia, Piptadenia*
Lythraceae	*Lafoensia*
Magnoliaceae	*Drimys*
Melastomataceae	*Miconia*
Meliaceae	*Cedrela, Guarea, Trichilia, Carapa*
Monimiaceae	*Mollinedia*
Moraceae	*Cecropia*
Myristicaceae	*Virola*
Myrtaceae	*Eugenia, Psidium*
Olacaceae	*Heisteria*
Proteaceae	*Roupala*
Rhamnaceae	*Rhamnus*
Rosaceae	*Licania*
Rubiaceae	*Genipa, Psychotria*
Rutaceae	*Esenbeckia, Fagara, Helietta, Pilocarpus*
Sapindaceae	*Paulinia*
Verbenaceae	*Cytharexylum, Vitexagnus*
Vochysiaceae	*Qualea*

Source: Abstract Books of Brazilian National Congress of Botany, Plant Physiology and Medicinal Plants, period 1994–1997.

18.3 Why are Brazilian forests endangered?

The area covered by original forests in the south of Brazil, which includes the States of Paraná, Santa Catarina and Rio Grande do Sul, was estimated to originally be ca. 35.3 million hectares but the intensive deforestation over the years has reduced this value to 11 per cent. The Atlantic Forest used to cover 12 per cent of the Brazilian territory, but by 1990 only 8.8 per cent of continuous remnants of the original cover were detected and a very recent survey has shown that this value is currently 7.2 per cent. This ecosystem is therefore one of the most threatened tropical ecosystems in the World (CIMA, 1991; Fonseca, 1985; Fundação SOS Mata Atlântica, 1992).

18.3.1   Abiotic factors

The main reasons why the Atlantic and Araucaria Forests have become so endangered were initially the intensive extraction of timber trees without replanting and later the unrestricted expansion of the sugar cane and cocoa cultures in the north-east of the country, the expansion of coffee and banana cultures in the south, the expansion of pasture areas and the increasing urban and tourism pressures. Over 70 per cent of the Brazilian population is concentrated in the Atlantic Coast of Brazil and the uncontrolled urbanization of the capitals and big cities like Rio de Janeiro and São Paulo has contributed significantly to the reduction and fragmentation of the original forest. Even now, the deforestation is continuing at high rates (Fundação SOS Mata Atlântica, 1992; Fundação SOS Mata Atlântica and INPE, 1993; Viana *et al.*, 1997).

18.3.2   Biotic factors

The main biotic factors which can interfere with the natural regeneration and availability of seedlings of tree species for replanting are erratic flowering, poor seed quality and, in many cases, limited seed viability. As a consequence of the intense fragmentation and genetic erosion of the populations, certain tree species and animal species may have been reduced to levels that interfere with their reproductive biology, affecting for instance the seed setting and quality (Viana *et al.*, 1997; Whitmore, 1997). Seed quality is one of the main factors which interferes with the maximum longevity of seeds in storage (Wang *et al.*, 1993). The slow growth of many species, their susceptibility to diseases and attack by insects and lack of knowledge about their ecology are considered the main limitations to commercial production and artificial regeneration, especially in the case of many tree species with valuable timber or medicinal importance such as *Cedrela fissilis, Copaifera trapezifolia, Ocotea odorifera, Ocotea catharinensis, Ocotea porosa, Tabebuia alba, Tabebuia cassionoides, Amburana cearensis* (Leguminosae-Papilionoideae), *Myrocarpus frondosus, Diatenopteryx sorbifolia* (Sapindaceae), *Ruprechtia laxiflora* (Polygonaceae), *Virola oleifera* (Carvalho, 1994) and many other endangered species. These are listed in Table 18.2.

18.4   Endangered tree species of the Brazilian forests

The list of Brazilian trees in Table 18.2 was compiled on the basis of the official lists of endangered species elaborated in the last decade by such organizations as the Brazilian Society of Botany. The trees are classified into three categories: E (threatened with extinction), R (rare) and V (vulnerable to extinction). These species became endangered for the reasons discussed in Sections 18.2 and 18.3. The strategies which have so far been proposed for the possible conservation of these species are through botanical gardens (*Cariniana ianeirensis, Vouacapoua americana*) and the creation of protected areas, biological reserves and national parks. For instance, in the Atlantic Forest there are only ca. 279 Units for Environmental Conservation covering ca. 2500 ha (Fundação SOS Mata Atlântica, 1992). Other proposed strategies for conservation are the establishment of seed collection programmes and propagation schemes for the enrichment of depleted forests or recuperation of degraded areas (*Ocotea catharinensis, Ocotea porosa, Ocotea odorifera* and *Pouteria psammophila)* and in some cases cultivation for industrial uses (e.g. *Pilocarpus* spp.) (Sociedade Botânica do Brasil, 1992).

Table 18.2 Some Brazilian tree species on the verge of extinction

Tree species	Family	Risk	Habitat	Uses
Aniba rosaeodora	Lauraceae	E	Ama	Linalol
Araucaria angustifolia	Araucariaceae	V	Ara	Wood, food
Bauhinia forficata	Leguminosae-Caesalpinoideae	V*	Ara	Medicinal compounds
Bertholletia excelsa	Lecithidaceae	V	Ama	Food (nut)
Bowdichia nitida	Leguminosae-Faboideae	V	Ama	Wood
Brosimum glazioui	Moraceae	R	Atl	Wood, aromatics
Caesalpina echinata	Leguminosae-Caesalpinioideae	E	Atl	Wood
Cariniana ianeirensis	Lecythidaceae	R	Atl	Wood
Dicypellium caryophyllatum	Lauraceae	V	Ama	Eugenol
Euxylophora paraensis	Rutaceae	V	Ama, Atl	Wood
Hirtella insignis	Chrysobalanaceae	E	Atl	Ornamental
Hirtella santosii	Chrysobalanaceae	E	Atl	Ornamental
Maytenus ilicifolia	Celastraceae	R*	Ara	Medicinal
Myracrodruon urundeuva	Anacardiaceae	V	Atl, Ama	Wood
Mollinedia gilgiana	Monimiaceae	R	Atl	Not known
Mollinedia lamprophylla	Monimiaceae	E	Atl	Ornamental
Mollinedia longicuspidata	Monimiaceae	R	Atl	Not known
Mollinedia stenophylla	Monimiaceae	E	Atl	Not known
Ocotea basicordatifolia	Lauraceae	R	Atl	Ornamental
Ocotea catharinensis	Lauraceae	V	Atl	Wood
Ocotea cymbarum	Lauraceae	V	Ama	Wood, oil
Ocotea porosa	Lauraceae	V	Atl, Ara	Wood, ornamental
Ocotea odorifera	Lauraceae	E	Atl, Ara	Wood, safrol, Metileugenol
Pilocarpus microphyllus	Rutaceae	E	Ama	Pylocarpine
Pithecellobium racemosum	Leguminosae-Mimosoideae	V	Ama	Wood
Pouteria psammophila var. xestophylla	Sapotaceae	V	Atl	Ornamental
Swietenia macrophylla King	Meliaceae	E	Ama	Wood
Torresea acreana Duke	Leguminosae-Papilionideae	V	Ama	Wood
Vouacapoua americana Aubl.	Leguminosae-Caesalpinioideae	E	Ama	Wood

Source: Sociedade Botânica do Brasil (1992) with Official List of Endangered Species.
* Paraná (1995).
Ama, Amazonian Forest; Ara, Araucaria Forest; Atl, Atlantic Forest; E, extinction; V, vulnerable; R, rare.

However, in the case of most of the endangered trees listed in Table 18.2, no strategies have yet been actually adopted and therefore there is a great urgency for implementation of appropriate conservation programmes in these cases. For example, no strategy has been adopted as yet for the conservation of *Aniba roseodora, Bowdichia nitida, Dicypellium caryophyllatum, Euxylophora paraensis, Ocotea cymbarum, Torresea acreana, Pilocarpus microphyllus, Pouteria psammophila* and *Ocotea porosa.* Examples of species which occur in Botanical Gardens or in areas where National Parks and Biological Reserves were already created are: *Brosimum glazioui, Caesalpina echinata, Cariniana ianeirensis, Mollinedia gilgiana, Mollinedia longicuspidata, Ocotea basicordatifolia, Ocotea catharinensis, Ocotea odorifera* and *Swietenia macrophylla* (Sociedade Botânica do Brasil, 1992).

In view of the fact that the Atlantic and Araucaria Forests are still undergoing drastic levels of degradation in such a way that whole ecosystems are endangered, it will only be a question of time before the species which are not considered endangered are included on the list.

18.5 Current and potential uses of biotechnology for *in situ* and *ex situ* conservation management of Brazilian forest species

With the emergence of biotechnological tools such as genetic fingerprinting and *in vitro* regeneration and propagation systems, much relevant information on the genetic characteristics of several important Brazilian species is being generated (as seen in Table 18.3). These developments will be useful for the implementation of *in situ* and *ex situ* conservation strategies, including possibilities for *ex situ* conservation using tissue culture banks of native tree species, especially in those cases where adult tree numbers are low and the natural process of seedling regeneration of the forests has been impeded.

18.5.1 Defining genetic diversity and differences between populations of flora and fauna

In view of their long regeneration and breeding times, research around the world has been actively pursued over the last two decades for the development of genetic markers for forest trees. Isozyme markers have been applied extensively during the past 15 years and have contributed significantly to tree conservation and breeding programmes. In the last five years, DNA-based genetic markers have been developed, most notably restriction fragment length polymorphisms (RFLPs) and polymerase chain reaction (PCR) based markers which have now been made even more sophisticated with the emergence of techniques such as amplified restriction fragment polymorphisms (AFLPs) which have already been demonstrated to be powerful tools in the breeding of short-rotation field crops. These methods have great potential to overcome some of the limitations of isozymes for tree genetics applications since they can be independent of many confounding environmental and biological influences which makes interpretation of biochemical markers difficult (Neale *et al.*, 1992).

Concerning Brazilian tree species, the two most relevant purposes of the application of molecular genetic marker tools are the assessments of levels of genetic variation, gene flow and degrees of genetic differentiation among tree populations. A recent International Foundation for Science Workshop held in Florianopolis in September 1997 considered the implications of these research tools for the conservation and management of Brazilian forest trees in the future (IFS, 1998).

Table 18.3 Biotechnology of Brazilian tree species

Tree species	Family	Isozyme	Molecular markers	Micropropagation	Somatic embryo	Embryo culture	Callus/cell culture	Goals**
Araucaria angustifolia	Araucariaceae	–	X	X	X	–	–	(1) (9)
Aspidosperma polyneuron Muell. Arg.	Apocynaceae	X	–	X	X	–	–	(1) (9)
Bauhinia forficata Link	Leguminosae	–	–	–	–	–	X	(6)
Bertholletia excelsa H. B.K.	Lecythidaceae	X	X	X	–	–	X	(1) (5) (9)
Bixa orellana L.	Lecythidaceae	X	X	–	–	–	–	(1)
Bowdichia virgilioides Kunth	Leguminosae	X	–	X	–	–	–	(9)
Caesalpina echinata Lam.	Leguminosae	–	X	–	–	–	–	(1)
Cariniana legalis	Lecythidaceae	X	–	–	–	–	–	(1)
Cedrela fissilis Velloso	Meliaceae	X	–	X	–	–	X	(1) (2) (4) (6) (7) (9)
Chlorophora tinctoria (L.) Gaudichaud	Moraceae	–	–	X	–	–	–	(9)
Chorisia speciosa St. Hil.	Bombacaceae	–	–	–	–	–	X	(8)
Copaifera langsdorffii Desf.	Leguminoseae	–	X	–	–	–	X	(1) (6)
Cordia thichotoma Vellozo	Boraginaceae	X	–	X	–	–	X	(1) (6) (9)
Cordia verbenaceae	Boraginaceae	–	–	X	–	–	X	(6) (9)
Dalbergia miscolobium Benth.	Fabaceae	–	–	X	–	–	–	(8) (10)
Dalbergia nigra	Fabaceae	–	–	X	–	–	–	(9)
Didymopanax morototoni	Araliaceae	–	–	X	X	–	–	(9)
Elaeis guineensis	Palmae	X	X	–	–	X	–	(1) (3)
Eugenia klotzchiana	Myrtaceae	–	X	–	–	–	–	(1)
Euterpe edulis Martius	Palmae	X	X	–	–	X	–	(1) (10)
Euterpe oleracea	Palmae	–	–	–	–	X	–	(10)
Feijoa sellowiana Berg.	Myrtaceae	X	X	X	–	–	–	(1) (9)
Ilex dumosa Reiss	Aquifoliaceae	–	–	–	–	–	X	(6)
Ilex paraguariensis St. Hil.	Aquifoliaceae	X	–	–	–	–	X	(1) (6)
Inga affinis	Leguminosae	–	–	–	–	–	X	(8)

Species	Family	1	2	3	4	5	Studies
Joannesia princeps	Euphorbiaceae	X	—	—	—	—	(1)
Machaerium angustifolium	Leguminosae	—	X	—	—	—	(1)
Maytenus aquifolium	Celastraceae	—	X	—	—	X	(6)
Maytenus ilicifolia Mart.	Celastraceae	—	X	—	X	—	(1) (9)
Miconia spp.	Melastomataceae	—	—	—	X	—	(9)
Mimosa scabrella	Mimosaceae	X	—	—	—	—	(1)
Ocotea catharinensis	Lauraceae	—	—	X	—	X	(6) (8) (10)
Ocotea porosa Nees	Lauraceae	—	—	—	X	—	(9)
Ocotea odorifera	Lauraceae	—	—	X	X	—	(8) (10)
Pilocarpus jaborandi Holmes	Rutaceae	—	—	—	—	X	(6)
Pilocarpus pennatifolius	Rutaceae	—	—	—	—	X	(6)
Psychotria carthagenensis	Rubiaceae	—	—	—	—	X	(6)
Psychotria ipecacuanha	Rubiaceae	—	—	—	—	X	(6)
Psychotria mai	Rubiaceae	—	—	—	—	X	(6)
Qualea grandiflora Mart.	Vochysiaceae	—	—	—	X	—	(9)
Qualea multiflora Mart.	Vochysiaceae	—	—	—	X	—	(9)
Qualea parviflora Mart.	Vochysiaceae	—	—	—	X	—	(9)
Rapanea ferruginea	Myrsinaceae	—	—	—	X	—	(9)
Rauvolfia ligustrina	Apocynaceae	—	—	—	—	X	(6)
Rauvolfia sellowii	Apocynaceae	—	—	—	—	X	(6)
Rollinia mucosa	Annonaceae	—	—	—	X	X	(6) (9)
Rudgea jasminoides	Rubiaceae	—	—	—	—	X	(6)
Senna multijuga	Leguminosae	—	X	—	—	—	(1)
Stryphnodendron polyphyllum	Leguminosae	—	X	—	—	—	(1)
Tabebuia avellanedae	Bignoniaceae	—	—	—	X	—	(8) (10)
Theobroma cacao L.	Sterculiaceae	—	—	—	X	—	(9)

Source: Abstract Books of Brazilian National Congress of Botany, Plant Physiology and Latin American Network on Plant Biotechnology, period 1995–1997.

'—' Indicates no information available; 'X' indicates that studies have been initiated and/or reported; *indicates that only Brazilian population studies were considered.

**(1) Genetic variability; (2) genetic mapping; (3) conservation; (4) genetic improvement; (5) gene regulation; (6) secondary metabolites; (7) regeneration; (8) seed propagation; (9) clonal propagation; (10) *ex situ* conservation; (11) mechanisms of genetic resistance.

The distribution of genetic variability in plants is influenced by the effective population size, geographic distribution, the primary mode of reproduction, the mating system, pollination and seed dispersal mechanisms, and the community type in which the species most commonly occurs (Hamrick, 1989). Bawa *et al.* (1990) present an important discussion on how the reproductive ecology of tropical plants affects the genetic variability of populations.

As described by Bawa (1992) and Bawa *et al.* (1990) for tropical species in Central America and Asia, the main population genetic characteristics of the tree species of natural forests in Brazil are as follows: they have a wide variety of breeding systems but display high outcrossing rates, are almost exclusively zoophilous and typically have low population densities. A vast majority are cross-fertilized by virtue of being either self-incompatible, dioecious or even monoecious with differences in the maturation of male and female flowers. Table 18.4 presents some information about pollination, seed dispersal, geographical range, density, and breeding systems of some Brazilian trees.

An additional fact to consider is that demographic fragmentation can modify pollen and seed dispersal patterns, influencing the ongoing levels and kinds of gene flow. The genetic consequences of habitat fragmentation depend critically upon whether or not there is dispersal between habitat islands and this parameter can be affected by the size and the distance between the fragments and mainly by the level of extinction of the species responsible for dispersal.

Species displaying small-sized populations are the most susceptible to the risks of inbreeding depression, demographic stochasticity and extinction, since the movement of pollen and seeds is affected. Such aspects are very important in the Araucaria and Atlantic Forests of Brazil, which are extremely fragmented. Fragments are in many cases extremely small and support many species responsible for dispersion of the trees. These in turn are also threatened with extinction and this eventual outcome could result in the development of different patterns of genetic variation within and between tree populations.

Use of biochemical fingerprinting techniques

It is only in the last decade that biochemical markers, in the form of allozymes, have been used to study mating systems and the genetic structure of native tree species in Brazil.

The rates of outcrossing can be quantified by the use of genetic markers and generating an estimation of the parameter 't', which can vary from 0 (no outcrossing) to 1 (100 per cent outcrossing). The most commonly used measures of intrapopulational genetic variation are: the percentage of polymorphic loci per population (P), the mean number of alleles per locus (A), and the average heterozygosity per individual (H). The most common measurement of population differentiation based on allozyme data is the ratio of allelic diversity among populations (G_{ST}), as defined by Nei's genetic diversity statistics (Nei, 1973).

Levels of outcrossing and intrapopulation allozyme variation have been calculated for some Brazilian species. As shown in Table 18.5, the outcrossing rate of many Brazilian species is high. The exception is *Myracrodruon urundeuva* – Anacardiaceae (Syn: *Astronium urundeuva*), a dioecious species, which displays high inbreeding coefficients (F=0.517 and 0.520) in two populations studied. Studies carried out with other tropical tree species have shown 't' values varying from 0.21 to 1.00 (Bawa, 1992).

Several studies on Brazilian tree species have indicated high levels of genetic variability (Table 18.5) with values of polymorphic locus (P), average heterozygosity (H) and average number of alleles per locus (A) higher than those related for tropical trees by

Table 18.4 Characteristics that influence the distribution of genetic variation in some Brazilian natural tree species

Species	Habitat†	Tax. status	Sexual systems	Geographic range	Density indiv./ha	Mating system	Pollination	Dispersion
Myracrodruon urundeuva	Ama	Angyos.	Dioecious	Widespread		Outcrossed	Animals	
Hevea brasiliensis	Ama	Angyos.	Monoecious	Regional	4–6	Outcrossed	Animals/water	
Copaifera langsdorffii	Ama	Angyos.	Monoecious	Widespread			Insects	Animals
Araucaria angustifolia	Ara	Gymno.	Dioecious	Regional	5–25	Outcrossed	Wind	Animals
Bowdichia virgilioides	Ama, Atl	Angyos.	Monoecious	Widespread	0.03		Insects	Wind
Cariniana legalis	Ama, Atl	Angyos.	Monoecious	Widespread	0.80		Insects	Wind
Cordia trichotoma	Atl	Angyos.	Monoecious	Widespread	0.03		Insects	Wind
Joannesia princeps	Atl	Angyos.	Dioecious	Regional	2.50	Outcrossed	Insects	Rodents
Euterpe edulis	Atl	Angyos.	Monoecious	Widespread	543		Insects	Animals
Ilex paraguariensis	Ara	Angyos.	Dioecious	Regional		Outcrossed	Insects	Birds

†Ama, Amazonian Forest; Ara, Araucaria Forest; Atl, Atlantic Forest.

Table 18.5 Genetic parameters within and among Brazilian natural tree populations estimated by allozyme variation

Species	NP	NL	NI	F	t	A	P	H	H_T	H_S	G_{ST}	References
Myracrodruon urundeuva	2	3	2798	0.51	0.49	3.24	68.28	0.15	0.45	0.44	0.04	Moraes (1992)
Bertholletia excelsa	–	–	–	–	0.95	–	–	–	–	–	–	O'Malley et al. (1988)
Hevea brasiliensis	2	4	3811	0.22	0.65	–	–	–	0.34	0.34	0.00	Paiva (1992)
*Bowdichia virgilioides**	1	19	56	–	–	2.25	0.28	0.13	–	–	–	Harrit (1991)
*Cariniana legalis**	1	20	92	–	–	2.68	0.30	0.14	–	–	–	Harrit (1991)
*Cordia trichotoma**	1	17	67	–	–	2.40	0.40	0.18	–	–	–	Harrit (1991)
*Joannesia princeps**	1	22	22	–	–	2.40	0.27	0.14	–	–	–	Harrit (1991)
Ilex paraguariensis	3	–	–	–	–	2.0	–	0.50	–	–	–	Winge et al. (1995)
*Aspidosderma polyneuron**	2	8	116	–	0.82	2.0	0.50	0.24	–	–	0.06	Maltez and Kageyama (1997)
*Cedrela fissilis**	2	4	720	–	–	2.4	–	0.31	–	–	–	Gandara and Kageyama (1997)

'–' = no value estimated or available.

NP, number of populations; NL, number of loci; NI, total number of individuals; F, endogamy coefficient; t, outcrossing rate; A, average number of alleles per polymorphic locus; P, percentage of loci polymorphic within each population; H, observed heterozygosity; H_T, total allelic diversity; H_S, mean allelic diversity within population; G_{ST}, the ratio of the allelic diversity among populations to total allelic diversity endangered.

* endangered

Hamrick and Godt (1989) (an average of 2.0 alleles per locus and a percentage of polymorphic loci of 53 per cent) and by Loveless and Hamrick (1987) (66 per cent polymorphic locus (P), an average heterozygosity (H) of 0.240 and an average number of alleles per locus (A) of 2.02).

Results of studies on Brazilian tree species (Table 18.5) have confirmed the data published by Hamrick (1989) for dioecious tropical species pollinated by animals, and show that there is a large genetic variation within populations (H_S), and a small genetic variation between populations (G_{ST}). This trend is also now being confirmed by the results emerging from more recent biochemical and molecular genetic studies on both Brazilian and South American tree species (IFS, 1998).

Development and deployment of molecular auditing techniques

RAPD analysis of nuclear DNA has been the most widely used DNA marker to date for studies on genetic variation in Brazilian tree populations (Table 18.3). These studies are still at a very preliminary stage. As far as the authors are aware, organelle DNA studies are just beginning in only two Brazilian tree species. Genetic variation within and between *Copaifera langsdorffii* populations are being determined using chloroplast DNA (cpDNA), aiming at the identification of regions of high genetic variability for the establishment of a genetic reserve for the species (Clampi *et al.*, 1997). Other studies have been initiated on *Araucaria angustifolia*, using cpDNA to evaluate the potential use of this technique in determining the levels of gene flow and the genetic structure of tree populations distributed in several remnants fragments of the *Araucaria* forest.

For some tree species which are important economically on the international markets, such as *Swietenia macrophylla*, *Hevea brasiliensis* and *Theobroma cacao* results of detailed studies on DNA molecular markers such as RFLPs and RAPDs are published, i.e. Newton *et al.* (1993) and Chalmers *et al.* (1994) for *Swietenia*, Lerceteau *et al.* (1997), Laurent *et al.* (1993a, 1993b, 1994), N'Goran *et al.* (1994), Ronning and Schnell (1994) for *Hevea* and Besse *et al.* (1993) for *Theobroma*. The methodologies developed for similar genera/species can now be used as models for investigations with Brazilian native trees.

The strategy to foster the application of molecular markers to important Brazilian native trees should be based firstly on the identification of the tree species by grouping them according to their levels of over-exploitation and actual utilization; and secondly, there must be areas identified for which the application of isozymes and DNA-based markers can be used to greatest effect.

In the past, many tree species were intensively exploited and this led to the exhaustion of tree populations and to a drastic reduction in timber production. The tree species which fall into this category are usually slow growing, have constraints in their multiplication by conventional methods and their seeds have short viabilities and cannot be stored for long periods (i.e. they are recalcitrant). These characteristics impeded the development of techniques for cultivation and genetic improvement and therefore there is an urgent need to conserve these species. Trees of this type are in great risk of extinction, because they are either rare, endangered or vulnerable species and the main current priority in such cases is to develop models for conservation of the remnant populations, e.g. *Araucaria angustifolia*, *Ocotea porosa*, *Ocotea catharinensis*, *Ocotea odorifera*, *Aspidosperma olivaceum*, *Caesalpinia echinata*, *Dalbergia nigra*, *Cedrela fissilis* and *Cariniana legalis* which are over-exploited trees from the south-east and south of Brazil. There is, therefore, an urgent need for the application of molecular markers to assess the genetic variability within and between populations of these tree species to support future conservation programmes. As most of these trees occur in remnants of natural forests distributed in several fragments, there will be a need to use molecular markers which will allow the determination of the gene flow, and the model of genetic diversity analysis shall consider the effect of the fragment component. Very few species in this group have yet been studied from the point of view of genetic variability using markers, either isozymes or DNA.

A very low percentage of native tree species are utilized now for wood in Brazil. However, natural populations of a few species are still over-exploited commercially on a large scale for solid wood products or veneer. In most cases there are plantations in other parts of the world; nevertheless, the Brazilian natural populations are still coming under strong pressure. Some of these species are vulnerable or even in danger of extinction and the examples of timber producing trees in this category are mainly from the Amazon Forest such as: *Virola surinamensis, Swietenia macrophylla, Carapa guianensis, Cordia goeldiana, Diplotropis* spp., *Torresea acreana, Cedrela odorata, Vouacapoua americana, Euxylophora paraensis, Calophyllum brasiliense, Carapa guianensis, Ocotea cymbarum, Bowdichia nitida* and *Aspidosperma* spp. For most of them, if not all, there is an urgent need to develop conservation programmes and to stimulate plantation as a strategy for conservation by substitution of the reliance on delicate natural forests. The application of DNA and isozyme markers to these species will be fundamental to raising information on genetic diversity within and between natural populations of these important trees.

Selective logging acts as a source of dysgenic selection, whereby the best genotypes (in terms of growth or form) are selectively depleted in their most favoured genotypes which in turn can lead to genetic erosion. Such genetic impoverishment of *Swietenia macrophylla* and *Cedrela* species has already been shown to have occurred in Brazil and Central America, where trees with good shape are now only rarely encountered in isolated areas. In the case of both species, it is likely that the poorly formed trees left by loggers are those with little resistance to *Hypsipyla grandella*, since frequent attacks by shoot borers affect the shapes of the trees. One important application of RAPDs is the characterization and quantification of genetic variation in plants and the identification of naturally existing genotypes resistant to pest attack. In this context, this technique could be used to select for borer-resistant tree genotypes and could contribute significantly to defining strategies for conservation and genetic improvement. On a longer term basis, integrated studies on molecular markers and tissue culture of the Brazilian trees will no doubt contribute to the development of resistance via genetic engineering approaches whereby foreign pest resistant genes could be introduced into these trees to enable the development of durable forms of resistance.

The Brazilian native trees which are used for non-timber production such as indigenous fruit trees, rubber trees and others belong to natural populations and these are experiencing genetic erosion due to the drastic destruction of the natural forests. Generally their plantations have restricted genetic constitutions due to the need for use of vegetative propagation techniques to conserve desirable phenotypes demanded by the market. In these cases, the deployment of DNA-based markers will be useful to assess the level of genetic variability and the application of quality linked traits (QLTs) and marker assisted selection techniques will be valuable for shortening and increasing the efficiency of genetic improvement programmes. The molecular markers will be especially useful in the identification of sites of genetic diversity and in the establishment of the most appropriate sustainable conservation strategies. Examples of some important Amazon native fruit tree species which could benefit from the application of this technology in genetic improvement programmes are *Euterpe oleracea, Orbignya speciosa, Bertholletia excelsa, Theobroma grandiflorum* and *Paulinia cupana*.

Particularly for medicinal tree species, it is very important to intensify the application of molecular approaches to study the available genetic variability and to identify specific genes controlling secondary metabolite production. The first steps in this direction will include the purification and characterization of key enzymes of biosynthetic pathways, the isolation of cDNA clones, gene sequence analysis and gene synthesis.

Clearly, DNA polymorphisms have enhanced a wide array of methodologies available to forestry geneticists. It now appears possible to choose appropriate genetic markers to answer specific questions. However, these technologies will only be valuable if there is a concomitant increase in the social and political awareness about the importance of conserving natural resources in Brazil.

18.5.2 Uses of *in vitro culture techniques for propagation and conservation*

In a general way, studies on the reproductive biology of Brazilian native trees are not sufficient alone to provide the basis for the establishment of consistent conservation strategies. Significantly, attempts to apply biotechnological approaches for propagation and conservation of Brazilian forest tree species have increased in the last few years (Table 18.3).

Embryo culture and somatic embryogenesis

The development of *in vitro* culture systems of zygotic embryos, embryonic axes or excised embryonic shoot apical meristems for tropical species presenting recalcitrant seeds has proved to be a relevant goal for it allows the production of entire seedlings which can then be micropropagated to generate plant materials for conservation, genetic improvement and plant production programmes for reforestation. The application of these techniques associated with *in vitro* culture techniques, such as growth restriction and cryopreservation, is the basis for the establishment of short, medium and long-term conservation strategies. Several models of embryonic axes and zygotic embryo culture systems aiming at *in vitro* conservation have already been developed for tropical tree species with recalcitrant seeds, such as *Araucaria hunsteinii* (Pritchard and Prendergast, 1986), *Hevea brasiliensis* (Normah *et al.*, 1986), *Theobroma cacao* (Pence, 1991), *Cocos nucifera* (Assy-Bah and Engelmann 1992a, 1992b, 1993), *Elaeis guineensis* (Dumet *et al.*, 1993), *Coffea canephora* (Abdelnour-Esquivel *et al.*, 1992) and *Swietenia macrophylla* (Normah and Marzalina, 1996). Examples of similar model systems developed for temperate forest tree species are those established for *Quercus* spp. and *Castanea* spp. by Fu *et al.* (1993) and Pence (1990).

A considerable number of the Brazilian native forest tree species produce recalcitrant seeds and their seed longevity cannot therefore be extended for long periods of time. Therefore, there is a great potential and a considerable need for the application of the *in vitro* zygotic embryo culture techniques to ensure their conservation, epecially concerning the tree species on the verge of extinction (Table 18.2). In this context, the model systems mentioned above will provide a very important basis for some of the species considered. As examples of economically important species which will benefit from the application of these techniques, those belonging to the families Araucariaceae (*Araucaria angustifolia*), Lauraceae (*Aniba roseodora, Dicypellium caryophyllatum, Ocotea cymbarum, Nectandra lanceolata, Ocotea odorifera, Ocotea porosa, Ocotea catharinensis, Ocotea puberula*), Palmae (*Euterpe edulis, Euterpe oleracea, Orbignya speciosa*) and Meliaceae (*Cabralea canjerana, Carapa guianensis*) can be considered appropriate. An additional number of species from other families may also be considered, such as *Theobroma grandiflorum, Paulinia cupana, Aspidosperma polyneuron, Centrolobium microchaete, Chlorophora tinctoria, Talauma ovata* and *Virola oleifera*. So far, only a few attempts to develop *in vitro* zygotic embryo culture systems for these species have been reported for *Euterpe oleracea, Ocotea odorifera* and *Ocotea catharinensis* (Viana and Mantell, 1998). Moreover, the establishment of zygotic embryo culture techniques aiming at the rescue and *in vitro* culture of immature embryos is a potent tool to be used in hybridization programmes of industrial tree species where there are restrictions for the zygotic embryo development *in vivo*. For instance, these techniques have been applied to rescue hybrids of *Pinus lambertiana* and *P. armandii, Populus trichocarpa* and *P. deltoides* and *Eucalyptus pellita* and *E. cloeziana* (Haynes, 1994). Concerning the Brazilian tree species, a great potential for the application of this technology will emerge parallel to the outcomes of the studies on reproductive biology which will elucidate the nature of the barriers to the fertilization for hybrid production.

Only a few examples of somatic embryogenesis systems have been described in the literature for Brazilian trees and these are for *Euterpe edulis* (Palmae) (Guerra and Handro, 1991), *Ocotea catharinensis* (Lauraceae) (Moura-Costa *et al.*, 1993; Viana, 1998; Viana and Mantell, 1998) and *Feijoa sellowiana* (Canhoto and Cruz, 1996). Other important species such as *Didymopanax morototoni* (Araliaceae), *Aspidosperma polyneuron* (Apocynaceae), *Araucaria angustifolia* (Araucariaceae) and *Ocotea odorifera* are currently being studied concerning their somatic embryogenesis potential. These systems, once optimized, will become model systems for each one of the families. For instance, the high frequency somatic embryogenesis system established recently for *Ocotea catharinensis* can hopefully be used for rescuing and conserving other trees in the Lauraceae.

In view of the low number of Brazilian tree species which have been studied concerning their potential for somatic embryogenesis, it is imperative to establish their natural level of polyembryony and to employ this technology to other important species of the Meliaceae (*Carapa guianensis, Swietenia macrophylla, Cedrela fissilis, Cedrela odorata, Cabralea canjerana*), Sterculiaceae (*Theobroma cacao, Theobroma grandiflorum*), Lecythidaceae (*Bertholettia excelsa, Cariniana ianeirensis, Cariniana legalis, Cariniana estrellensis*), Leguminosae-Caesalpinioideae (*Caesalpinia echinata, Vouacapoua americana*) and Rutaceae (*Pilocarpus jaborandi, Pilocarpus microphyllus*).

Shoot culture and micrografting

Application of micropropagation technology to elite clones of the Brazilian forest tree species is a powerful tool for genetic improvement programmes and for the large scale propagation industries, especially in those cases where there are limitations in the conventional methods of propagation by cuttings. It is predicted that this technology will have a profound impact on conservation, for it will allow the rapid production of plants to meet the demand for reforestation and enable the enrichment of forests which have ecologically and economically important species which have the potential to be exploited on a sustainable basis (also see Marzalina and Krishnapillay, Chapter 17, this volume). Currently, the forest plantation programmes with native species are carried out using seedlings and there is a strong request for seedlings of the endangered species, which are not available in most cases. Micropropagation systems have now been described for members of the Lecythidaceae, Araucariaceae, Moraceae, Leguminosae-Faboideae, Meliaceae, Boraginaceae, Araliaceae, Celastraceae, Melastomataceae, Lauraceae, Vochysiaceae, Annonaceae, Fabaceae, Bignoniaceae and Sterculiaceae (Table 18.3). Not all of the tree species mentioned belong to the Amazon, Atlantic or Araucaria forests but the micropropagation systems established will function as models for the propagation of other species in the same genus present in these ecosystems.

One of the main constraints to the application of micropropagation technology to forestry is *in vitro* recalcitrance due to the expression of the adult phase. In some cases it has been shown that it is possible to revert adult shoot meristems utilizing serial *in vitro* micrografting techniques. The technique has been applied successfully to rejuvenate mature shoot tips of *Dalbergia miscolobium* (Fabaceae) (Teixeira *et al.*, 1995).

Apart from this, the combined techniques of micrografting and micropropagation offer potent tools for the genetic improvement programmes of the Brazilian native trees, and most especially for the transfer of disease resistance and other targeted traits of interest. One potential application of the *in vitro* micrografting technology is in the development of insect resistance in members of the Meliaceae family such as *Cedrela fissilis*, *Cedrela odorata* and *Swietenia macrophylla*. The development of consistent regeneration and genetic transformation systems via callus culture will be a valuable tool for programmes aimed at the development of insect resistance in susceptible tree species such as those belonging to the Meliaceae family. Callus culture and regeneration systems have been developed for a few species in Table 18.3, including *Cedrela fissilis* (Moreno *et al.*, 1995).

Secondary product biosynthesis in callus and cell culture

In general, the phytochemical studies (see also Schumacher, Chapter 9, this volume) of tree species which belong to endangered ecosystems are severely constrained by the lack of readily available plant material. For instance, some of the species of medicinal importance in families listed in Table 18.1, such as Lauraceae, Monimiaceae, Rutaceae, Leguminosae, Apocynaceae, Bignoniaceae and Magnoliaceae are already rare and/or endangered. Therefore, the development of tissue culture systems for the Brazilian native trees is an urgent goal since their availability will foster studies on the regulation of biosynthesis and isolation of synthesized biologically active compounds of interest. One example where the development of tissue culture systems can foster phytochemical studies is with *Ocotea catharinensis*. The genus *Ocotea* is important not only for the production of essential oils but also for the production of lignans and neolignans with cytotoxic effects against tumour cells. Phytochemical studies carried out on the bark, wood and leaves of *Ocotea catharinensis* has led to the isolation of 15 novel neolignans. Further studies on the somatic embryogenesis system of this species at different stages of development showed that the biosynthesis of neolignans is related to cell differentiation and one new neolignan, as yet not described for *Ocotea catharinensis*, was detected in dehydrated mature embryoids (Lordello, 1996).

The employment of the cell, shoot and root culture technologies, integrated with *in vitro* conservation techniques, will undoubtedly contribute to the sustainable management and conservation programmes of the Brazilian forests. In this context, an increasing number of callus culture and cell suspensions aiming at secondary metabolite production have been developed for species of the genus *Bauhinia*, *Cedrela*, *Ilex*, *Maytenus*, *Ocotea*, *Pilocarpus*, *Psychotria*, *Rauvolfia*, *Rollinia* and *Rudgea*, but it is urgent that this technology is applied to a broader range of tree species, especially to the rare ones whose phyto-chemical studies have been restricted in the past by the lack of plant material.

Growth limitation and cryopreservation

The *in vitro* conservation techniques of growth limitation and cryopreservation have not been developed to any large extent for Brazilian trees as compared to field cultivated species such as *Manihot, Ipomoea batatas, Dioscorea, Rubus, Asparagus, Vanilla planifolia, Stevia, Mentha piperita, Fragaria × ananassa* and *Origanum vulgare*. There is now a pressing urgency for the application of conservation biotechnology to the tree species already endangered and to those known to have recalcitrant seeds or seeds whose storage life is short. Research in this field can benefit a great deal from the *in vitro* conservation research on tropical tree species carried out in countries such as Malaysia (see Marzalina and Krishnapillay, Chapter 17, this volume), India (see Mandal, Chapter 14, this volume), Indonesia, South Africa and Japan. Examples of species of interest which have been studied and could be used as models are *Rauvolfia serpentina* (Imelda, 1996), *Trichilia dregeana* (Berjak *et al.*, 1996), *Cedrela odorata* and *Jacaranda mimosaefolia* (Maruyama *et al.*, 1997) and those mentioned in the subsection on 'Embryo culture and somatic enbryogenesis'.

18.6   Conclusions and future prospects

Conservation programmes have been established for just a few Brazilian tree species. Of the 100 most important Brazilian forest trees, there are conservation programmes *in situ* (protected areas and managed resource areas) and/or *ex situ* (conservation stands, plantations) for *Amburana cearensis, Araucaria angustifolia, Balfourodendron riede-lianum, Chlorophora tinctoria, Cordia trichotoma, Didymopanax morototoni, Dipteryx alata, Enterolobium contortisiliquum, Gallesia gorarema, Genipa americana, Hymenaea courbaril, Myracrodruon urundeuva, Parapiptadenia rigida, Pterogyne nitens, Tabebuia impetiginosa, Talauma ovata* and *Zeyheria tuberculosa*. There is an urgent need for the establishment of conservation programmes for the following species: *Caesalpinia echinata, Calophyllum brasiliense, Cariniana strellensis, Cedrela fissilis, Centrolobium microchaete, Myracrodruon balansae, Ocotea odorifera, Ocotea porosa, Ocotea catharinensis* and *Peltophorum dubium* (Carvalho, 1994).

Biotechnology is being applied increasingly to the native Brazilian tree species al-though, to date, the application of molecular markers has been restricted to a few cases where studies have been carried out to assess genetic variability within and between tree populations. A greater number of tree species and populations now need to be assessed since many species of economical, ecological and medicinal importance still remain completely unknown from the genetic point of view. The lack of information in this field not only causes restrictions to the development of *in situ* and *ex situ* conservation strat-egies but it also impedes the sustainable utilization of resources based on the knowledge of the genetics of the tree populations to be explored.

The establishment of a consistent strategy on biodiversity conservation of native tree species in Brazil requires efforts to generate more information on the reproductive bio-logy and ecology of native tree species and to integrate this information with the studies in genetic variability and molecular markers. Concomitantly, it is imperative that studies on the fauna genetic structure must be carried out, due to the extreme importance of animals and insects in the pollination and seed dispersion of the Brazilian tree species.

The development of somatic embryogenesis systems and artificial seeds for the Brazilian native forest trees is imperative to circumvent the erratic seed production and the availability of seedlings for replanting in those cases where seed storage cannot be extended and no other alternative is available for a large scale production. The integration of this technology with molecular characterization of genetic diversity and conservation biotechnology will allow the large scale multiplication of selected tree genotypes and the establishment of consistent *ex situ* germplasm conservation strategies through tissue culture and DNA banks.

The increasing availability (for several species) of consistent systems of zygotic embryo culture, somatic embryogenesis, micropropagation, callus culture and cell suspensions will be the basis for appropriate integration of *in vitro* with conventional approaches and methods for short, medium and long-term germplasm conservation of native trees in the future.

Acknowledgements

We would like to thank the International Foundation for Science (IFS), Stockholm, Sweden (for grants D/1265-1, D/1265-2, D/1265-3 to AMV and for workshop fellowships to AMV and MCM), Conselho Nacional de Desenvolvimento Científico e Tecnológico (CNPq, Brasil), Fundo Nacional do Meio Ambiente (FNMA, Brasil) for research support and the British Council and CAPES (Brazil) for sponsoring valuable academic links between Brazilian and UK universities which assisted our research efforts. The valuable assistance, critical advice and stimulating discussions provided to us during the elaboration of this chapter by Dr João de Deus Medeiros (UFSC, Florianópolis, Brasil), Dr Eric Gomes Schaitza and Dr Patricia Povoa de Mattos (CNPF, EMBRAPA) are gratefully acknowledged.

References

ABDELNOUR-ESQUIVEL, A., VILLALOBOS, V. and ENGELMANN, F., 1992, Cryopreservation of zygotic embryos of *Coffea* spp., *Cryo-Letters*, **13**, 297–302.

ASSY-BAH, B. and ENGELMANN, F., 1992a, Cryopreservation of immature embryos of coconut (*Cocos nucifera* L.), *Cryo-Letters*, **13**, 67–74.

ASSY-BAH, B. and ENGELMANN, F., 1992b, Cryopreservation of mature embryos of coconut (*Cocos nucifera* L.) and subsequent regeneration of plantlets, *Cryo-Letters*, **13**, 117–126.

ASSY-BAH, B. and ENGELMANN, F., 1993, Medium-term conservation of mature embryos of coconut, *Plant Cell, Tissue and Organ Culture*, **33**, 19–24.

BAWA, K.S., 1992, Mating systems, genetic differentiation and speciation in tropical rain forest plants, *Biotropica*, **24**, 250–255.

BAWA, K.S., ASHTON, P.S. and MOHD NOR, S., 1990, Reproductive ecology of tropical forest plants: management issues, in BAWA, K.S. and HADLEY, M. (Eds), *Reproductive Ecology of Tropical Forest Plants, Unesco Man and the Biosphere Series*, Vol. 7, pp. 3–13, New Jersey: Parthenon Publishers.

BERJAK, P., MYCOCK, D.J., WESLEY-SMITH, J., DUMET, D. and WATT, M.P., 1996, Strategies for *in vitro* conservation of hydrated germplasm, in NORMAH, M.N., NARIMAH, M.K. and CLYDE, M.M. (Eds), *In Vitro Conservation of Plant Genetic Resources*, pp. 19–52, Kuala Lumpur: Percetakan Watan Sdn. Bhd.

BESSE, P., LEBRUN, P., SEGUIN, M. and LANAUD, C., 1993, DNA fingerprints in *Hevea brasiliensis* (rubber tree) using human minisatellite probes, *Heredity*, **70**, 237–244.

CANHOTO, J.M. and CRUZ, G.S., 1996, *Feijoa sellowiana* Berg. (Pineapple Guava), in BAJAJ, Y.P.S. (Ed.), *Biotechnology in Agriculture and Forestry*, Vol. 35, pp. 155–171, Berlin: Springer-Verlag.

CARVALHO, P.E.R., 1994, *Espécies florestais brasileiras: potencialidades e uso da madeira*, Brasília: EMBRAPA-CNPF/SPI.

CHALMERS, K.J., NEWTON, A.C., WAUGH, R., WILSON, J. and POWELL, W., 1994, Evaluation of the extent of genetic variation in mahoganies (Meliaceae) using RAPD markers, *Theoretical Applied Genetics*, **89**, 504–508.

CHELIAK, W.M., YEH, F.C.H. and PITEL, J.A., 1987, Use of electrophoresis in tree improvement programs, *Forestry Chronicle*, 63, 89–96.

CIMA, 1991, *Relatório da comissão interministerial sobre desenvolvimento e meio ambiente*, Brasília, Brazil.

CLAMPI, A.Y., BRONDANI, R.V. and GRATTAPAGLIA, D., 1997, 'Variabilidade em sequências de DNA de cloroplasto em populações de copaíba (*Copaifera langsdorffii*)', presentation at II Encontro Brasileiro de Biotecnologia Vegetal, November, Gramado, Abstract 225.

DUMET, D., ENGELMANN, F., CHABRILLANGE, N. and DUVAL, Y., 1993, Cryopreservation of oil palm (*Elaeis guineensis* Jacq.) somatic embryos involving a desiccation step, *Plant Cell Reports*, **12**, 352–355.

FONSECA, G.A.B., 1985, The vanishing Brazilian Atlantic forest, *Biological Conservation*, **34**, 17–34.

FU, J.R., XIA, Q.H. and TANG, L.F., 1993, Effects of desiccation on excised embryonic axes of three recalcitrant seeds and studies on cryopreservation, *Seed Science and Technology*, **21**, 85–95.

FUNDAÇÃO SOS MATA ATLÂNTICA, 1992, *Dossiê SOS Mata Atlântica*, São Paulo: Estudo do Projeto e Edições Ltda.

FUNDAÇÃO SOS MATA ATLÂNTICA and INPE, 1993, *Evolução dos remanescentes florestais e ecossistemas associados do domínio da Mata Atlântica*, São Paulo: Fundação SOS Mata Atlântica, Instituto de Pesquisas Espaciais.

GALVÃO, F., KUNIYOSHI, Y.S. and RODERJAN, C.V., 1989, Levantamento fitossociológico das principais associações arbóreas da Floresta Nacional de Irati-PR, *Floresta*, **19**, 30–49.

GANDARA, F.B. and KAGEYAMA, P.Y., 1997, Alta taxa de cruzamento em uma espécie arbórea tropical de baixa densidade: *Cedrela fissilis* Vell. (Meliaceae), *Revista Brasileira de Genética*, **20**, 328.

GUERRA, M.P. and HANDRO, W., 1991, Somatic embryogenesis in tissue cultures of *Euterpe edulis* Mart. (Palmae), in AHUYA, R. (Ed.), *Woody Plant Biotechnology*, pp. 189–196, New York: Plenum Press.

HAMRICK, J.L., 1989, Isozymes and the analysis of genetic structure in plant populations, in SOLTIS, D.E. and SOLTIS, P.S. (Eds), *Isozymes in Plant Biology*, pp. 87–105, Portland: Dioscorides Press.

HAMRICK, J.L. and GODT, M.J., 1989, Allozyme diversity in plant species, in BROWN, A.H.D., CLEGG, M.T., KAHLER, A.L. and WEIR, B.S. (Eds), *Plant Population Genetics, Breeding, and Genetic Resources*, pp. 43–63, Sunderland: Sinauer Associates Inc.

HARRIT, M.M., 1991, Ecology and genetic variation of four hardwoods of Brazil's Atlantic Forest region, PhD thesis, North Carolina State University, USA.

HAYNES, R., 1994, Biotechnology in forest tree improvement with special reference to developing countries, *FAO Forestry Paper*, No. 118.

IFS 1998, *Recent Advances in Biotechnology for Tree Conservation and Management*, MANTELL, S.H., BRUNS, S., TRAGARDH, C. and VIANA, A.M. (Eds), International Foundation for Science, Stockholm, Sweden.

IMELDA, M., 1996, Current Status on *in vitro* plant conservation in Indonesia, in NORMAH, M.N., NARIMAH, M.K. and CLYDE, M.M. (Eds), *In Vitro Conservation of Plant Genetic Resources*, pp. 263–270, Kuala Lumpur: Percetakan Watan Sdn. Bhd.

KLEIN, R.M., 1963, Observações e considerações sobre a vegetação do planalto nordeste catarinense, *Sellowia*, **15**, 39–54.

LAURENT, V., RISTERUCCI, A.M. and LANAUD, C., 1993a, Chloroplast and mitochondrial DNA diversity in *Theobroma cacao*, *Theoretical and Applied Genetics*, **87**, 81–88.

LAURENT, V., RISTERUCCI, A.M. and LANAUD, C., 1993b, Variability for nuclear ribosomal genes within *Theobroma cacao*, *Heredity*, **71**, 96–103.

LAURENT, V., RISTERUCCI, A.M. and LANAUD, C., 1994, Genetic diversity in cocoa revealed by cDNA probes, *Theoretical Applied Genetics*, **88**, 193–198.

LERCETEAU, E., ROBERT, T., PETIARD, V. and CROUZILLAT, D., 1997, Evaluation of the extent of genetic variability among *Theobroma cacao* accessions using RAPD and RFLP markers, *Theoretical Applied Genetics*, **95**, 10–19.

LISBOA, P.L.B., TEREZO, E.F.M. and SILVA, J.C.A., 1991, Madeiras amazonicas: considerações sobre exploração, extinção de espécies e conservação, *Boletim de Museo do Para Emilio Goeldi, Series Botanicas*, **7**, 521–542.

LONGHI, S.J., 1980, A estrutura de uma floresta natural de *Araucaria angustifolia* (Bert.) O.Ktze no sul do Brasil, MSc Thesis, Universidade Federal do Paraná, Brazil.

LORDELLO, A.L.L., 1996, Constituintes Químicos de Folhas e de Cultura de Células e Tecidos de *Ocotea catharinensis* Mez. (Lauraceae), PhD Thesis, Universidade de São Paulo, Brazil.

LOVELESS, M.D. and HAMRICK, J.L., 1987, Distribución de la variación en especies de árboles tropicales, *Revista de Biologia Tropical*, **35**, 165–175.

MAACK, R., 1981, *Geografia física do Estado do Paraná*, 2nd edn, Rio de Janeiro: Livraria José Olympio Editora S.A.

MALTEZ, H.M. and KAGEYAMA, P.Y., 1997, Variabilidade alozímica entre e dentro de duas populações naturais de peroba rosa (*Aspidosperma polyneuron* Muell. Arg. – Apocynaceae), *Revista Brasileira de Genética*, **20**, 329.

MARUYAMA, E., KINOSHITA, I., ISHII, K. and OHBA, K., 1997. Germplasm conservation of the tropical forest trees, *Cedrela odorata* L., *Guazuma crinita* Mart., and *Jacaranda momisaefolia* D. Don, by shoot tip encapsulation in calcium-alginate and storage at 12–25°C, *Plant Cell Reports*, **16**, 393–396.

MORAES, M.L.T., 1992, Variabilidade genética por isoenzimas e caracteres quantitativos em duas populações naturais de aroeira *Myracrodum urundeuva* F.F. & M.F. Allemão – Anacardiaceae (Syn: *Astronium urundeuva* (Fr. Allemão) Engler), MSc thesis, Universidade de São Paulo, Brazil.

MORENO, F., MANTELL, S.H. and VIANA, A.M., 1995, 'Micropropagation of *Cedrela fissilis* Velloso (Meliaceae)', presentation at Redbio 95 Segundo Encuentro Latino Americano de Biotecnologia Vegetal, Puerto Iguazu, June, Abstract A-88.

MORI, S.S., 1989, Eastern extra-Amazonian Brazil, in CAMPBELL, D.G. and HAMMOND, H.D. (Eds), *Floristic Inventory of Tropical Countries: The Status of Plant Systematics, Collections, and Vegetation, plus Recommendations for the Future*, pp. 428–454, New York: New York Botanical Garden.

MORI, S.A., BOOM, B.M. and PRANCE, G.T., 1981, Distribution patterns and conservation of eastern Brazilian coastal forest tree species, *Brittonia*, **33**, 233–245.

MOURA-COSTA, P.H., VIANA, A.M. and MANTELL, S.H., 1993, *In vitro* plantlet regeneration of *Ocotea catharinensis*, an endangered Brazilian hardwood forest tree, *Plant Cell Tissue and Organ Culture*, **35**, 279–286.

MYERS, N., 1983, Tropical moist forests: over-exploited and under-utilised systems, *Forest Ecology and Management*, **6**, 59–79.

MYERS, N., 1988, Tropical forests: much more than stocks of wood, *Journal of Tropical Ecology*, **4**, 209–221.

NEALE, D.B., DEVEY, M.E., JERMSTAD, K.D., AHUJA, M.R., ALOSI, M.C. and MARSHALL, K.A., 1992, Use of DNA markers in forest tree improvement research, *New Forests*, **6**, 391–407.

NEI, M., 1973, Analysis of gene diversity in subdivided populations, *Proceedings of the National Academic of Sciences of the United States of America*, **70**, 3321–3323.

NEWTON, A.C., LEAKEY, R.R.B. and MESEN, J.F., 1993, Genetic variation in mahoganies: its importance, capture and utilization, *Biodiversity and Conservation*, **2**, 114–126.

N'GORAN, J.A.K., LAURENT, V., RISTERUCCI, A.M. and LANAUD, C., 1994, Comparative genetic diversity studies of *Theobroma cacao* L. using RFLP and RAPD markers, *Heredity*, **73**, 589–597.

NICKLES, D.J., 1980, Realised and potential gains from using and conserving genetic resources of *Araucaria*, in *IUFRO Forestry Problems of the Genus Araucaria*, pp. 87–95, Curitiba: FUPEF.

NORMAH, M.N. and MARZALINA, M., 1996, Achievements and prospects of *in vitro* conservation for tree germplasm, in NORMAH, M.N., NARIMAH, M.K. and CLYDE, M. (Eds), *In Vitro Conservation of Plant Genetic Resources*, pp. 253–261, Kuala Lumpur: Percetakan Watan Sdn. Bhd.

NORMAH, M.N., CHIN, H.F. and HOR, Y.L., 1986, Desiccation and cryopreservation of embryonic axes of *Hevea brasiliensis* Muell.-Arg., *Pertanika*, **9**, 299–303.

OLIVEIRA, Y.M.M. and ROTTA, E., 1982, Levantamento da estrutura horizontal de uma mata de araucaria no primeiro planalto paranaense, *Boletim de Pesquisa Florestal*, **4**, 1–46.

O'MALLEY, D.M., BUCKLEY, D.P., PRANCE, G.T. and BAWA, K.S., 1988, Genetics of Brazil nut (*Bertholletia excelsa* Humb. & Bonpl.: Lecythidaceae). 2. Mating system, *Theoretical Applied Genetics*, **76**, 929–932.

PAIVA, J.R., 1992, Variabilidade genética em populações naturais de seringueira (*Hevea brasiliensis* (Willd) ex Adr. De Juss) Muell. Arg.), PhD thesis, Universidade de São Paulo.

PARANÁ, 1995, *Lista vermelha de plantas ameaçadas de extinção no Estado do Paraná*, Curitiba: Secretaria de Estado do Meio Ambiente.

PENCE, V.C., 1990, Cryostorage of embryo axes of several large seeded temperate tree species, *Cryobiology*, **27**, 212–218.

PENCE, V.C., 1991, Cryostorage of immature embryos of *Theobroma cacao*, *Plant Cell Reports*, **10**, 144–147.

PRANCE, G.T., 1973, Phytogeographic support for the theory of Pleistocene forest refuges in the Amazon Basin, based on evidence from distribution patterns in Caryocaraceae, Chrysobalanaceae, Dichapetalaceae and Lecythidaceae, *Acta Amazonia*, **3**, 5–28.

PRANCE, G.T. (Ed.), 1982, Forest refuges: evidence from woody angiosperms, in PRANCE, G.T. (Ed.), *Biological Diversification in the Tropics*, pp. 137–158, New York: Columbia University Press.

PRANCE, G.T., 1985, The changing forests, in PRANCE, G.T. and LOVEJOY, T.E. (Eds), *Key Environments: Amazonia*, pp. 146–165, Oxford: Pergamon Press.

PRITCHARD, H.W. and PRENDERGAST, F.G., 1986, Effects of desiccation and cryopreservation on the *in vitro* viability of embryos of the recalcitrant seed species *Araucaria hunsteinii*, *Journal of Experimental Botany*, **37**, 1388–1397.

REITZ, R., KLEIN, R.M. and REIS, A., 1978, Projeto Madeira de Santa Catarina, *Sellowia*, **28**, 1–320.

REITZ, R., KLEIN, R.M. and REIS, A., 1979, *Madeiras do Brasil*, Florianópolis: Editora Lunardelli.

RONNING, C.M. and SCHNELL, R.J., 1994, Allozyme diversity in a germplasm collection of *Theobroma cacao* L., *Journal of Heredity*, **85**, 291–295.

SILVA, S., 1996, *Frutas no Brasil*, São Paulo: Editora das Artes.

SOCIEDADE BOTÂNICA DO BRASIL, 1992, *Centuria Plantarum Brasiliensium Exstintionis Minitata*, Brazil: Sociedade Botânica do Brasil.

STOCKEY, R.A., 1982, The Araucariaceae: an evolutionary perspective, *Review of Paleobotany and Palynology*, **37**, 133–154.

TEIXEIRA, J.B., LEMOS, J.I., NETO, L.M. and GONDIM, M.T.P., 1995, 'Micrografting of Jacaranda Tree (*Dalbergia miscolobium* Benth.)', presentation at Redbio 95 Segundo Encuentro Latino Americano de Biotecnologia Vegetal, Puerto Iguazu, June, Abstract A-134.

TESKE, M. and TRENTINI, A.M.M., 1994, *Herbarium Compêndio de Fitoterapia*, Curitiba: Herbarium Laboratório Botânico.

VIANA, A.M., 1998, Somatic embryogenesis in *Ocotea catharinensis* Mez. (Lauraceae), in MANTELL, S.H., BRUNS, S., TRAGARDH, C. and VIANA, A.M. (Eds), *Recent Advances in Biotechnology for Conservation and Management*, Stockholm: International Foundation for Science.

VIANA, A.M. and MANTELL, S.H., 1998, Somatic embryogenesis of *Ocotea catharinensis* an endangered tree of the Mata Atlantica (S. Brazil), in JAIN, M.J., GUPTA, P.K. and NEWTON, R.J. (Eds), *Somatic Embryogenesis in Woody Plants*, Volume 4, Dordrecht: Kluwer Academic Publishers.

VIANA, V.M., TABANEZ, A.J. and BATISTA, J.L., 1997, Dynamics and restoration of forest fragments in the Brazilian Atlantic Moist Forest, in LAURENCE, W.F. and BIERREGAARD JR, R.O. (Eds), *Tropical Forest Remnants, Ecology, Management and Conservation of Fragmented Communities*, pp. 351–365, Chicago: University of Chicago Press.

WANG, B.S.P, CHAREST, P.J. and DOWNIE, B., 1993, *Ex situ* storage of seeds, pollen and *in vitro* cultures of perennial woody plant species, *FAO Forestry Paper*, No. 113.

WHITMORE, T.C., 1997, Tropical forest disturbance, disappearance, and species loss, in LAURENCE, W.F. and BIERREGAARD JR, R.O. (Eds), *Tropical Forest Remnants, Ecology, Management and Conservation of Fragmented Communities*, pp. 3–12, Chicago: University of Chicago Press.

WINGE, H., WOLLHEIM, C., CAVALLI-MOLINA, S., ASSMANN, E.M., BASSANI, K.L.L., AMARAL, M.B. *et al.*, 1995, Variabilidade genética em populações nativas de erva-mate e a implantação de bancos de germoplasma, in WINGE, H., FERREIRA, A.G., MARIATH, J.E.A. and TARASCONI, I.C. (Eds), *Erva-mate, biologia e cultura no Cone-Sul*, pp. 323–345, Porto Alegre: Editora da UFRGS.

Index

abiotic elicitors 125
abiotic stress 182, 187
abscisic acid (ABA) 47, 85, 159, 168, 238–9
Acanthaceae 229
achira 165–6, 169, 175
acid phosphate (Ac-P) 189, 258
acquisition (plant germplasm) 5–6, 27–36, 43, 84
adventitious shoots 42, 49, 50, 90–1, 147, 236, 273
adventitive regeneration 185–6
Advisory Committee on Genetic Manipulation (UK) 161
Africa (tropical root and tuber crops) 179–202
Agavaceae 229
Agricultural Research Service (US) 28–9
agriculture 25–9, 158, 181
Agriculture and Agri-Food Canada 158
agro-ecosystems 26
Agrobacterium 50, 132, 134
ajipa 165
alcohol dehydrogenase (ADH) 188, 189
algae (conservation strategies) 111–21
Algensammlung am Institut fur Botanik (ASIB) 114
alkaloids, indole 129–31, 134
alleles 16, 17, 27, 189, 286, 288–9
Alliums 212–14, 216, 217, 222
allozymes 12–14, 16–19, 286
almond 213, 219
Alocasia 213, 215, 217
Aloeaceae 229
Alstroemeriaceae 229
Amaryllidaceae 229, 236, 253
Amazon Forest 277–80, 282, 286–7, 290–1, 293
American Type Culture Collection (ATCC) 114, 116
amino acids 13, 52, 158
amplified fragment length polymorphism (AFLP) 13–17, 101, 188, 190–2, 222, 283
Andean root and tuber crops (ARTCs) 165–6, 168–70, 172, 175

aneuploidy 99
Animal and Plant Health Inspection Service (US) 66
anthocyanins 126–7
anti-stress agents 85
antibiotics 125, 167, 239
antibodies 68–72, 76, 197
antiviral chemicals 45, 74, 76–8
apex necrosis 50
Apiaceae 229
apical meristems 74–8, 99, 195, 235–6, 273, 291
apical shoot tips 173, 174
Apocynaceae 129
arabis mosaic virus 75–6
Araliaceae 229
Araucaria Forest 277–9, 281–2, 286–7, 289, 293
Aristolochiaceae 229
arracha 165–6, 169–70, 175
arthropods 74
Artocarpus heterophyllus 218
Asclepiadaceae 229
aseptic cultures (establishment) 166–7
Asia and Pacific Plant Protection Commission (APPPC) 65
assessment of plant diversity 11–19
Asteraceae 229, 252–7 *passim*
Atlantic Forest 277–82, 286–7, 293
auditing germplasm collection 187–92
autopolyploidy 18
auxins 47, 48, 125, 184, 185
axillary buds 184, 195, 219, 236, 273
Azadirachta indica 218

bacteria 74, 78, 147–8, 167, 239
bacteriophage M13 101
bananas 53, 159, 212–14, 219–21
barley yellow dwarf virus (BYDV) 72
Begoniaceae 230
benzyl adenine purine 214
benzyl amino purine 140, 186, 195, 252
berberine 131, 134–5